France and Its Spaces of War

France and Its Spaces of War

Experience, Memory, Image

Edited by

Patricia M. E. Lorcin and Daniel Brewer

First published in 2009 by
PALGRAVE MACMILLAN®
in the United States—a division of St. Martin's Press LLC,
175 Fifth Avenue, New York, NY 10010.

Where this book is distributed in the UK, Europe and the rest of the world,
this is by Palgrave Macmillan, a division of Macmillan Publishers Limited,
registered in England, company number 785998, of Houndmills,
Basingstoke, Hampshire RG21 6XS.

Palgrave Macmillan is the global academic imprint of the above companies
and has companies and representatives throughout the world.

Palgrave® and Macmillan® are registered trademarks in the United States,
the United Kingdom, Europe and other countries.

ISBN: 978–0–230–61561–8

Library of Congress Cataloging-in-Publication Data

France and its spaces of war : experience, memory, image / Patricia M.E. Lorcin and Daniel Brewer, editors.
    p. cm.
    "The papers in this volume were first presented at a conference on Spaces of War in the French and Francophone World, held at the University of Minnesota in October 2006"—Introd.
    Includes bibliographical references and index.

    1. War and society—France—History—Congresses. 2. World War, 1914–1918—France—Influence—Congresses. 3. World War, 1939–1945—France—Influence—Congresses. 4. Collective memory—France—History—Congresses. 5. France—History, Military—Congresses. I. Lorcin, Patricia M. E. II. Brewer, Daniel.

DC363.F73 2009
303.6′60944—dc22   2009013782

A catalogue record of the book is available from the British Library.

Design by Newgen Imaging Systems (P) Ltd., Chennai, India.

First edition: October 2009

Contents

Figures

Acknowledgments

T he papers in this volume were first presented at a conference on Spaces of War in the French and Francophone World, held at the University of Minnesota in October 2006. Its success was due to the 93 active participants, the attendees, and the staff and graduate students who helped in its organization and ensured its smooth operation.

Our thanks, therefore, go to Laura Seifert and Alli Lindberg who did an extraordinary job handling the administration and paperwork necessary in preparation the conference, thus sparing us sleepless nights and guaranteeing a problem-free event. Evelyn Davidheiser and Klaas Van der Sanden provided much-appreciated support, while the graduate students of the Department of French and Italian and Department of History contributed their time on the two days of the conference.

A conference of this size would not have been possible without substantial financial support. We are especially grateful for the funding provided by the Gould Foundation in New York and by the University of Minnesota, in particular the Institute of Global Studies, the European Studies Consortium, the Center for German and European Studies, the Center for Genocide and Holocaust Studies through the Mark and Muriel Wexler Speakers Fund, the Department of History, and the Department of French and Italian. Finally, we should like to thank the participants themselves whose stimulating and on occasion provocative contributions made for an intellectually exciting two days.

War and Its Spaces: Introduction

Patricia M. E. Lorcin

War has scarred France territorially, ideologically, socially, and culturally in unique ways. It is closely connected to the emergence of France, first as a state, then as a nation, and finally as a republic. The Albigensian Crusade (1209–22) against what were perceived as the heresies of the south, the Hundred Years' War (1337–1453), which expelled the Plantagenets from all but a sliver of the continental mainland, and the Burgundian Wars (1469–77), which ended with the demise of the house of Burgundy, secured the French monarchy and established the approximate territory that would come to be known as *l'hexagone*. Internally, the wars of religion first anchored France in the Roman Catholic camp (1562–98) and then, during the 1789–99 revolution, cut its Catholic moorings to engage in a century-long secularizing process that made *laïcité* a tenet of the Republic and French-style democracy. In the modern period, all but the second of France's five republics have been born of or created in the aftermath of a war. Nor should it be forgotten that three of the Fifth Republic's presidents, Charles de Gaulle, François Mitterrand, and Jacques Chirac, served or were involved in the twentieth-century wars that were seared into the memories of the French people. It is no coincidence that during the tenure(s) of each of these three men the wars in which they were involved, the two World Wars and the wars of decolonization, were scripted and rescripted onto French national memory.

The horrors of World War I prompted the right-wing activist and author Léon Daudet to publish *La Guerre totale* in 1918. It was, he wrote, a "new type of war," which he had analyzed in order to "open the eyes of his fellow countrymen (*concitoyens*)" as to the "ferocious" totality of waging war against the Germans.[1] Although Daudet's analysis was essentially an anti-German exposé, he introduced a concept that led the way to broader analyses of war and its impact. Daudet's was the first in a series of popular and scholarly works at the time that emphasized the multiple and overlapping spaces of civilian and military engagement and war's all encompassing nature, a theme that runs through many of the articles in the volume.[2] The inter-war period saw the development of a number of intellectual movements that started the shift away from accounts of battles, strategies, and generals to a more holistic, interdisciplinary approach. The establishment of the Annales School in 1929, with its emphasis on an interdisciplinary methodology and the *longue durée*, was a significant scholarly force in encouraging the move toward an examination of the multiple social spaces of war. In the past three decades, scholars interested in issues of women and gender have added new spaces from which to view wars' experiences, while literary scholars and cultural historians have examined the spaces of imagination and memory or the voids of obfuscation and obliteration.[3] The concept of total war has recently been revisited in a project initiated by Roger Chickering and Stig Förster to unravel the nature of its totality.[4] Initially conceived as a comparative study of American and German experiences from the Civil War to the Second World War, Chickering and Stig opened the door to a wider debate. The latest round was initiated with the publication in 2007 of David Bell's *The First Total War: Napoleon's Europe and the Birth of Warfare as We Know It,* which pushed the concept back to the Revolutionary period.[5]

By emphasizing war's all-embracing dimension, however, the concept obscures the internal ideological divisions that are often brought to the fore by war and its aftermath. Dubbed the Franco-French wars as an indication of their bitterness, inherent violence, and trauma, these divisions have periodically flared up with particular vehemence throughout France's history, starting with the earliest wars mentioned above and continuing well into the twentieth century. World War II and the Algerian War of Independence greatly exacerbated Franco-French divisions, leading to veiled responses in the post-war periods. De Gaulle, a veteran of World War I and leader of the Free French in World War II, was intent on papering over the splits caused during the German occupation. He encouraged the concept of unity by stressing the role of French resistance and de-emphasizing collaboration. Vichy remained in French scholarly shadows until the advent of François Mitterrand, the very personification of wartime ambiguity. Mitterrand served under Vichy and was awarded the *Francisque,* the regime's highest civilian decoration, although toward the end of the war he switched camps and joined the resistance. During his presidency (1981–95), the lid of the Vichy caldron, which had first been lifted in 1972 by Robert Paxton's *Vichy France: Old Guard, New Order,* flew off.[6] The trials for war crimes of Klaus Barbie, Paul Touvier, Maurice Papon and the indictment of Mitterrand's friend, René Bousquet, obliged the French to confront their recent murky past and the shadow it cast over the present, and prompted scholars interested in war to focus more sharply on issues of violence, trauma, and memory.[7] The historian Henry Rousso dubbed these interlocking spaces of war *The Vichy Syndrome.*[8]

When Jacques Chirac took over from Mitterrand in 1995, one of his first presidential acts was to acknowledge France's collaborative role in the deportation of the Jews. He did so on the 53rd anniversary of the *rafle du Vel d'Hiv* (the round-up of the Jews at the Velodrome d'Hiver).[9] A month earlier, in June 1995, he had ordered the resumption of nuclear tests in the Pacific, ending the moratorium imposed by Mitterrand in 1992.[10] He justified the tests—six in all—by emphasizing their need for scientific research. These two gestures re-inscribed French spaces of war, the first by recalling what previous regimes had "forgotten" and the second by the scientific obfuscation of France's Cold War policy of nuclear deterrence. As the ghosts of Vichy were beginning to be laid to rest, those of the Algerian War of Independence (1954–62) started to re-emerge from the limbo to which they had been relegated. Although the war never ceased to be the focus of scholarly and literary attention, rather like René Magritte's painting of two lovers, whose inability to communicate was symbolized by veil-enveloped heads, the face of the war remained shrouded in a veil of silence in official circles; a "war without a name."[11] The growing scholarly and public spaces of memory, trauma, and violence surrounding Vichy instigated a similar movement for recognition of the personal, social, cultural, and political costs of the Algerian War in both France and Algeria. During Chirac's first mandate, the clamor for official recognition of the war's horrific dimensions and their legacy grew louder, leading to a series of acknowledgments on the part of the French state of its role in their perpetration.[12] On December 5, 2002, in the first year of his second presidential mandate and 40 years after the war's end, Chirac inaugurated the Algerian War memorial on the Quai Branly; its three electronic columns dwarfed by the Eiffel Tower, thus creating a new spatial relationship between the state and its symbols of commemoration and culture.

The ongoing processes of these recent re-inscriptions may account for the fact that the focus of the majority of the presentations at the 2006 conference on "Spaces of War" from which the articles in this volume have been drawn, was on the wars of the twentieth century. The proximity of these wars and their relevance to the present may be due to this preponderance, but the methodological shift in writing about war, from an emphasis on diplomacy, military operations, and personalities, to an exploration of the social, cultural, and commemorative dimensions, opened up a range of research possibilities that make twentieth-century wars a more readily accessible terrain on which to work.[13] Nonetheless, as the articles by Christy Pichichero and Michael Wolfe in this volume demonstrate, the newer methodological approaches to war are by no means restricted to the modern period.

We chose to remain with the conference's concept of space rather than chronology or theme as a way of dividing up the volume. The three sections—experience, memory, and imagination—suggest both individual and collective spaces, highlighting as well the continuities and changes between the different periods and their wars. The most radical changes in wars over time have been in the technologies and strategies of war, the constants are violence, dislocation, and trauma. But, as the articles in this volume demonstrate, the use of space as a conceptual starting point removes the constraints of a linear, temporal progression of change or continuity and allows us to think in terms of overlap, disparity, and concentricity. All wars reconfigure traditional spaces and relationships; the disparity lies in the way this reconfiguration takes place.

The articles in the first section, on the experiences of war, explore the relationship between the civilian and military spheres and the ways in which they impinge on one another. In pre-nineteenth century France, as Pichichero points out, the chivalric codes of honor allowed the aristocracy to move seamlessly "between the boudoir and battlefield," thus eliminating the distinctions between civilian and military "strategies" in the upper levels of society. Class distinctions, however, meant that officers held the rank and file in contempt, a situation that critics of the military saw as responsible for their poor performance in the eighteenth-century wars. Enlightenment reformers therefore sought to improve military morale and performance by advocating the need for officers and men to espouse values of liberty, equality, and fraternity. The moralizing of Enlightenment reformers may have aimed to "civilize" military behavior within the French ranks, but as Jennifer Sessions shows in her article, the civilizing ideals of France were seriously challenged in Algeria. French soldiers trampled over the spatial boundaries between military and civilian by committing uncalled-for atrocities, as in the case of the *enfumade* (death by smoke asphyxiation) of civilians in the caves of the Dahra Mountains. The symbolism, which was certainly lost on the military at the time, cannot be so for a post-colonial audience. The colonizing forces' smothering of Algerians sheltering in caves, archetypical symbol of the womb and hence of life, was not just a denial of French values of humanity; it represented the human annihilation-ist impulse that undergirds the waging of war. Soldiers can be "barbarized" by the violence of war with civilians bearing the brunt of that barbarity.

The overlap of military and civilian spaces forms an altogether different relationship in Martha Hanna's article of rural society under duress during World War I. She high-lights how the civilian space—as anxiety and image of home and hearth—encroaches onto the mental space of peasant soldiers at the front, affecting their actions in and reac-tions to the progress of the war. The reconfiguration of traditional relations between geographical and psychological space is also developed in Nicole Dombrowski Risser's chapter on civilian safe havens during World War II. Aerial bombing destroyed the distinctive spaces of home front and battlefront, as the home lost its quality of shelter and was sucked into the atrocity of battle. In a particularly telling passage Dombrowski Risser reappraises the symbolism of the cave, "man's most primitive dwelling." The civilian escape to shelter from the destructive force of modern technology was, she sug-gests, one of the "de-civilizing" processes engendered by war. Loss, as a space rather than a cost, extends beyond casualty figures and material destruction to include moral and ideological impoverishment, a reversal of the course of humanization.

Scholarship on gender and war has greatly enhanced our understanding of the dif-ferent spaces inhabited by men and women during wartime. Wars' experiences blur the boundaries of gender and unsettle the way gender is imagined. Women move into men's roles when their menfolk are away; men are "effeminized" at the front, a two-edged denigration used either to spur perceived slackers into action or to cre-ate a detrimental image of the enemy.[14] But war also sharpens gender tensions. The effeminization of eighteenth-century officers, as Pichichero points out, was attributed in part to the detrimental influence of elite women whose frivolity degraded "both French national character and French men." Similarly, Hanna informs us, in World War I peasant women, who bore the weight of working the land while their men were at the front, were blamed for the lack of food. Anxieties about female infidelity and the symbolic emasculation and powerlessness that war produced created gendered antago-nisms. War also shattered comfortable (and comforting) perceptions of women. The image of nurses as "ministering angels" was often undermined by male resentment at being "patched up" to be sent back to the front.

The ambiguous spaces of war extend well beyond gender, however. Patrice Arnaud's article highlights the mixed feelings of French conscripts toward allied troops in World War II. If the bombing of the French fleet at Mers-el-Kebir by the British in the early months of the war resuscitated sentiments of *Albion perfide*, the allied bombing sur-rounding the Normandy landings, which killed thousands of civilians, did nothing to dissipate them. The ambiguity of feeling was probably also filtered through lingering feelings of humiliation at the collapse of France and the hardships of German occupa-tion. Ambiguity is inherent in any occupied territory as resistance and collaboration can take many divergent or intersecting paths. In the post-Occupation period these ambiguities are often intentionally erased to accommodate national or regional exi-gencies. A case in point, as Bronson Long demonstrates, was the Francophilia in the post-World War II Saar region, which re-inscribed the Saarlanders' actions during the war from participants in Nazi activities to victims of it.

Long's article on wartime experiences of occupation and their aftermath is the bridge to the second section of the volume, which explores the different ways in which war is written onto individual and collective memory. The inscription of war on the body of the individual or the nation is examined in the first three articles of the section. If bodily wounds are the most obvious evidence of war's horrors, the reshaping of individual and national identity is the inevitable corollary. For a member of the sixteenth-century "warrior class," the inability to function as a soldier due to physical impairment meant a loss of one's defining identity. The nobleman Blaise de Monluc re-inscribed his wounds in the text of his memoirs as "badges of honor," as much to emphasize his manliness as to curry favor with the crown to obtain compensation for what he deemed was his due for the injuries he had sustained. The horrific war injuries of the sixteenth century bore a resemblance to those of the *gueules cassées* of the First World War who, like Monluc, had the remembrance of war carved onto their bodies but unlike him were better able to obtain compensation and recognition from the state.[15] The shaping of a veteran's identity by pain, whether it was the long-term scarring of the body or the searing of the mind due to war's traumas, is a constant over time.[16] If, on the individual level, it is the scarred body and mind that memorialize war, at the national level memorialization is of the dead, symbols of the nation's sacrifice and valor, rather than the impaired who serve as reminders of the barbarity and futility of war.

Memorials, in the form of commemorative holidays, statues, stained glass windows, ossuaries, chapels, or well-ordered cemeteries, mark the body and soul of the nation, shaping its post-war identity. Whereas the rituals of commemoration are often shaped by the politics and exigencies of a particular era, memorials symbolize the duty of the civilian to remember and the desire of the state to educate. The articles by Alexandre Neiss, Christopher Fischer, and Julien Fragnon develop the reasons and politics behind commemoration at the regional and national level. Neiss demonstrates that positioning of memorials in carefully selected spaces was a political act, whereas the actual representations on the memorials were highly symbolic, responding to the impact or extent of the area's involvement in the war. Memorials are often constructed long after the wars in question have ended, either as an overtly political act or, as Neiss points out, because they could be erected only after the devastated areas could be rebuilt.[17] Differences in regional and national conceptions of how war was remembered created conflicting spaces of memorialization. In Alsace-Lorraine, as Fischer explains, Alsatians had a different understanding of the war shaped by the ambiguities of an identity that was rooted in both French and German cultural practices. Fischer's article highlights the useful distinction between commemorative space (memorials) and commemorative time (holidays) and draws attention to the fact that commemoration and mourning were often difficult to reconcile.

Language is one of the defining features of commemoration. Both Fischer and Fragnon illustrate how the language of commemoration is influenced, not just by the political rhetoric of the moment, but also by differing approaches to and experiences of the wars in question. War's memory, Fragnon argues, is a political object in so far as its discourses are used for mobilization and identification, thus shaping the "national habitus." Linguistic eloquence serves commemoration well, but language can also be used to obfuscate, distort, or eradicate what is deemed best forgotten. The themes of eradication and distortion of wars' memorials and memory run through the last two articles in the section. Kirrily Freeman focuses on empty statuary pedestals as spaces of absence. The World War II destruction of bronze statues for use in the manufacture of munitions engendered mixed public reaction: relative acquiescence in Paris, where their removal symbolized the demise of the Third Republic; condemnation in the provinces, where a sense of loss was compounded by an affront to the patriotism the statues had helped to foster. Freeman also demonstrates the shift in symbolic perception from wartime to post-Liberation when the empty pedestals became reminders of the occupation and its divisive impact on the French people. If, as Freeman suggests, the decapitated pedestals symbolized the decapitation of the Third Republic, one cannot help but reflect on Vichy's avenging mimicry of the decapitating zeal of French Revolution: the Republic vanquished at last, or so its adherents thought.

Like Long's article at the end of the first section of the volume, Richard Golsan's contribution bridges the second section with the themes that connect the third. The article explores the literary dimensions of the Vichy Syndrome and sketches the parameters of forgetting and remembering. Unlike the ongoing commemoration and remembrance of World War I, in France the remembrance of World War II and indeed of the wars of decolonization that followed, in particular the Algerian War of Independence, has been muted. Even their twenty-first century acknowledgments and memorials have been geared to lay the memory of these wars to rest once and for all. Commemoration is thus a necessary step in the process of forgetting. The trauma of war cannot be laid to rest that easily, however, and wars have always haunted the imagination.

As the articles in the final section of the volume attest, the way war is imagined depends not only on the war but also at what stage, during or after the war, the imagining takes place. Describing war has always been difficult, as its violence and trauma go well beyond the linguistic frame. The divisions between participants and non-participants, soldiers and civilians, men and women, create distinctive voices and gazes to say nothing of proprietorial conflicts about who can or cannot accurately represent war. Jan Bethke Elshtain, for example, has argued that irony, a voice that was so frequently used by men to describe war, is largely absent in women's writing about war.[18] Indeed, until the second half of the twentieth century, women were considered best equipped to concern themselves with the home front. Battle and home fronts also shape the genres used to represent war. Military artists and, more recently, photographers capture the images of war in action, whereas at the battlefront the fractured existence of active service, where time is at a premium, is often best captured by the conciseness of poetry. The imagination of the home front is altogether different. Artistic creation, whether through painting, poetry, or prose, is after the fact. The raw sentiments that emerge in the heat of battle are tempered by distance and political, cultural, and social agendas come into play. The blurring of the lines between fact and fiction adds to the epistemological disjuncture between the battlefront and the home front.

War literature, in all its forms, creates multiple spaces, some of which become gendered, as Libby Murphy demonstrates in the case of World War I, an extension of the gendering of the two fronts. Popular fiction with its sentimentality and themes of resentment and revenge was perceived as emotional and inaccurate, triggering anxieties about the emasculation of French literature. Only facts, unemotional and direct, could truly represent the culture of war. But facts are also suppressed or distorted for purposes of propaganda. The silencing effect of war on art is broached in Béatrice Vernier-Larochette's analysis of Fernand Léger's art and writing. The lack of explicit expressions of Léger's experiences of the atrocities of battle were, she suggests, sublimated by his desire to identify with his dead comrades through his art by equating the armed struggle in which they had been lost with the creative struggle of his post-war existence. The projection of the armed struggles of war onto peacetime activities is often long-lived, as Annette Becker demonstrates in her discussion of the debates surrounding the 1993 commissioning of a new World War I monument in Biron. Conflicts about memorialisation and memory ripple out from the war's epicentre, creating spaces that are shaped by the political and social rhetoric of the moment.

Imagining war can transpose spaces or create overlaps, as the articles of both Floréal Jiminez Aguilera and Paul Shue demonstrate. Jean-Jacques Annaud's film, La Victoire en Chantant, the subject of Jiminez Auilera's contribution, is a satire about colonization during World War I. Annaud shifts the action of the war to Africa, thus removing the horrors of the Western and Eastern fronts. Not only is the war perceived as more anodyne, but it also becomes more ambiguous as the excesses of colonization mask the enmity between the warring factions. Indeed at the end of the film the French and Germans appear to have more in common than the French and the British, a reflection no doubt of the political situation at the time the film was made, when Franco-German relations were at their harmonious height.[19] The overlapping of wars' space and their ambiguity is highlighted in Shue's article on right-wing writers' use of the Spanish Civil War (1936–39) to work through their own ideas on the Popular Front and French democracy. Envisaging Franco's Nationalists as latter-day Chouans, the contributors to L'Action française and Je suis partout sought to convince their readers that Franco was restoring the sort of order that was necessary in France to combat communism, anarchism, and the excesses of democracy.

The masked spaces of war are the subject of the last two chapters of the volume. Propaganda, the subject of Angélique Durand's contribution, is of course the most familiar of these masks, but Durand argues that there is much more to propaganda than its overt and most familiar manifestations. The parameters of propaganda are much greater, encompassing nearly every aspect of media and cultural representation. It is apt that the volume ends with Steven Ungar's discussion of two recent representations of twentieth-century wars, the films La Trahison (The Betrayal) and Indigènes (literally "natives" but translated as Days of Glory). The role of colonial troops in the world wars and the wars of decolonization, in particular the Algerian War of Independence, may be emerging from the occluded spaces in which they have hitherto been situated but, as Ungar suggests in the case of the two films he discusses, competing memories mask wars' spaces as often as they illuminate them.

The diversity of the articles in this volume reflect how the concept of space can complicate an analysis of war and its impact and demonstrate that the wars in which France has been engaged have had a profound impact on the social, cultural, and political development of the nation.

Notes

1. Léon Daudet, La Guerre totale (Paris: Nouvelle librairie nationale, 1918), 243.
2. Ibid. See also Henri Lémery, De la Guerre totale à la paix mutilée (Paris: Alcan, 1931) and Gal Erich von Ludendorff, La Guerre totale (Paris: Flammarion, 1936).

3. For women and gender see Miriam Cooke and Angela Woollacott, *Gendering War Talk* (Princeton: Princeton University Press, 1993): Margaret H. Darrow, *French Women and the First World War: War Stories of the Home Front* (New York: Berg, 2000); Jean Gallagher, *The World Wars through the Female Gaze* (Carbondale: Southern Illinois University Press, 1998); Susan R. Grayzel, *Women's Identities at War: Gender, Motherhood and Politics in Britain and France During the First World War* (Chapel Hill: University of North Carolina Press, 1999); Margaret R. Higonnet, ed., *Behind the Lines: Gender and the Two World Wars* (New Haven: Yale University Press, 1987) and her *Lines of Fire: Women Writers of World War I* (New York: Plume, 1999); Margaret R. Higonnet, Mary Borden, and Ellen Newbold La Motte, eds., *Nurses at the Front: Writing the Wounds of the Great War* (Boston: Northeastern University Press, 2001); Miranda Pollard, *Reign of Virtue: Mobilizing Gender in Vichy France* (Chicago: University of Chicago Press, 1998); Margaret Collins Weitz, *Sisters in the Resistance: How Women Fought to Free France, 1940–1945* (Somerset, NJ: J. Wiley, 1995); Marie-Louise Jacotey, *Femmes aux armées: de 1792 à 1815* (Langres: D. Guéniot, 1999).

   For wars' legacies see Leonard V. Smith, *The Embattled Self: French Soldiers' Testimony of the Great War* (Ithaca; Cornell University Press, 2007); Matthew Cambell, Jacqueline M. Labbe, and Sally Shuttleworth, eds., *Memory and Memorials 1789–1914* (New York: Routledge, 2000); Philip Dine, *Images of the Algerian War: French Fiction and Film, 1954–1992* (Oxford: Clarendon Press, 1994); Martin Evans and Kenneth Lunn, *War and Memory in the Twentieth Century* (Oxford: Berg, 1997); Sarah Farmer, *Martyred Village: Commemorating the 1944 Massacre at Oradour-Sur-Glane* (Berkeley: University of California Press, 1994); Daniel J. Sherman, *The Construction of Memory in Interwar France* (Chicago: University of Chicago Press, 1999); Debra Kelly, *Remembering and Representing the Experience of War in Twentieth-Century France: Committing to Memory*, vol. 20 (Lewiston: E. Mellen, 2000).

4. Roger Chickering and Stig Förster, *The Shadows of Total War: Europe, East Asia, and the United States, 1919–1939* (New York: Cambridge University Press, 2003).

5. David Avrom Bell, *The First Total War: Napoleon's Europe and the Birth of Warfare as We Know It* (New York: Houghton Mifflin, 2007). For the debate raised by Bell, see Michale Broers, "The Concept Of 'Total War' In the Revolutionary-Napoleonic Period," *War in History* 15:3 (2008): 247–68. See also the H-France Forum, 2:3 (summer 2007), with reviews by Jeremy Popkin, Annie Jourdan, Jeremy Black, and Howard Brown with a response by Bell.

6. Robert O. Paxton, *Vichy France: Old Guard and New Order, 1940–1944*, 1st ed. (New York: Knopf, 1972). Among the many books that appeared in the Mitterrand years, see Sherman, *The Construction of Memory in Interwar France.*

7. Both Papon and Bousquet were associates and friends of Mitterrand. Bousquet was assassinated and was therefore never tried.

8. For some recent works on war and memory see William & Kidd and Brian Murdoch, eds., *Memory and Memorials: The Commemorative Century* (London: Ashgate, 2005); Henry Rousso, *The Haunting Past: History, Memory, and Justice in Contemporary France* (Philadelphia: University of Pennsylvania Press, 2002) and his *Le Syndrome de Vichy: 1944–198—* (Paris: Seuil, 1987).

9. For extracts of his speech, see "L'Occupation et la Republique," under "Société," *Le Monde*, December 6, 1997.

10. He later made a commitment to sign the Comprehensive Test Ban Treaty.

11. The literature on the Algerian War is vast; there are over 2,200 entries in WorldCat on the period of the war. For a bibliography of forty years of literature about the war, see Benjamin Stora, *Le Dictionnaire des livres de la guerre d'Algérie: romans, nouvelles, poésie, photos, histoire, essais, récits historiques, témoignages, biographies, mémoires, autobiographies: 1955–1995* (Paris: Harmattan, 1996). The Algerian war was recognized as a war only in 1999. Formerly it was regarded as "an operation to maintain order." The designation "war without a name" was used by John E. Talbott, *The War without a Name: France in*

Algeria, 1954–1962, 1st ed. (New York: Knopf, 1980) and Patrick Rotman and Bertrand Tavernier, eds., *La Guerre sans nom: les appelés d'Algérie, 1954–1962* (Paris: Editions du Seuil, 1992). See also Patricia M. E. Lorcin, ed., *Algeria and France, 1800–2000: Identity, Memory, Nostalgia* (Syracuse: Syracuse University Press, 2006). Tavernier made a documentary, *La Guerre sans nom* (1992), based on his book with Rotman. Magritte painted *Les Amants* in 1928.

12. The recognition of the war in 1999 was followed on by an acknowledgment of the massacre of October 16, 1961. Bertrand Delanoë, then mayor of Paris, placed a memorial plaque on the St. Michel bridge on October 17, 2001. The same year the publication of two memoirs exposed the (self-satisfied) role of torturer and the plight of the tortured. Claude Pedroncini, Guy Carlier, *Les Troupes coloniales dans la Grande Guerre* (Paris: IHCC-CNSV, Economica, 1997). For more details on these events see Lorcin, "Introduction" and Joshua Cole, "The Memory of Police Violence in Paris, October 1961m" both in Lorcin, ed. *Algeria and France 1800–2000: Identity, Memory, Nostalgia* (Syracuse: Syracuse University Press, 2006).

13. Military history has not been eclipsed altogether, however. See Robert A. Doughty, *Pyrrhic Victory: French Strategy and Operations in the Great War* (Cambridge, MA: Harvard University Press, 2005); Jonathan R. Dull, *The French Navy and the Seven Years' War* (Lincoln: University of Nebraska Press, 2005); John A. Lynn, *The Wars of Louis XIV, 1667–1714* (New York: Longman, 1999); and John H. Gill, *1809 Thunder on the Danube: Napoleon's Defeat of the Habsburgs*, vol. 1: *Abensberg* (London: Frontline Books, 2008).

14. Leo Braudy, *From Chivalry to Terrorism: War and the Changing Nature of Masculinity* (New York: Alfred A. Knopf, 2003); Stefan Dudink, Karen Hagemann, and John Tosh, eds., *Masculinities in Politics and War: Gendering Modern History* (New York: Manchester University Press, 2004). For a review of recent books on the subject see Robert A. Nye, "Western Masculinities in War and Peace," *American Historical Review* 112:2 (2007): 417–38.

15. See Rebecca Scales, "Radio Broadcasting, Disabled Veterans, and the Politics of National Recovery in Interwar France," *French Historical Studies* 31:4 (2008):643–78; Roxanne Panchasi, "Reconstructions: Prosthetics and the Rehabilitation of the Male Body in World War I France," *Differences* 7 (1995):109–41. Antoine Prost, *Les Anciens Combattants et la société française: 1914–1939*, 3 vols. (Paris: Presses de la Fondation nationale des sciences politiques, 1977). Sophie Delaporte, *Les Gueules cassées: les blessés de la face de la Grande Guerre* (Paris: Editions Noêsis, 1996).

16. Mark S. Micale and Paul Lerner, eds., *Traumatic Pasts: History, Psychiatry and Trauma in the Modern Age, 1870–1930* (Cambridge: Cambridge University Press, 2001); Alec G. Hargreaves, *Memory, Empire, and Postcolonialism: Legacies of French Colonialism (After the Empire)* (Lanham, MD: Lexington Books, 2005); Gregory Mathew Thomas, "Post-Traumatic Nation: Medical Manifestations of Psychological Trauma in Interwar France" (Thesis, University of California, 2005).

17. The inauguration of the Franco-Prussian war memorial in Limoges in 1899 and the memorialization of the village of Ouradour-sur-Glane in the 1950s are good examples of the politicization of commemoration. For analyses of these two memorializations, see John M. Merriman, *The Red City: Limoges and the French Nineteenth Century* (New York: Oxford University Press, 1985), 156; and Sarah Farmer, *Martyred Village: Commemorating the 1944 Massacre at Oradour-Sur-Glane* (Berkeley: University of California Press, 1999).

18. Quoted by Margaret R. Higonnet, "Not So Quiet in No-Woman's Land," in *Gendering War Talk*, Miriam Cooke and Angela Woollacott, ed. (Princeton: Princeton University Press, 1993), 205.

19. See Hélène Miard-Delacroix, *Partenaires de choix?: le Chancelier Helmut Schmidt et la France, 1974–1982* (New York: Peter Lang, 1993); Haig Simonian, *The Privileged Partnership: Franco-German Relations in the European Community, 1969–1984* (Oxford: Clarendon Press: New York, 1985).

Part I

War Experienced

2

Moralizing War: Military Enlightenment in Eighteenth-Century France

Christy Pichichero

Though history remembers him better for his many condemnations of war, eighteenth-century French *philosophe* Voltaire penned works celebrating warfare such as the "*Poème de Fontenoy*" of 1745 commemorating the French victory at the Battle of Fontenoy during the War of Austrian Succession (1741–48), one of the only great French military victories of the eighteenth century prior to the Revolution. In his eulogy of the battle written in classical *alexandrins*, Voltaire focuses his praise on Louis XV and the noble officers of the army, making multiple references to the king and Maréchal Maurice de Saxe in addition to painstakingly naming no less than 51 aristocratic officers of the French army in the 348-line poem. In particular, Voltaire glorifies what he sees as the dual identity of these military men, marveling at their transformation from gentlemanly courtiers to fierce warriors of the battlefield:

> Comment ces courtisans doux, enjoués, aimables,
> Sont-ils dans les combats des lions indomptables?
> Quel assemblage heureux de grâces, de valeur!
>
> [How is it that these gentle, jocular, amiable courtiers
> Become indomitable lions in combat?
> What a happy assemblage of graces and valor!]

Voltaire's wonder regarding this metamorphosis both reveals the central tenet of early modern aristocratic culture of war in France and hints at the deterioration of this culture that this article proposes to explore.

The above passage and Voltaire's incessant name-dropping throughout the poem demonstrate the strongly aristocratic ethos that guided pre-Revolutionary warfare, an arena in which princes and noblemen who made up the officer corps of the armed forces battled for prestige and power among themselves. "War was waged in the way that a pair of duelists carried out their pedantic struggle," Carl von Clausewitz would later assert; "one battled with moderation and consideration, according to the conventional proprieties.... War was caused by nothing more than a diplomatic caprice, and the spirit of such a thing could hardly prevail over the goal of military honor."[1] Indeed, writers of this period frequently used the metaphor of the duel to describe their experience at war. During the early modern period, David A. Bell explains, both war and duels "followed intricate sets of rules and involved scrupulous attention to appearance, gesture, movement, and expression. Both demanded a high degree of physical courage. Both, of course, were socially acceptable arenas for the taking of human life.... Dueling and warfare came together in the (originally medieval) practice of single combat: officers form the opposing sides fighting while their men looked on, as spectators."[2] Centered on a code of honor and a strong sense of theatricality, the culture of war in early modern France shared many elements with the elite culture of *mondanité* (worldliness) practiced at court and in the upper-crust of Parisian society.[3] Here, noble officers "passed easily from the theater of the aristocracy that was the royal court, with its intrigues and scandals and seductions, to the theater of the aristocracy that was the military campaign, where they could find more of the same. In each arena, they were expected to show the same grace, coolness, and splendor."[4]

However, Voltaire's sincere astonishment at the notion of such a "happy union of graces and valor" reflects the deteriorating foundations of this culture of the "courtier-warrior," which was indeed disintegrating throughout the age of Enlightenment toward a devastating outcome. From the beginning of the eighteenth century until the Revolution, the French army experienced a disastrous decline in combat effectiveness that initiated a general military decay and resulted in a number of grave losses in men, equipment, and territory during the War of the Spanish Succession (1701–14), the War of the Austrian Succession, and the French and Indian/Seven Years' War (1754–63).[5] This crisis spurred generations of military thinkers—largely French noblemen who formed the officer corps and military administration—to undertake a vast inquiry into almost all areas of the armed forces.[6] Some reform-minded thinkers scrutinized the structures of the military system (tactics, administration, organization, discipline, and command), suggesting changes in policy, practice, and infrastructure, while other reformers theorized about much less tangible subjects, bringing increasing attention to cultural, moral, and educational issues. Many of the latter thinkers believed that the crisis of military performance was closely connected to the disintegration of the traditional "courtier-warrior" military identity and culture within the aristocratic officer corps. They deemed that those whom Voltaire characterized as "gentle, jocular, affable courtiers" were in fact no longer transforming into "indomitable lions in battle," and that what had been a "happy union of graces and valor" in the space of war had collapsed into a pathetic spectacle of powdered wig-clad fops, timorous to the point of impotence on the battlefield. Many examples fueled this opinion as time and again the French royal army was stifled by inglorious mistakes such as the humiliating tactical defeat at Dettingen in June 1743 during the War of the Austrian Succession, a battle at which French troops—even the prestigious *Gardes françaises*—took off in panicked flight in the face of the enemy.

The abundance of similar tales of cowardice and defeat convinced military thinkers that the aristocratic culture of war had eroded, its traditional martial values having been replaced by corrupt ones from the highly effeminized, materialistic elite society in Paris and Versailles. As we shall see, these reformers argued that not only had virtuous, chivalric values disappeared, but the balance between noble *courtoisie* and wartime belligerence had fallen off kilter, officers privileging their worldly personas and pursuits over their military ones. This cultural deterioration and lopsidedness were seen to have corrupted the meaning of key military concepts of the traditional aristocratic culture of war (honor, glory, merit) and perpetuated social discord as well as practical problems within the French army. We will see how a number of thinkers of the mid-eighteenth century sought to amend this situation by transforming the predominant figure of the "courtier-warrior" into a *"philosophe-militaire."* In this attempt, thinkers introduced a new conceptual and linguistic lexicon based upon a fusion of natural law, theories of sociability, and practices of etiquette originating in the salon milieu. This amalgam reflected contemporary moral philosophy that posited reason-based natural sociability, which thinkers believed would help the army plant the roots of a culture of functional social equality and essential human bonds amongst military men. Contrary to what such historians as André Corvisier and David D. Bien have claimed, these philosophically-minded thinkers strongly advocated socially democratizing forces as the key element to strengthening the French army and envisioned it as an ideal civil society in which all men are united by their common humanity and psychological constitution. Rather than see the army as a space for technical and administrative reform, these military thinkers viewed it as a space for moral improvement in which the strength and applicability of Enlightenment visions of human society could be put to the test.

The Dysfunctions of Eighteenth-Century Military Culture

In eighteenth-century France, noblemen of the sword were expected to move seamlessly between the battlefield and the enclaves of polite society in Paris and Versailles, exhibiting civility as well as gentlemanly martial values in both spheres. For the *noblesse d'épée* of Enlightenment France there was no conceptual or linguistic indication for distinguishing civilian and military realms, making it to some extent unthinkable that a tension between these two spheres could exist. However, as the eighteenth century progressed, military documentation shows that thinkers perceived a growing and deeply problematic tension between the life of the noble officer on duty during the warring seasons of spring and summer and the life he led at home during the fall and winter months. According to reform-minded military thinkers, officers old and young had either forgotten or not yet had the opportunity to practice the art of war, making them insufficiently bellicose on a culture level and frightfully unprepared for military action on a practical level. Many of these officers, especially the younger ones, it was argued, had also been morally corrupted by spending all of their time frequenting the elite, urban society of *le monde* at the court of Versailles as well as the salons and *hôtels* of Paris. Speaking of young nobles destined for the military's officer corps, M. de Saint Hilaire complained that:

> We've settled for teaching them some Latin and Humanities and then we thrust them into the corrupt *monde* at the tender age of 14 or 15 without having given them the time to arm themselves with proper protection. Their fathers are already corrupted themselves, driven by the glitter of false vanity and have, from the moment their sons enter *le monde*, already procured important positions for them through credit, intrigues, and money.[7]

In addition to being characterized by a dangerous "false vanity" and power struggles involving "credit, intrigues, and money," le monde and its cultural code of mondanité were seen as highly effeminized, at times to a fault. Intellectuals of the eighteenth century concurred that la mondanité, whose codes emphasized superficial forms of sociability and politeness accompanied by frivolous gaiety (légèreté), was indeed dominated by the tastes, attitudes, and actions of women. In period writings on gender and French mores (mœurs), women were thought to have held a strong influence over this realm because these moral and civil qualities combined with the penchant toward luxury (ornate and exorbitantly expensive clothing, meals, pastimes, and material accoutrements) were believed to be innate to the female sex.[8] François Ignace d'Espiard de la Borde's L'Esprit des nations of 1752 conveys the ambivalence of many male intellectuals regarding the penetrating influence that women were seen to hold over men of le monde and French mores more largely. On the one hand, he declares that "the Frenchman owes the amiable qualities which distinguish him from the other peoples to interchange with women," while he later admits bitterly, "foreigners say that in France, men are not men enough."[9]

It was not just foreigners who thought French men were not "men enough." Many eighteenth-century military thinkers also believed that French men—particularly those of the army—had fallen prey to the cultural influence of women to the detriment of their military science and masculine belligerence. They had grown overly effeminate in their mores, polite to the point of ridicule, and decadent in their taste for sensual and material pleasures. In treatises, memoirs, and letters addressing this cultural conundrum and its dangerous consequences in the space of war, military thinkers wrote of linguistic and social problems that they saw to be directly related to the "civilian" lives of noble officers. Ethically and linguistically, the system of values and language of aristocratic service were tainted by those of le monde, creating semantic corruptions of traditional key concepts of the noble warrior ethos and culture of military service. Socially, officers had become absorbed by the superficial culture of gossip and petty competition that marked interactions of elite society and polluted social exchange in the military.

With regard to the values and language of military service, reformers complained of semantic erosion in the understanding and enactment of central terms in the noble tradition of military service, such as honor, glory, and merit. M. de Lamée, a noble lieutenant in the army, explains the troublesome and backward concepts of honor and glory that many young men assimilated in their "civilian" lives and then brought with them to the army to no good end:

> Few people know the true point of honor. Prejudices too often prescribe its laws and give birth to ridiculous delicacies that betray the smallness of one's esprit rather than the purity and grandeur of one's sentiments. When a young man leaves his home for the military, he is overwhelmed with advice. Everyone wants to have their part in his glory and his education.... They want to make him honest and good (brave)? Instead they make him quarrelsome. They want to inspire gentleness in him? They make him into a coward.... Many young men join a regiment, and their ignorance causes them to make stupid mistakes. They believe that they must stand by these mistakes as a point of honor.... They will always stand behind what they've said and done. Their glory depends upon it—or so they think.[10]

Lamée tacitly portrays the meanings of honor and glory exhibited by these young officers, meanings that came from the men's social and familial upbringings, as quite

opposite to ones more fitting to the military sphere. Rather than a purity and grandeur of sentiment, a polite and measured speech and reasoning, an esprit that is open
to learning and recognizing one's errors, honor, and glory are understood by these
youngsters to mean a combination of stubbornness, aloofness, opinionatedness,
undeterred pride, and uninhibited verbosity. As Paul-Henri Thiry, baron d'Holbach,
would relay in his treatise of moral philosophy *La Morale universelle, ou les devoirs
de l'homme fondés sur la nature* (1776), the meaning of honor had deteriorated to the
point of becoming pathological, degrading the lofty sentiment into "the fear of being
despised because one knows that one is, in truth, despicable."[11]

More grave than the degradation of honor and glory was that of merit, a central
term of the older tradition of military service.[12] As Jay Smith illustrates in his study of
the culture of military service in early modern France, the virtue and generosity that a
nobleman displayed toward the king through military service illustrated his "merit," a
word that signaled both his deservedness and the obligation to recompense him for his
virtuous actions. Merit represented a reciprocal, mutualistic system involving gallant,
courageous self-sacrifice performed in order to be recognized by the king's "sovereign
gaze" and subsequently rewarded both monetarily and symbolically.[13] According to an
officer named Lagarrigue, this older meaning of merit had disappeared from the army
by the mid-eighteenth century, having been replaced by one that could be traced back
to the overly effeminate mores and extravagant taste of the culture of *la mondanité*:

> We have adopted a kind of merit that has made its way into the troops, one that nourishes
> laxity and smothers the love of duty. This "merit" is luxury. A well-powdered, perfumed,
> and dressed officer dolled up with tassels and embroidery, though foolish, disorderly,
> and without the least application to his profession, has a greater chance of receiving the
> king's graces than an officer who is modestly dressed, who gives all of his attention to
> fulfilling the functions of his rank, who speaks little but to the point, and who always has
> 30 louis d'or in his purse to repair the losses that could occur in his unit.[14]

Lagarrigue was not alone in observing that luxury and its varied material representations were the merits toward which officers aspired, supplanting a concept of merit
based on a sense of moral justness or professional duty. An anonymous writer of the
1730s was of the same mind, remarking that "these days one only speaks of handsome
regiments—we are all preoccupied with reputations of grand nobility, good looks,
cleanliness, and stylishness of clothing...not at all with moral uprightness."[15]

The result of this semantic erosion of key terms and concepts in the traditional
paradigm of noble military service became drastically palpable in certain battle situations. Speaking about his experience in the War of the Spanish Succession, an anonymous writer brings up what he calls "a very detrimental abuse": "most infantry officers
use the king's soldiers as valets when they cannot afford to hire one for wages, such
that quite often you will find 50 men missing in each battalion during a battle, each
officer using one soldier and sometimes two to tend to his equipage in the baggage
train....This abuse is even more detrimental in an army of 80 battalions in which
upwards of 4,000 soldiers will never see battle...and in a word are rendered useless."[16]
By removing trained soldiers from the battlefront to serve as valets and by putting
their money into ornate silver dishes, embroidered coats, and silk stockings, a great
number of officers prioritized their personal reputations far more than the military
prowess and effectiveness of the French army.

These priorities were seen to manifest themselves socially, a second problem pinpointed by military thinkers, both in officers' poor treatment of their troops as well as in the vicious gossiping and competition that thrived amongst noble officers and entire regiments. In *La Morale universelle*, D'Holbach decries "the revolting spectacle" of "commanders who, by their luxury, liberality, and sumptuous meals, starve the camp," allowing "a mob of do-nothing servants to swim in abundance while the exhausted soldier lacks even the most basic necessities."[17] In this vein, another writer remarks that:

> The troops have nothing left but courage. Their officers are without dedication to their jobs and without emulation....They neglect to take care of their troops, nor do they apply themselves to the daily tasks of military service. They have forgotten the significance of obedience and do not bother to make sure that they are obeyed. What is more, often they do not deserve to be obeyed because of the lack of care they have for their troops.[18]

Soldiers often fled the battlefield in fright, deserted from the army, or died of sickness, isolation, boredom, or unhappiness, all of which was only further exacerbated by the degrading treatment by their officers. As David A. Bell affirms, officers of the French army "saw soldiers in general as beneath contempt" and "usually failed to credit their social inferiors with the same values" of personal or family honor, glory, or merit.[19] An anonymous writer tells of the detrimental psychological effects of this social prejudice, relaying that, for soldiers, "the incertitude of their future, the contempt heaped upon them, plunge them into a despair that breaks them down in sickness or forces them to desert the army...such bad treatment destroys the little good will that they could have had. What service can we expect from such soldiers?"[20]

Addressing a similarly nefarious social trend related to the prejudices of *le monde*, war minister Marc-Pierre de Voyer de Paulmy, comte d'Argenson (1696–1764) slings disdainful aspersions toward those who perpetuated vicious competition between men of the same unit. He decries the misguided people:

> who are delighted when officers of their unit are officially reprimanded, people who, instead of taking care to hide such dishonor, take great pleasure in spreading word of it, not only within their regiment but also to the public. Is it not shameful that a unit composed of allegedly honest people desire to destroy one another?...Every day we see entire Regiments behave as if they were enemies. Where do such ideas come from? From whence does such a lack of *société* emerge?[21]

According to d'Argenson, officers and whole regiments were more concerned with destroying one another's reputations than with destroying enemy armies, a manifestation of the backward and self-sabotaging elements of French military culture at the time. The comte de Bombelles similarly observes that the soldiers from different provinces of France "live together with great difficulty," since "a Gascon has no reason to make himself feel fondness towards a Norman nor to give him the helping hand that he would need, not even just advice."[22] In both the officer and soldier corps, tensions between veterans, new recruits, and recent appointments from court ran high, the veterans proud and war-hardened and the latter groups naïve at best, and at worst exhibiting the effeminate worldly character and behaviors that certain military men despised.

What were the solutions to this failing culture of the "courtier-warrior" and the related military crisis? How could the army and the spaces of war that it inhabited purge these unwanted, deeply destructive cultural influences of *le monde*? For some, the solution was a function of social exclusivity within the officer corps: the bourgeois, the recently ennobled, and other affluent urbanite nobles (as opposed to the nobles of the sword with long ancestral histories of service) were to be systematically prevented from attaining leadership positions in the army via the creation of strict social criteria for entrance into the officer corps.[23] For others, the solution lay in creating a wholly separate and isolated military realm with its own doctrine and educational methodology, the latter focusing on mathematics and analytical skills rather than history, classics, and debate.[24] Still others had a different solution and a somewhat counter-intuitive one at that. Instead of suggesting social exclusion or military isolation as the path out of crisis, some thinkers turned back toward *la mondanité* and its female influences to find attitudes and corresponding behaviors that could ameliorate rather than exasperate the army's crisis in culture and combat effectiveness. Seeking to appropriate and redirect cultural ideas and practices that were already in place, these thinkers saw positive potential in ideologies and practices related to natural law, theories of sociability, and the moral and behavioral system of salon etiquette that had combined to develop a new kind of moral philosophy based upon society (*la société*) during the Enlightenment.

Imagining Enlightened Military Society

In his *Réflexions diverses* of 1665, aristocratic officer La Rochefoucauld dedicated his second meditation to society. "It would be useless to say how necessary society is to mankind," he declared, defining society as "the particular commerce that honest people (*honnêtes gens*) must have together."[25] He detailed the personal attributes that were constitutive of society—an open mind, kindness, politeness, humanity, confidence, and the ability to speak sincerely.[26] He also stipulated that each person in such a society should be free, and that although certain inequalities (in birth or personal qualities) between these persons may exist, these inequalities should not be palpable and certainly not demonstrated in any abusive manner. La Rochefoucauld's idea of society, one that military thinkers of the eighteenth century would emulate, implied a moral and social code amalgamating functional equality, polite behavior, and the importance of interior personal qualities. The background for such thinking about society was largely developed in the space and discourses of the seventeenth-century salon milieu in France. As Carolyn Lougee Chappell tells us, "the ideology of the salons rested on [the] substitution of behavior for birth. The quality perennially cited as the earmark of '*belles gens*' was '*esprit*': wit, urbanity, the ability to converse and participate in all the pleasures of society."[27] The salon was a place of cultural and social assimilation, in which it was thought that women served as the key catalyst to men's learning of proper comportment and becoming "honnête hommes." As seventeenth-century "feminist" scholar of manners Poulain de la Barre observed about men of his age, "if they wish to center in the '*Monde*' and play well their role in it, they are obliged to go to the school of ladies in order to learn there the politeness, affability, and all the exterior graces which today make up the essence of '*honnêtes gens*.'"[28]

The key to this cultural and social binding was thus learning and practicing the moral and behavioral tenets of sociability. Daniel Gordon has attributed five modes to sociability in seventeenth- and eighteenth-century France, three of which are particularly useful to us in understanding the military's adoption of this ideology. First, sociability took the form of socialization and education, initiating people into the cultural norms of a specific space. Second, it was seen as the love of exchange that held pleasure as its principle rather than utility, though in the case of military sociability, exchange was suggested to be both pleasurable and useful.[29] Third, sociability was seen to be constitutive of bonds among strangers in spaces of *mixité* that included men and women, nobles and bourgeois alike.[30] The language of sociability that included such words and concepts as society, humanity, reciprocity, mutualism (*mutualité*), politeness, honesty, and esprit, was as much a practical language as a philosophical one that was utilized and reflected upon throughout the Enlightenment.

Military thinkers of the eighteenth century integrated and expanded upon this vocabulary and socio-moralistic ideology in their ideas for reform. References to the salon and the influence of women were largely implicit in these ideas, but nevertheless quite discernable, especially for the noble or royal reader to whom these eighteenth-century military treatises were directed. First, reformers spoke of the primacy of polite behavior and "noble" morals, emphasizing that one's upright inner qualities were to be privileged over gracefulness and good looks. As Lamée avers, "the substance of an officer is in his *esprit*, his heart, and his sentiments. A pleasing countenance, charm, and graces are supplements to these other perfections. The first ones keep him from committing errors; the second ones make him commit new ones every day." He goes on to explain how open-mindedness, moral uprightness, and sociability link up with military professionalism to form the essential qualities of the ideal officer: "having an *esprit* that is flexible, sociable, and even a bit ingratiating; being polite and obliging without insipidness; loving virtue; savoring one's job (*faire son métier par goût*) without inhibition; the desire to instruct; these are the qualities that form an officer. Possessing these qualities is what we call having *l'esprit du métier*."[31] Having the *esprit du métier*—the spirit of the military profession—meant being an *honnête homme* and, as Lamée further argues, being a kind of *philosophe* in pursuit of truth. He insisted, "we could not form a good officer without forming an *honnête homme*. The lieutenant-colonels must then apply themselves to inspiring the officers they command with the love of truth, horror for even the least artifice, and a taste for honor and virtue."[32] Lamée's words actually prefigure the *Encyclopédie*'s definition of the *philosophe* as an "*honnête homme* who acts through reason, joining an *esprit* of reflection and justice with sociable qualities and mores."[33] The ideal French army, according to Lamée, was to be composed not of "courtier-warriors," but rather of *philosophes-militaires*.

The comte d'Argenson likewise upheld that embracing such philosophical, sociable ways of thinking and being could be highly productive in improving military service at large. "Be polite, generous, compassionate," offers d'Argenson, "the well-being of military service, the pleasure of society, and tranquility of the soul—all of this will be found in this happy way of thinking."[34] D'Argenson believed that a generous and compassionate politeness was vital to eliminating the counter-productive forces of gossip and competition within the army because it would help reflect and reveal the deeper importance of social unity among human beings. "For the betterment of military service," he argues, "it is of utmost importance to establish a perfect union within a unit. The foundation of this union is humanity." A central term of liberal Enlightenment thought, humanity not only signified a common human nature, but implied a moral dimension. As defined in the *Encyclopédie* as well as in many treatises of moral philosophy written during the eighteenth century, humanity was a sentiment of benevolence for all men that is born of compassionate sensibility (*sensibilité*); it was a feeling that not only made one share in the pain of others and wish to alleviate it, but spurred one to seek moral and social justice.[35] Sensibility, in addition to mothering humanity, gave birth to generosity and worked in service of merit, according to the chevalier de Jaucourt.[36] D'Argenson's vocabulary of military betterment unites politeness with moral philosophy, a tie that is further iterated through his usage of the term society.

"People must be reasonable, establish social bonds, and live together as friends and as comrades according to the tenets of good society (*la bonne société*)."[37] D'Argenson's argument merits closer inspection. His first exhortation is to "be reasonable," suggesting a voluntarism by which each man applies his faculty of reason to debunk his social presuppositions and prejudices in order to reveal commonalities between men that run deeper than their surface-level regional, cultural, and class differences. In the tradition of Samuel Pufendorf's theory of natural law developed in *De Jure Naturae et Gentium* (1672), a work translated into French by Barbeyrec and published in five editions between 1706 and 1734, d'Argenson upheld that rational choice would beget social bonds.[38] D'Argenson then specifies that these social bonds of friendship and camaraderie were to function according to the tenets of good society, a variation of the term society that, as historian Antoine Lilti explains, indicated "the ensemble of salons and those who frequent them."[39] To sum up d'Argenson's proposition: men of the French army are to apply their faculty of reason, thereby realizing their common humanity and natural mutual benevolence, which in turn would permit them to form social bonds that should function according to the tenets of salon etiquette. All men of d'Argenson's ideal army, from the common soldier to the highest ranked officer, are equal in their humanity as well as in their capacity and duty to form social bonds by rational choice and subsequently to treat one another with the polite manners and respect that one would in the salon milieu.

Lamée joined d'Argenson in his idea of transforming the army into a space of social egalitarianism based upon principles of moral philosophy and salon-style sociability, maintaining a similar vision and process of social bonding. "The military profession calls men forth from the farthest provinces of the Kingdom," Lamée explains, "the faces, mores, sentiments, and even the names of those with whom we are going to live are unfamiliar to us. This idea alone is surprising at first; it is too difficult to grasp. Our *esprit* brings this idea closer to us, familiarizes us with it, teaching us that humanity submits all men to the same laws and that men cannot remain strangers amongst themselves."[40] For Lamée, men of the army realize their absolute equality and natural sociability, rendering their bond through society an inexorable human inclination. Lamée reiterates the egalitarian principle behind his thesis, proclaiming: "man was created to think, to have ideas, to communicate ideas and to receive the impressions of the ideas that others create. *L'Esprit* is the link of society, and all men are brought together in society." Accordingly, and again intermingling discourses of natural law and sociability in Pufendorfian fashion:

> the power of personal interest and the love of pleasure make necessary what we call life's commerce (*le commerce de la vie*) between men…it is a mutual deference that unites men and that serves as a foundation for society. Man must be useful to man, and the

reciprocal communication of our ideas is as necessary to us in rendering our existence useful as the regular movement of the planets is indispensable to conserve the harmony of the world.[41]

A mutual deference and need, as well as the desire to be useful to one another, were to be found in the heart and mind of every man. What is more, as in a salon, reciprocal communication was a prime goal of coming together, as were the pleasures of companionship and the building of knowledge through discussion and collective reason. Such natural forces and practices of communication were useful for the betterment of military service and learning, said d'Argenson, since it was "through such society [that] one finds opportunities to discuss the military profession and, during these little conversations, discover things about it of which one was quite unaware."[42] The trust formed between military men who converse and interact according to the standards of salon etiquette would increase the ability to make internal decisions independent of higher military and administrative authority, thus creating a functional autonomy that would be useful in many situations. D'Argenson confirmed that "this society gives you confidence in one another, so that if ever some unhappy affair were to occur in a unit, no one outside of the unit would know what is happening therein and there would never be any need for a decision from outside, which is extremely useful for the well-being of the service."[43] In addition to this useful application of society, human equality, combined with utilitarian mutual deference, fostered subordination, an institution of central importance for the army as well as society at large. Lamée argued that when human equality is the basis upon which military society is founded, men feel more inclined to see the organizational and functional utility of subordination. Being aware of their status as essentially equal to all other men in the army and then witnessing those around them—subordinates, equals, and superiors—agreeing to fall into the hierarchy of subordination would depersonalize and thus mitigate the initially unpleasant feeling of submitting oneself to orders.[44]

The moral responsibility in the army was to become that of politeness, profound human bonds, and equality. In the minds of these military thinkers, establishing this kind of social unity would finally optimize the army's functioning, sweeping away the cultural crisis marked by semantic errors, intra-army competition, social abuses, and general lack of dedication to the service. The French army's combat effectiveness would reach its pinnacle and rehabilitate France's martial glory to the heights it had attained in its past. "It would be of infinite benefit," proclaimed d'Argenson, "if all of the King's troops could live together in perfect union. If we could just get to that state, we would be invincible."[45]

Military Enlightenment and the Moralization of War

The writings of d'Argenson and Lamée illustrate that for eighteenth-century military thinkers, men of the French army should adopt the ideologies and practices of an enlightened *honnête homme* or *philosophe-militaire*, exercising a type of moral philosophy based upon the ideals of salon-style sociability. These theorists pointed toward the necessity of reordering moral priorities and of each individual's taking a philosophical perspective in their regard for fellow army men. Recognizing both common human traits and needs, they argued, would lead men to discover their essential equality and then to join together in partnership, subordination, and polite comportment. The positing of man's proactive sociability not only reflected developments in natural law theory advanced by Pufendorf, and later d'Holbach, Diderot, and the Encyclopedists, it also reveals how the discourse on military reform and professionalization both conjoined and contributed to Enlightenment moral philosophy's championing of liberty, equality, fraternity, and politeness. The ideas that these military thinkers explored were part of a moral philosophy in which natural law and the rules of salon etiquette combined to give a particular account of human nature and society. D'Holbach's *Morale universelle* of 1776 represents this trend in that, as Daniel Gordon puts it, his "maxims on this subject [of "la politesse mondaine"] echoed the conventional precepts of salon etiquette, though he tried to extend their field of application beyond the scope of the salons and into the whole sphere of human interdependence."[46] Here it is important to note that the social egalitarianism that d'Holbach, d'Argenson, Lamée, and others championed was underscored by a sharp distinction between the legal institutions of the state apparatus and its ethical and convivial structures. These thinkers did not advocate political democracy while supporting a greater degree of social egalitarianism. Rather conciliatory in nature, "the purpose of this language was to define a sphere of practice that was based on the egalitarian premises of natural law, yet was compatible with the hierarchical legal foundation of the French regime."[47] True revolution was yet to come.[48]

This narrative of military thought and reform also offers a new perspective on cultural history and gender during the Enlightenment. Historians have long focused on antifeminist perspectives regarding French mores and national character that developed in the latter half of the eighteenth century. These scholars have shown how in the context of military and political decay that characterized Louis XV's and XVI's reigns, many adopted a highly misogynistic, "caustic Rousseauian view of history, held up the ideal of '*patrie*' over civilization, and started to think of impaired national virility as an urgent problem" that could only be remedied by annihilating female cultural influence and replacing it with a classical republicanism that was strictly male in character.[49] The archival sources analyzed in this article, however, show a contrary line of thinking by which military decay could be ameliorated by embracing and actualizing female cultural influences in the realms of philosophical and practical sociability. Though these military thinkers were not acting on the part of feminism, their practical openness toward what were considered to be feminine qualities and cultural influences complicates the misogynist interpretation of discourses on military reform and on cultural history and gender during the French Enlightenment, indicating the need for further study of this subject.

Also revealed in these ideas for renewing military culture are strong notions of personal and moral autonomy. Far different from Michel Foucault's historical account of military men functioning as "docile bodies" policed by a coercive, panoptic machine of absolute surveillance, these reformers viewed humans at all ranks of the army as possessing a far higher degree of agency and autonomy.[50] The soldiers and officers these thinkers envisioned were not automatons eating, sleeping, and shooting smoothbore muskets into enemy lines without a thought or an emotion. Rather, the ideal army imagined by these military thinkers was composed of autonomous, passionate, and compassionate self-policing men who spontaneously adhered to the theoretical and practical tenets of a morally enlightened realm. This idea relates directly to what J. B. Schneewind has called "the invention of autonomy" that came about during the eighteenth century and involved a "new outlook...centered on the belief that all normal individuals are equally able to live together in a morality of self-governance."[51]

There are obviously deep ironies though in the method of improving army cohesion and performance that we are considering. First, one wonders how the natural human sociability that was seen to be capable of uniting men from all walks of life in the French army would not have also united the men of the French army with those of enemy armies, eliminating the desire and/or capacity to wage war altogether. Since this peaceful utopian result was clearly not the goal of these reformers who in fact sought to bolster the French army's ability to wage victorious wars, we are faced with the second great irony. This theorized benevolence, fellowship, and politeness between men of the royal army was ultimately directed toward the goal of becoming a more efficient killing machine. "If all of the King's troops could live together in perfect union," pleaded d'Argenson, "if we could just get to that state, we would be invincible."

At a general cultural level, when we consider the increasing humanitarianism, cosmopolitanism, and anti-war sentiments of the Enlightenment era, we can see that the efforts of these reformers reflected a desire to revitalize the esteem given to war as a legitimate human practice and to soldiers as righteous men fighting for the preservation of *la patrie* and the greater good. War, as they portrayed it, could still give occasion to human dignity, and the army could be a sphere in which Enlightenment theories of society and its socio-moralistic tenets could be put into action. In a longer narrative of French military culture from the Renaissance to the Napoleonic age, the eighteenth century marks the period when war was divested of divine qualities and was understood to be a space of human agency and self-jurisdiction ripe for such experimentation. Left in the wake of a receding influence of God, Saint-Hilaire argued that the space of war offered man the opportunity to prove his dignity and agency through his conduct in armed conflict, which was seen to be an inevitable part of human life:

> I think that no one can reasonably contest that in this mortal life, empires form, maintain themselves, expand and are overthrown by the processes and the results of armed conflict. It seems that God has left mankind a sort of agency (*franc-arbitre*), that is to say the power to carry arms in dignity or indignity....In order to achieve the former, it is necessary to study, to observe oneself, [to have] continual dedication, precise obedience, courage, patience, valor, science, in a word, to possess the virtues of civil life.[52]

For the Enlightenment *philosophe* as for the enlightened *philosophe-militaire*, the "virtues of civil life" and society had become, as Du Marsais proclaimed it in the famous entry *Philosophe* of the *Encyclopédie*, divinity on earth.

Notes

1. Carl von Clausewitz, "Bekenntnisdenkschrift," *Schriften, Aufsätze, Studien, Briefe*, Werner Hahlweg, ed. (Göttingen: Vandenhoek and Ruprecht, 1966), 1:750. Translated by David A. Bell in *The First Total War: Napoleon's Europe and the Birth of War as We Know It* (New York: Houghton Mifflin, 2007), 241.
2. Bell, *The First Total War*, 35.
3. For a recent work on *le monde* and its corresponding cultural code of *mondanité*, see Antoine Lilti, *Le Monde des salons* (Paris: Fayard, 2005).
4. Bell, *The First Total War*, 24.
5. During the War of the Spanish Succession (1701–14), the French army experienced a series of increasingly bloody military defeats between 1704 and 1709 that ousted France from the

Low Countries, Italy, and Spain, and left the royal coffers and army on the verge of ruin. Save for a handful of battle victories, bungling and ineffectiveness also severely hampered French military performance in the War of Austrian Succession (1741–48). Later, during the French and Indian/Seven Years' War (1754–63), France endured tremendous losses in Europe and in the Americas. These wars would virtually annihilate France's naval power and eliminate her colonial presence in the Americas (all that remained were French Guyana, Saint-Domingue and Saint Pierre, and Miquelon).

6. See Louis Tuetey, *Les Officiers sous l'ancien régime: nobles et roturiers* (Paris: Plon-Norrit, 1908); André Corvisier, *L'Armée française: de la fin du XVIIème siècle au ministère de Choiseul: le soldat* (Paris: PUF, 1964); Émile-Guillaume Léonard, *L'Armée et ses problèmes au XVIIIème siècle* (Paris: Plon 1958). In addition to these older works, see more recent research by David D. Bien, "The Army in the French Enlightenment," *Past and Present*, 85 (1979): 68–98; Jay M. Smith, *The Culture of Merit: Nobility, Royal Service, and the Making of Absolute Monarchy in France 1600–1789* (Ann Arbor: Michigan University Press, 1996); and Rafe Blaufarb, *The French Army 1750–1820: Careers, Talent, Merit* (Manchester: Manchester University Press, 2002).

7. Service Historique de l'Armée de Terre (SHAT) series 1M, no. 1701, M. de Saint-Hilaire, *Traitté de la guerre où il est parlé des moyens de rédiger les troupes et y restablir l'ancienne et bonne discipline* [1712]. This and all other translations from early-modern French into English of archival SHAT documents were made by the author unless otherwise specified.

8. For a discussion of gender and French mores, see David Bell, *The Cult of the Nation in France* (Cambridge, MA: Harvard, 2001), ch. 5.

9. Translated in Bell, *The Cult of the Nation in France*, 149.

10. SHAT series 1M, no. 1703, M. de Lamée, *Essay sur l'art militaire dessein de l'ouvrage* [1742].

11. From the Paul Henri Thiry, baron d'Holbach, *La Morale universelle ou, Les devoirs de l'homme fondés sur sa nature* (Amsterdam: M.M. Rey, 1776), quoted in Bell, *The First Total War*, 69.

12. See Jay M. Smith, *The Culture of Merit*.

13. Smith, 27.

14. SHAT series 1M, no. 1702, Lagarrigue, "Mémoire" [1733–36].

15. SHAT series 1M, no. 1702, anonymous, untitled [1736].

16. SHAT series 1M, no. 1701, anonymous, *Mémoire pour l'infanterie en cas de guerre* [1720].

17. Holbach, *La Morale universelle*, part II, 127.

18. SHAT series 1M, no. 1702, anonymous, untitled [1736].

19. Bell, *The First Total War*, 36.

20. SHAT series 1M, no. 1703, anonymous, "Mémoire concernant les premières opérations à faire à la paix, tant dans le corps de la vielle infanterie que dans celuy des milices afin de pouvoir tirer de l'un et de l'autre des avantages considérables pour le service du Roy," also entitled "Mémoire contenant les moyens de porter le corps de l'infanterie françoise au plus haut point de perfection" [May 24, 1748].

21. SHAT series 1M, no. 1703, le comte d'Argenson, "Raisonnemens sur ce que le lieutenant Colonel d'un Regimen, commandant de corps, ou capitaine doit observer pour le bien du service, avec un détail de l'état de l'officier, de sa vie et de sa conduite" [~1743–50].

22. SHAT series 1M, no. 1708, Dossier Bombelles, "Mémoire contenant les moyens de remedier aux défauts qui se trouvent dans le corps de l'infanterie françoise et de le porter au plus haut poin de perfection" [1756].

23. Rafe Blaufarb summarizes the suggested changes to the military system in *The French Army 1750–1820: Careers, Talent, Merit*, chap 1.

24. On this process of cultural separation between the army and society at large, see David D. Bien, "The Army in the French Enlightenment," *Past and Present*, 85 (1979): 68–98; and

"Military Education in 18th Century France: Technical and Non-Technical Determinants," with commentary by John Shy, Thomas P. Hughes, and Gunther E. Rothenberg, in Monte D. Wright and Lawrence J. Paszek, eds., *Science, Technology, and Warfare: The Proceedings of the Third Military History Symposium United States Air Force Academy 8–9 May 1969* (Washington: Office of Air Force History, 1969): 51–84.

25. La Rochefoucauld, *Maximes et réflexions diverses* (Paris: Gallimard , 1976), 163. My translation.

26. La Rochefoucauld, 163–66.

27. Carolyn Lougee, *Le Paradis des femmes: Women, Salons, and Social Stratification in Seventeenth-Century France* (Princeton: Princeton University Press, 1976), 52.

28. Quoted in Lougee, 54.

29. In *The Republic of Letters: A Cultural History of the French Enlightenment* (Ithaca: Cornell University Press, 1995), Dena Goodman argues that utility did become a parameter in sociability of salons of the eighteenth century in which female heads of salons (or *salonnières*) promoted the agendas in human progress of the Enlightenment and the Republic of Letters (53–54).

30. Daniel Gordon, *Citizens Without Sovereignty: Equality and Sociability in French Thought, 1670–1789* (Princeton: Princeton University Press, 1994), 38.

31. SHAT series 1M, no. 1703, M. de Lamée, *Essay sur l'art militaire dessein de l'ouvrage* [1742].

32. SHAT series 1M, no. 1702, Lagarrigue, "Mémoire" [1733–36].

33. César Chéneaux Du Marsais, "Philosophe," *Encyclopédie ou dictionnaire raisonné des sciences, des arts et des métiers, par une Société de Gens de lettres* (Paris: 1751–72), my translation. For more on the term "honnête homme" and its connection to the mid-eighteenth-century definition of the "philosophe" in the *Encyclopédie*, see Daniel Brewer "Constructing Philosophers," in *Using the Encyclopédie: Ways of Knowing, Ways of Reading*, Daniel Brewer and Julie Candler Hayes, eds. (*SVEC* [Studies on Voltaire and the Eighteenth Century], 2002:05): 21–35.

34. SHAT series 1M, no. 1703, le comte d'Argenson, "Raisonnements…"

35. Denis Diderot, "*Humanité*," *Encyclopédie*. In particular, "*l'humanité*" causes men to seek the abolition of slavery, superstition, vice, and unhappiness.

36. Chevalier Louis de Jaucourt, "*Sensibilité*," *Encyclopédie*.

37. SHAT series 1M, no. 1703, le comte d'Argenson, "Raisonnemens…"

38. See Samuel, Baron von Pufendorf, *De Jure Naturae et Gentium* book 2, chapter 3 (Lund, Sweden: 1672); and Gordon, *Citizens Without Sovereignty*, 62–63, for a discussion of the work and its relation to other theories of natural law and sociability.

39. Lilti, *Le Monde des salons*, 86. My translation.

40. SHAT series 1M, no. 1703, M. de Lamée, "Essay sur l'art militaire dessein de l'ouvrage" [~1742].

41. *Ibid.*

42. SHAT series 1M, no. 1703, le comte d'Argenson, "Raisonnemens…."

43. *Ibid.*

44. *Ibid.* Interestingly, in Diderot and d'Alembert's *Encyclopédie*, the entry "*société*" lists subordination as "the link of society." In both Lamée's definition and this one, essential human equality was the precondition for subordination, which itself was a practical organizing principle rather than a reflection of a profound inequality between men. Keith Baker has connected this theory of natural equality and functional inequality to Claude Buffier's *Traité de la société civile* (1726), see "Enlightenment and the Institution of Society Notes for a Conceptual History," in Willem Melching and Wyger Velema, eds., *Main Trends in Cultural History* (Amsterdam: Rodopi, 1994), 95–120.

45. SHAT series 1M, no. 1703, le comte d'Argenson, "Raisonnemens…."

46. Gordon, *Citizens Without Sovereignty*, 68.

47. Gordon, 69.

48. Though a few reforms reflective of the socially egalitarian thinking examined in this chapter were implemented prior to 1789, the French Revolution was the catalyst of larger-scale changes in the armed forces as in the rest of society. For more on social egalitarianism, military psychology, and reform during the Revolutionary and Napoleonic eras, see Christy Pichichero, "Le Soldat Sensible: Military Psychology and Social Egalitarianism in the Enlightenment French Army," *French Historical Studies*, 31:4 (2008): 553–80.

49. Bell, *The Cult of the Nation in France*, 150 and ch. 5.

50. See Michel Foucault, *Discipline and Punish: The Birth of the Prison*, Robert Hurley, trans. (New York: Vintage, 1995), in particular *Part Three: Discipline*.

51. Quoted in Lynn Hunt, *Inventing Human Rights: A History* (New York: Norton, 2007), 28. The assumption of personal and moral autonomy was also fundamental to "inventing human rights," a process that Lynn Hunt traces through the eighteenth century into the American and French Revolutions.

52. SHAT series 1M, no. 1701, M. de Saint-Hilaire, *Traitté de la guerre où il est parlé des moyens de rédiger les troupes et y restablir l'ancienne et bonne discipline* [1712]. In *The First Total War*, David Bell traces the process by which war was no longer thought to be an inevitable part of human society.

3

"Unfortunate Necessities": Violence and Civilization in the Conquest of Algeria

Jennifer E. Sessions

War was central to the history of the modern French empire. The army and navy acted as key agents of colonial conquest and administration, while strategic positioning was a central motive for expansion overseas. Empire entered the metropolitan consciousness most materially during times of war. With the notable exception of the wars of decolonization, however, warfare and the particular characteristics of colonial war have remained strikingly marginal in investigations of French imperialism. The extraordinary violence of the drawn-out conquest of Algeria, where France fought continuously for nearly half a century after the invasion of 1830, never prompted the kind of sustained legal or philosophical argument that characterized the Enlightenment or contemporary British inquiries into colonial abuses.[1] Even Alexis de Tocqueville, the only nineteenth-century French thinker to draw serious scholarly attention for his considerations of empire, is universally reproached for his "failure" to confront rigorously the terrible conflict that followed the French invasion of Algiers in 1830.[2] The question of colonial violence likewise provoked little serious reflection among French military men who, like subsequent historians, focused on strategic and tactical matters, rather than on moral ones.[3]

The conquest of Algeria did nonetheless give rise to significant debate, which sheds important light on French understandings of colonial war. In the mid-nineteenth century, as in scholarship on Western colonial warfare, the key problem was whether colonies constituted a separate space of war governed by distinct codes of military conduct.[4] Scholars have tended to view this issue through the lens of race and to emphasize the ways that dehumanizing racial stereotypes authorized forms of violence, including torture, summary execution, collective punishment, and mass killing, considered unacceptable in "normal" conflicts.[5] In Algeria, representations of Arabs as nomadic, devious, bellicose, and fanatical helped to justify French atrocities, especially after Governor-General Thomas-Robert Bugeaud instituted a strategy of total war to "pacify" the new colony in the 1840s.[6] French officers, lawmakers, and observers maintained that Algerians' supposedly "savage" and "barbaric" nature exempted them from restrictions governing warfare between "civilized" nations. But the primary concern in French discourse, especially during Bugeaud's tenure as governor-general (1841–47), was not the victims of colonial warfare, but its potential ramifications for the conquerors. Bugeaud's strategic thinking, military responses to his tactics, and public controversy over the most notorious event of the period, the massacre of the Ouled Riah tribe in 1845, reveals a widespread conviction that colonial warfare's greatest danger lay in its potential to barbarize French soldiers and even France itself.

War and Civilization

Fears of military barbarization had particular resonance in mid-nineteenth century France, where attitudes toward war were deeply ambivalent. The intellectual and political leaders of Louis-Philippe's July Monarchy shared a positivist worldview that identified civilization with the triumph of reason in individual, social, and political life. They expected war to vanish as civility, trade, and the rule of law spread inexorably over the globe.[7] Although positivists accepted the necessity of defensive war, they believed that the progress of civilization would eliminate even this type of conflict. "If all governments and all peoples convert to contempt for force," wrote journalist Joseph Lingay, "none will become a conqueror, none will therefore need to defend itself. Not just conquest, but war itself will disappear from the face of the civilized world!"[8]

This understanding defined the "civilizing mission" that was so central to modern French colonial ideology. Positivists believed France to have the most advanced civilization of its day and thus a particular duty "to extend the glorious empire of reason."[9] By the mid-1840s, proponents of the Algerian conquest had begun to use the phrase *mission civilisatrice* to describe their self-imposed obligation to "rais[e] the beacon of civilization on barbarian shores." This meant, according to Lingay, "not France conquering Africa by arms," but "transporting to Africa her justice, her mores, her trade, all her attractions."[10] The spread of French civilization was assumed, by definition, to be a nonviolent process.

The new century did not entirely disdain the spirit of conquest, however. Nineteenth-century Frenchmen still associated national honor with military glory. To the question "Must we, in order to encourage the growth of industry and commerce, repress national pride and weaken the warrior spirit?" the left-leaning *Siècle* newspaper answered emphatically, no.[11] Contemporary French culture, from royal propaganda to popular prints, Romantic literature, and political philosophy, was suffused with paeans to Napoleon Bonaparte and the sublime glories of combat.[12] These celebrations of military might have shaped a second ideological justification for the conquest of Algeria, which was as militaristic as the civilizing mission was presumed to be peaceable. From this perspective, conquest was "an imperious glory that must be obeyed" and war one of "the most noble human things."[13]

The French War in Algeria

The war in Algeria challenged both pacifist and militarist notions of French civilization. By the end of the first decade of the conquest, French policymakers had decided upon the full occupation and colonization of the former Ottoman regency. The appointment of Bugeaud as governor-general in December 1840 marked a corresponding shift to a military strategy of "total war," which was designed to end the tenacious resistance of emir Abd-el-Kader. During the 1830s, the French had leveled accusations of barbarity primarily against the enemy. In the 1840s, the unconventional tactics and institution-alized atrocities of total war called into question their own civilization.

Bugeaud's strategy was predicated on the idea that essential differences between Algerian and European societies made war in North Africa fundamentally different from war in Europe. "In Europe," he explained repeatedly to the Chamber of Deputies,

> you take the enemy's capitals, you cut off the main roads and navigable rivers, and when you hold the country's guts, you make him capitulate. But in Africa there are no capitals, no towns, no villages, no farms. There is only one vulnerable interest, which is the agri-cultural interest spread over the whole surface of the country.... That is why this war is different from all others.[14]

Civilians and the country's economic resources were thus legitimate military targets. This reasoning was buttressed by the belief, widespread in the Armée d'Afrique, that Algerians themselves made no distinction between combatants and noncombatants, and that support from this militarized population constituted Abd-el-Kader's greatest strength. Arabs, Bugeaud wrote, were "without a doubt, the most bellicose [population] in the world. Among this people, all men are warriors from their earliest adolescence to their last days."[15] Combined with the guerrilla tactics adopted by Abd-el-Kader and his allies, these convictions underpinned the decision to make the destruction of civil-ian assets France's primary war aim.

Abandoning the earlier system of fortified outposts and expeditionary columns, Bugeaud sent mobile "flying" columns out to pursue Abd-el-Kader's forces, to punish rebellious tribes, and to terrorize the population into submission. The primary weapon in this campaign was the *razzia*, a scorched-earth raid designed to destroy the ene-my's economic resources and to feed French troops. The term itself was Arabic, but the Armée d'Afrique's *razzia* bore little resemblance to the relatively bloodless North African raiding tradition for which it was named.[16] A letter from Lieutenant-Colonel Lucien de Montagnac to his uncle vividly describes what Montagnac called a "perfect" *razzia*:

> Once the tribe's location is known, we each charge, dispersing in all directions. We reach the tents, whose inhabitants, awoken by the soldiers' approach, emerge pell-mell with their animals, wives, and children; all these people flee in all directions; gunshots ring out from all sides at these miserable, defenseless, surprised people; men, women, chil-dren are pursued, quickly surrounded and assembled by the soldiers who gather them up. The stampeding cattle, sheep, goats, and horses are soon rounded up.... Then we set fire to everything we can't carry away, while the beasts and people are taken to the convoy.[17]

French forces in Algeria had raided for supplies since the early 1830s,[18] but the *razzia* had a purely destructive purpose that contravened existing French conventions of war. Army regulations mandated respect for the persons and property of civilians in occu-pied countries, and destroying crops in the field was expressly forbidden by French mil-itary law, which made it a capital offense to destroy or pillage civilian property without written orders.[19] These standards led Bugeaud's predecessor, Count Valée, to express horror in 1839, when provincial commanders first began to carry out *razzia*-style raids. "I will never be disposed to approve," he wrote "the sorts of expeditions whose results can be explained thusly, 'Only four *Douars* [encampments]...were laid waste and burned, etc., etc.'":

> Such punishment of the innocent along with the guilty, these wars of savages, are not in our manners and can only corrupt the French who witness them. Vigorously fighting men under arms, who attack us or resist our operations, giving no quarter, taking no prisoners, killing men under arms, this I can understand and if it is necessary, the circumstances will excuse whatever may be cruel about this manner of making war. But Razzias, *pillaging*, fire, these are nothing but detestable.[20]

Bugeaud had no such qualms, and in 1840 he ordered commanders throughout the Armée d'Afrique to adopt the practice. Officers' reports, letters, and memoirs reflect the extent to which "cut," "burn," "devastate," "ravage" became the order of the day, as French columns cut swathes of destruction across the Algerian landscape. Officers soon coined the verb *razzier* to describe the routine activity of burning grain silos, trees, villages, and whatever crops and animals they could not carry away.[21] Regulations according the troops one-third to one-half of the booty captured and regularizing its distribution, although ostensibly intended "to attenuate its sad effects," actually exacerbated the *razzia*'s violence by giving soldiers further incentive to pillage.[22]

Other horrific practices proliferated too, all aimed at terrorizing Algerian civilians. Although there had been summary executions and collective punishments from the earliest days of the war, exemplary violence became standard procedure under Bugeaud, and French soldiers, in Montagnac's words, "[took] on a savagery that would make the hair stand up on the head of an honest bourgeois."[23] The heads of dead Algerian soldiers, for instance, were displayed on bayonet points, flagstaffs, saddles, or camp walls as a warning to local populations. When Abd-el-Kader's powerful *khalifa* Sidi Embarek was killed in 1844, his head was sent to General Lamorcière, who ordered it paraded along the road from Oran to Sidi Embarek's seat at Milianah, where it was exposed to the public for three days.[24] Cash rewards were distributed to soldiers for each pair of ears taken from the enemy, and one commander ordered that any man to bring in a living prisoner be beaten.[25] Officers' accounts rarely mention rape specifically, but scattered references to "insulted women" leave no doubt that sexual violence was part of the French repertory of terror.[26]

The impact of Bugeaud's war on the Algerian population was devastating, as intended. Although the lack of reliable pre-conquest population data makes it impossible to determine how many died as a result of combat, disease, and starvation during this period, it is clear that the demographic effects were staggering.[27] Caught between French predations and pressure to join Abd-el-Kader's *djihad*, Algerian tribes faced not only battlefield casualties and executions, but also economic devastation as the French seized cattle, burned fodder, and destroyed grain stores. By the mid-1840s, famine was widespread throughout central and western Algeria.[28] Bugeaud's system also began to have the desired strategic effect. Abd-el-Kader's exhausted allies began to desert him and to request *aman* (pardon, protection) from the French. As Montagnac noted, even far-flung tribes, "seeing the punishments we inflicted on the recalcitrant, found it prudent to throw themselves at [our] feet...for fear of meeting the same fate."[29] Critical blows to Abd-el-Kader's regular forces, including the capture of his *smalah* (mobile tent capital) in 1843 and the defeat of his Moroccan allies at the

Battle of Isly the following year, forced the emir onto the defensive, where he largely remained until surrendering in 1847.

For the Armée d'Afrique, Bugeaud's command offered relatively improved conditions and opportunities for advancement, along with serious psychological pressures and new moral questions. *Razzias* and better planning largely resolved supply problems that had plagued the army in the 1830s. Officers newly-arrived from metropolitan garrisons were still appalled by the Armée d'Afrique's living conditions and ragged appearance, but food, bedding, and uniforms were better than they had been. Bugeaud freely exercised his belief in the value of decoration and promotion for improving morale, which made him popular with the rank-and-file and ambitious young officers.[30]

To many military men, however, the Algerian war conflicted with understandings of honorable warfare. General de Castellane, conservative peer and commander at Perpignan, where regiments destined for Algeria trained, became Bugeaud's most outspoken parliamentary critic. His voluminous correspondence with former subordinates encapsulates critical military responses to the conquest. One young infantry captain, for example, found that little in his training had prepared him for a "totally exceptional" conflict that "follow[ed] none of the prescribed rules for large or for small wars." In his first four months, Captain Cler had "sought in vain an occasion to fight," as his column "made war only on herds, homes, crops, and an infinitesimal fraction of the population who, disarmed and driven by hunger and misery, prefer surrender to combat."[31] If junior officers' expectations of glorious battle were disappointed by the realities of anti-guerrilla warfare, senior officers feared its brutalizing effects on the rank-and-file and its impact on military discipline. "I have often had to shudder," the future General Canrobert lamented, "at the profound demoralization that [the *razzia*] casts into the heart of the soldier who slits throats, steals, rapes and fights for himself under the eyes of officers often powerless to stop him."[32] Although ordinary soldiers left few traces of their own thoughts on the war, officers did note increasing their "repugnance" for *razzias*, which could only be overcome by increasing their allocation of booty.[33] Rampant alcoholism, indiscipline, and suicide in the Armée d'Afrique hint at the ways that soldiers tried to escape from the horrors of the war.[34]

When civilian authorities raised questions about the disparities between Bugeaud's unconventional tactics and French conventions of war, however, the general claimed that military necessity outweighed ethical concerns. "One does not make war with philanthropic sentiments," he told the Chamber of Deputies. "If you want the ends, you must accept the means."[35] And policymakers in Paris were, for the most part, willing to accept them. Bugeaud held onto his post until 1847, despite ongoing conflicts with Minister of War, Marshal Nicolas Soult, and the parliament continued to provide the funds and manpower he requested. The Armée d'Afrique swelled from 63,000 to 118,000 men during his tenure, accounting at its height for one-third of the entire French army.[36]

Censoring Barbarism

Rather than attempt to reign in the army's violence, Soult sought to hide it from the French public. The government had long made it standard practice to print North African campaign bulletins in metropolitan newspapers, and Bugeaud himself considered such publicity essential in generating political support for the conquest.[37] The governor-general's insistence that published reports accurately portray the Algerian war, however, put him at odds with the minister, who believed that its extreme violence had to be dissimulated from metropolitan readers. A devoted servant of the July Monarchy's party of order, Soult was particularly sensitive about the "unfortunate impression" that Bugeaud's unconventional warfare might make on public opinion.[38] "I do not think," he wrote to Bugeaud, "that we can expose [our strategy] to the examination and censure of a public unfamiliar with the nature of the very particular war we must make in Algeria and the entirely exceptional character of its inhabitants."[39] Employees in the Ministry of War, and often the minister himself, thus carefully redacted official reports to eliminate potentially damning information prior to publication.

Drafts prepared for the press are indicative of the lines Soult believed the public would be unwilling to see French soldiers cross. Besides intelligence data and political commentary, ministerial officials consistently deleted three types of information: French soldiers' suffering from harsh conditions or enemy reprisals; summary executions or the mutilation of Algerian bodies by French troops; and pillaging or destroying villages and crops, especially by burning. On the one hand, widespread allegations that ambitious officers in Algeria sought advancement at the expense of their men, neglecting soldiers' well-being and taking foolhardy risks, seem to have prompted Soult's work to conceal the hardships of the Armée d'Afrique. Accounts of French troops' hunger, cold, and fatigue, like Algerian cruelty to French prisoners, also conflicted with the optimistic picture of military success painted in government propaganda. Summary execution and mutilation, on the other hand, belied contemporaries' faith in a French civilization defined by the rule of law. Such acts violated French military jurisprudence, which required that civilians in occupied territories be tried by court martial and defined mutilating or killing disarmed and wounded enemy soldiers as murder.[40] Finally, pillaging and "devastation," while technically legalized by Bugeaud's written orders, nonetheless ran counter to the "civilized" values of economic prosperity and respect for private property. The particular effort to eliminate references to destruction by burning reflected associations in the contemporary imagination between arson and the uncivilized forces of brigandage, social vengeance, and political revolt, as well as the assumption that fire's indiscriminate, uncontrollable, and exclusively destructive nature made it an especially barbarous weapon.[41]

Despite Soult's constant vigilance, keeping the details of French violence from metropolitan audiences proved a Sisyphean task. Private correspondence traveled faster than official dispatches, and Bugeaud himself published reports before forwarding them to Paris. Metropolitan awareness of controversial events was often sparked by articles in the Algerian press or letters sent to legislators by friends and acquaintances within the Armée d'Afrique. The government attempted to stem the tide of unwanted news by tightening restrictions on newspapers published in Algiers, clamping down on soldiers' communications with journalists, and prosecuting metropolitan newspapers for libel.[42] These efforts were stymied, however, by Bugeaud's manipulation of the press, by the ministry's inability to control newspapers' correspondents, and by metropolitan juries' refusal to convict editors for criticizing the war.

The impossibility of concealing the extremes of colonial warfare became dramatically evident in the summer of 1845, when the French public learned that an expeditionary column had killed nearly 800 members of the Ouled (tribe) Riah in the Dahra mountains west of Algiers. What made the massacre of the Ouled Riah so horrifying to French readers was the manner of their death: asphyxiation by the smoke from bonfires set at the entrances of the cavern where they had sought refuge from French forces. The so-called *enfumade* (smoking) combined the features of summary execution, civilian punishment, and fire that Soult had rightly identified as especially shocking to metropolitan sensibilities, and it quickly became a symbol of the extremes of the North African war as a whole. The ensuing public outcry gave rise to the conquest period's most intensive discussion of the war and crystallized the terms of debate about its methods. As we will see, opponents and defenders of the Armée d'Afrique turned to the categories of civilization and barbarism to explain colonial warfare, its effects on French soldiers and, ultimately, its impact French society as a whole.

The Massacre of the Ouled Riah

The Ouled Riah, like most of the tribes in the Dahra mountains and Chélif River valley, had formally submitted to the French in the winter of 1842–43. When they joined a mahdist revolt led by the young *marabout* Bou Maza in the spring of 1845, Bugeaud sent Lieutenant-Colonel Aimable Pélissier to punish and disarm them.[43] As Pélissier approached in mid-June, conducting massive *razzias* against neighboring tribes along the way, over 800 Ouled Riah retreated into the Dahra foothills and took refuge with their herds in a complex of caverns known as Ghar el Frachich.[44] On June 18, Pélissier surrounded the cave and, when its occupants refused to surrender, ordered that wood piled at each of its entrances be set ablaze. The fire burned through the next two nights, with only a short pause on the morning of the nineteenth for fruitless negotiations. Fed continuously by French soldiers, the flames reached over 60 meters into the sky, sending thick columns of smoke towering over the cave's main entrance. When Pélissier finally ordered the inferno doused just before dawn on June 20, the detail sent to reconnoiter the cavern found fewer than 100 survivors.

The siege of Ghar el Frachich was neither the first nor the last use of *enfumade* in the Dahra region, which was riddled with caves that the local populations had long used as a sanctuary in times of emergency. A year earlier, Colonel Eugène Cavaignac had used the same method against some 50 members of the Sbéah tribe, who surrendered after one smoke-filled night. The incident was largely ignored in France, but it set a precedent that Bugeaud recommended that Pélissier follow if the Ouled Riah fled to their caves.[45] In August 1845, Colonel de Saint-Arnaud began an *enfumade* of the Mchaïas fraction of the Sbéah tribe, and, when they refused to surrender, he sealed off the entrances to the cavern, leaving 500 people trapped inside.[46] Colonel Canrobert executed a similar *emmurement* in the northern Dahra.[47]

Bugeaud responded with delight to Pélissier's action, which he expected to have significant strategic impact.[48] Soult's first reaction was to ensure that it remained secret. Pélissier's original report is not in the archives, but an excerpt from a letter to Saint-Arnaud, which Bugeaud forwarded to Paris, has been preserved and bears the unambiguous signs of the minister's reaction: Scrawled across the portion of the text describing the night of June 19 is the single word "BRUTE!!!" "TO HIDE," reads a notation across the top of the document.[49] Soult "refrained from inserting in the newspapers details of the rigors exercised by Colonel Pélissier against the Ouled-Riah," and provided the *Moniteur universel* with a heavily redacted report from which all reference to the Ouled Riah had been removed.[50]

The *enfumade* would not be silenced, however. Within days, Algiers' *Akhbar* newspaper had published a lengthy account of the massacre, which was reprinted by the Parisian *Journal des Débats* on July 10. These initial reports were followed over the next weeks by a series of eyewitness accounts that provided graphic details of the scene, which the official press had carefully avoided. Emphasizing the indescribable horror of the scene inside the cave, these dramatic narratives dwelled on the tangle of human and animal bodies covering the floor. Especially haunting were tableaux of family devotion, which they described in great detail: infants clasped to their mothers' breasts for comfort; small children enveloped in their mothers' clothing to protect them from the smoke; a man frozen in the act of protecting his wife and child from a smoke-maddened ox.[51] Such emotional descriptions were given even greater immediacy in a widely-circulated lithograph by noted illustrator Tony Johannot, whose imagined vision of the chaotic struggle inside the cavern featured poignant vignettes of desperate women clutching pleading children (figure 3.1).

Serious public outrage began as soon as the *Journal des Débats* confirmed rumors already circulating about the *enfumade*. On July 11, the Prince de Moskowa read the *Débats* report to the Chamber of Peers and demanded that the government deny or disavow "an act of inexplicable, indescribable cruelty."[52] Soult's disingenuous response—a claim of ignorance and agreement that the government would "disapprove" and even "deplore" such a thing, if true—helped to feed an outcry in the political press, which denounced a "lamentable," "terrible," "appalling," "horrible," and "deplorable" act.[53] The press fury was driven in part by political opportunism, and both left and right seized on the Dahra affair as a weapon for attacking the July Monarchy. Across the political spectrum, however, defenders and critics of the *enfumade* adopted the language of civilization in a debate that ultimately turned on the question of colonial war's exceptionalism and its barbarization of French soldiers.

Bugeaud's apology, drawn up in response to Soult's request for explanations comprehensible to the metropolitan public, provided the blueprint for Pélissier's few parliamentary and journalistic defenders. The massacre of the Ouled Riah, like other unconventional tactics, was a "cruel but inevitable" necessity in war against barbarians.[54] It was essential, Bugeaud claimed, for Pélissier to demonstrate that the Dahra's caverns offered no safe haven from French power. But a frontal assault on the cave would have cost too many French lives, and a pressing rendezvous with another column left no time for a lengthy blockade. When the Ouled Riah refused to negotiate an end to the siege, Pélissier had turned to the *enfumade* as a last resort. In the long run, the governor-general argued, the massacre would actually shorten the war and spare lives on both sides.

Articles from Algerian newspapers loyal to Bugeaud, inserted at Soult's urging in ministerial journals at home, reframed this argument as a question of relative barbarism and civilization. *La France algérienne* led the ferocious response to what it called a "paroxism of philanthropy for barbarians, and of anti-French cruelty towards our officers." It was "an act of the most complete unreason," the newspaper's editors claimed, "to ask that the army behave in a distant country as with civilized enemies."

To require that our soldiers, who have long witnessed acts of the most odious barbarism committed by the Arabs against their brothers in arms, observe their ferocity with calm and without being moved by a sentiment of legitimate vengeance, our regiments would have to be made up not of men but of gods.[55]

"Johannot pinx.

Colin sc.

Figure 3.1   Tony Johannot, "Les Grottes du Dahra," in P. Christian [pseud., Christian Pitois], *L'Afrique française, l'Empire de Maroc et les Déserts de Sahara: conquêtes, victoires et découvertes des Français, depuis la prise d'Alger jusqu'à nos jours.* Paris, 1846.

However awful the *enfumade*, "it loses much of its atrocity when contrasted with African morals and the inescapable necessities of the warfare that we are making with the natives."[56]

Such justifications of Pélissier's actions had two critical features, which also characterized broader defenses of Bugeaud's unconventional war. First, they made no effort to deny the barbarity of the *enfumade* or other unconventional tactics. Instead, they argued that war with an uncivilized enemy was by nature barbarous. "In Europe," Soult stated baldly, "such an action would be awful, detestable; in Africa, it is war itself."[57] Second, they assumed that if standards of conventional warfare had to be put aside in Algeria, the military code of honor did not. The enemy's barbarism, manifested by the Ouled Riah in their torture of French prisoners two years earlier, constituted an insult that required a response. Honorable French men could not be expected, Soult continued, to allow such offenses to pass. "You will never impose on a soldier conscious of his dignity," he declared, "enough abnegation that he will accept an insult without returning it."

From these claims of military and honorable obligation, it was only a short step to shifting responsibility for colonial violence onto its victims. Bugeaud articulated this conclusion in a private letter to Pélissier: "It is a cruel extremity to which these madmen have reduced you, but they have only their own blindness to blame."[58] If actions like the *razzia* or *enfumade* were, in Tocqueville's words, "unfortunate necessities which cannot be escaped by any people at war with Arabs,"[59] their authors could be portrayed as victims of Algerian barbarism themselves, sacrificing their own civilized status for the nation. It was this logic that underpinned the claims of Bugeaud and his supporters that the Armée d'Afrique deserved praise rather than blame for its patriotic selflessness.[60] Only men unable to comprehend the exceptional nature of the Algerian war, they maintained, could fail to recognize the necessity of harsh measures like the *enfumade* and the self-abnegation of the Armée d'Afrique in carrying them out.

Where defenders of Pélissier and of Bugeaud's tactics argued from a fundamental distinction between conventional and colonial war, critics of the *enfumade* departed from the assumption that there was no difference among wars waged by French armies. Characterizing the *enfumade* of the Ouled Riah as an act of barbarism unworthy of a civilized nation, critics on the left and the right argued that such actions jeopardized not only French military interests but also the legitimacy of the civilizing mission that justified French domination in Algeria. Moreover, intimate ties between the army and the nation, between the colony and the metropole, made extremes of colonial violence a direct threat to the integrity of French civilization as a whole.

Critics rejected the argument of military necessity on the grounds that Pélissier's column had faced neither imminent danger nor urgent time pressure. Pointing to its deliberate execution and the lack of confrontation with the Ouled Riah, they suggested that there had been time for further negotiation or a blockade. The absence of extenuating circumstances made the *enfumade* not an act of war, but a crime, "a deplorable act of premeditated murder." Repeated references to Pélissier and his men as *brûleurs*, *chauffeurs*, *brigands*, or *incendiaires* reinforced characterizations of the *enfumade* as a criminal action, rather than a military one. Such horrific means, furthermore, would not speed pacification, but rather prolong the war by sowing hatreds that would feed Algerian resistance for generations to come.[61]

As for Bugeaud's and Soult's appeals to honor, "such a doctrine is infamous," the liberal *Courrier français* retorted. To republicans and other left-leaning critics, the personal honor of Pélissier's men was subordinate to a national honor based on what they called the "principle of modern war": "never to do the enemy any harm that can be spared without deviating from the assigned goal of operations and without compromising success."[62] Honorable warfare, from this perspective, lay in humanizing combat as much as possible, minimizing its destructive force, and treating the defenseless—women, children, and disarmed enemy combatants—generously. "The noble and holy France of the nineteenth century," the *Courrier français* continued, "combats heroically on the battlefield, but does not massacre her conquered enemies, wages war with the sword, but not with faggots."[63] From this perspective, the barbarism of the *enfumade* was incompatible with French claims to epitomize modern civilization and undermined the legitimacy of the French civilizing mission in Algeria. Critical journalists compared Pélissier's behavior with the worst excesses of an earlier age, from the Inquisition and the Spanish conquest of the New World to the *chauffeurs* that terrorized the roads of eighteenth-century France, and with the irrational violence of uncivilized peoples. Thus, the *enfumade* had been "an act worthy of the Spanish adventurers of the sixteenth century conquering the New World," "a monstrous act...such as one may read of in the annals of savage tribes."[64]

Condemnations of the *enfumade* did not question the basic categories of civilization and barbarism, however, nor challenge the ideology of the civilizing mission. Indeed, the most vehement critics attacked colonial violence in its name. *Le National* spelled out the potential consequences of acts like Pélissier's: "If our mores, our ideas, our civilization don't demonstrate themselves to be superior, what right have we to conquer their country?"[65]

> We pretend to be better, that is more enlightened, more humane, more moral than the Arabs: that is our only title to the conquest. And yet! what do we show the Arabs? Men who borrow [Arab] mores, who burn crops, who destroy herds, who cut off heads, who suffocate by the hundreds women and children piled in a cave; who, instead of giving examples of humanity, have invented atrocious tortures for themselves.[66]

Le National's query raised the potential of colonial warfare to endanger not only French domination in Algeria, but also France's place in the ranks of civilized nations. Fears that actions like the *enfumade* would "earn our nation the contempt of civilized Europe" and "banish us from humanity"[67] drove a pervasive concern among critics about France's status in the eyes of Algerians and, especially, of Europe. This was evident from the very beginning of the Dahra affair, when Count de Montalembert demanded that the Chamber of Peers consider the effect that the *enfumade* would produce in England and elsewhere abroad.[68] French newspapers carefully tracked foreign coverage of the massacre and, when the European press joined the chorus of denunciation, even critics of the *enfumade* responded with charges of hypocrisy. English attacks, in particular, provoked lengthy reviews of atrocities in the British Empire, from the conquest of Ireland to the American Revolution, Wellington's retreat from Spain, the defeat of the Maori in New Zealand, and taxation in India.[69] By comparison with British imperial atrocities, French wartime excesses appeared "much more excusable."[70]

Yet even these self-exculpatory comparisons could not disguise the fundamental dangers of colonial violence to the positivist conception of civilization in which French self-understandings were so invested in the nineteenth century. The very possibility of barbarization suggested that the spread of civilized values was not, as positivists maintained, an ineluctable process. "Civilization would be a vain word," wrote *La Presse*, "if, when polite [*policées*] nations go to war with barbaric nations, the former find no other recourse than to regress into barbarism themselves."[71] The potential of colonial violence to reverse the progress of civilization constituted a threat that could be contained to neither the colonial theater nor the realm of the purely metaphysical. Numerous commentators warned that the "demoralization" of French troops in colonial warfare could impact metropolitan society directly when the demobilized soldier returned home, bringing with him "the deplorable habits he gets from the half-savage kind of war he is forced to make" in North Africa.[72] Algeria's relative proximity to French shores, just 40 hours distant by steamship, fed fears that the government could turn the Armée d'Afrique on French citizens at home, as indeed it did in the insurrections of February and June 1848. "They accustom our soldiers to barbarism in order to then launch them against the people of Paris," cautioned both republicans and legitimists.[73] It was these assumptions about the porousness of the boundaries between metropole and colony that gave such urgency to contemporary French fears about the barbarizing effects of the Algerian conquest.

Conclusion

What, in the end, came of the Dahra affair? Practically speaking, it had relatively little impact on Bugeaud's strategy or on the individuals responsible for the worst atrocities. Despite Soult's initial suggestion that public pressure might force Pélissier's recall,[74] the army took no disciplinary action against any of the officers known to have ordered *enfumades*. As a group, Pélissier, Cavaignac, and Saint-Arnaud went on to brilliant careers. Cavaignac served briefly as governor-general of Algeria and minister of war, before being named head of the Executive Power under the Second Republic. During the Second Empire, Pélissier and Saint-Arnaud took control of Algerian policy as governor-general and minister of war, respectively. Both were eventually promoted to marshal of France, the army's highest rank. If anything, the outcry over the *enfumade* of the Ouled Riah simply strengthened authorities' determination to hide the Armée d'Afrique's excesses, as Saint-Arnaud's *emmurement* of the Sbéah tribe in mid-August 1845 demonstrates. The Ministry of War falsified reports of the operation to characterize the siege as "less painful than that of the Ouled-Riah" and to claim that the Sbéah had capitulated. When the press aired rumors about the *emmurement* a few months later, Soult moved rapidly to quash the rumors.[75]

The massacre of the Ouled Riah was also quickly displaced from the headlines by allegations of disciplinary torture within the Armée d'Afrique. In the same week that the *enfumade* became public, the press reported on illegal corporal punishments being used in Algeria. Detailed descriptions of practices like the *silo* (imprisonment in open pits), *barre* (attachment of the feet to a raised bar), *crapaudine* (tying of crossed hands and feet behind the back), and *clou* (suspension of a prisoner tied in the *crapaudine*) initially added to outrage about the barbarity of the *enfumade*. By the end of July, however, the press had largely turned its energy to this "revolting abuse of power" over French soldiers.[76] Commentators found these punishments even more horrifying than the *razzia* or *enfumade* because their victims were not "barbarian"

Algerians, but fellow citizens carrying out their legal, patriotic duty. Disciplinary torture troubled French observers even more than the massacre of the Ouled Riah because it collapsed the last boundaries between French civilization and colonial barbarism.

As we have seen, it was this threat that weighed most heavily in French concerns about the conduct of the Algerian war. Historians of the Algerian conquest have interpreted attacks on Bugeaud's tactics as evidence of the Armée d'Afrique's "alienation" from metropolitan values.[77] What the Dahra affair shows, however, is that anxieties about barbarization in the 1840s stemmed as much from the perceived links between the Armée d'Afrique and metropolitan society as from the increasing distance between them. The army's behavior in Algeria was deemed particularly terrible because the boundaries between colony and metropole were understood to be fluid, or even nonexistent. The exceptional nature of colonial warfare was invoked to rationalize French excesses, but fears of barbarization reflected the conviction that metropole and colony constituted a single, intimately connected space.

Notes

1. Anthony Pagden, *Lords of All the World: Ideologies of Empire in Spain, Britain and France, c. 1500–c. 1800* (New Haven: Yale University Press, 1995), esp. 94–102; Jennifer Pitts, *A Turn to Empire: The Rise of Imperial Liberalism in Britain and France* (Princeton: Princeton University Press, 2005), ch. 3 and ch. 5, esp. 150–60.

2. Melvin Richter, "Tocqueville on Algeria," *The Review of Politics*, 25:3 (1963): 363; Cheryl Welch, "Colonial Violence and the Rhetoric of Evasion: Tocqueville on Algeria," *Political Theory*, 31:2 (2003): 236, 255, 256; Roger Boesch, "The Dark Side of Tocqueville: On War and Empire," *The Review of Politics* 67:4 (2005): 738–41, 746; Pitts, 239.

3. Jean Delmas, "L'Organisation militaire en France: les ministères, l'armée, 1815–1870," in *Histoire militaire de la France*, Jean Delmas, ed., vol. 2, *De 1715 à 1871* (Paris: Presses Universitaires de France, 1992), 455–56; Jean Gottmann, "Bugeaud, Galliéni, Lyautey: The Development of French Colonial Warfare," in *Makers of Modern Strategy: Military Thought from Machiavelli to Hitler*, Edward Mead Earle, ed. (Princeton: Princeton University Press, 1943), 234–59; Douglas Porch, "Bugeaud, Galliéni, Lyautey: The Development of French Colonial Warfare," in *Makers of Modern Strategy from Machiavelli to the Nuclear Age*, Peter Paret, ed. (Princeton: Princeton University Press, 1986), 376–407.

4. See most recently Raphaëlle Branche, *La Torture et l'armee pendant la guerre d'Algérie (1954–1962)* (Paris: Gallimard, 2001); Caroline Elkins, *Imperial Reckoning: The Untold Story of Britain's Gulag in Kenya* (New York: Henry Holt, 2005); Isabel Hull, *Absolute Destruction: Military Culture and the Practices of War in Imperial Germany* (Ithaca: Cornell University Press, 2005); Paul Kramer, *The Blood of Government: Race, Empire, the United States and the Philippines* (Chapel Hill, NC: University of North Carolina Press, 2006).

5. Olivier Le Cour Grandmaison, *Coloniser, exterminer: sur la guerre et l'état colonial* (Paris: Fayard, 2005); Elkins, 46–50; Kramer, ch. 2.

6. On these stereotypes, see Patricia Lorcin, *Imperial Identities: Stereotyping, Prejudice and Race in Colonial Algeria* (New York: I. B. Taurus, 1995).

7. Lucien Febvre, "Civilisation: Évolution d'un mot et d'un groupe d'idées," in *Civilisation, le mot et l'idée* (1930), Jean-Marie Tremblay, ed., http://classiques.uqac.ca/classiques/febvre_lucien/civilisation/civilisation.html (accessed 8 Dec 2008); Alice Conklin, *A Mission to Civilize: The Republican Idea of Empire in France and West Africa, 1895–1930* (Stanford: Stanford University Press, 1997), ch. 1.

8. *La France en Afrique* (Paris: Musée des Familles, 1846), 9.
9. *Histoire générale de la civilisation en France* (1839), quoted in Febvre, 57.
10. Lingay, 5, 12, 18.
11. *Le Siècle*, 20 Mar 1845.
12. Michael Marrinan, *Painting Politics for Louis-Philippe: Art and Ideology in Orleanist France, 1830–1848* (New Haven: Yale University Press, 1988), esp. part IV; Bernard Ménager, *Les Napoléon du peuple* (Paris: Aubier, 1988); Nancy Rosenblum, "Romantic Militarism, *Journal of the History of Ideas*, 43:2 (1982): 258–63.
13. Eugène Lerminier, "De la conservation d'Alger," *Revue des deux mondes*, 1 Jun 1836: 605, 612.
14. "Discours du maréchal Bugeaud à la Chambre des Députés, le 24 janvier 1845, à l'occasion de la discussion de l'adresse au Roi," in Thomas-Robert Bugeaud, *Par l'épée et par la charrue: écrits et discours de Bugeaud*, Paul Azan, ed. (Paris: Presses Universitaires de France, 1948), 193.
15. "L'Algérie. Des moyens de conserver et d'utiliser cette conquête," [1842], in Bugeaud, 268.
16. Douglas Porch, "French Colonial Forces on the Saharan Rim," in *The Military Between Cultures: Soldiers at the Interface*, James Bradford, ed. (College Station, TX: Texas A&M University Press, 1997), 166.
17. Letter to Bernard de Montagnac, 19 Dec 1841–2 Feb 1842, in Lucien-François de Montagnac, *Lettres d'un soldat. Algérie, 1837–1845* ([1885]; Vernon: Éditions Christian Destremau, 1998), 105.
18. On early raids, Service historique de la Défense (Vincennes), henceforth SHD 1H 22, doss. 1, General Voirol to Minister of War Soult, 3 and 11 Oct 1833.
19. "Ordonance du 3 mai 1832 sur le service des armées en campagne" and "Extrait du code des délits et des peines pour les troupes de la République du 21 brumaire an 5 (11 novembre 1796)" in Louis Durat-Lasalle, *Droit de législation des armées de terre et de mer* [...] 10 vols. (Paris: n.p., 1842–1857), 6:285, 9:21–24.
20. SHD 1H 62, doss. 3, General Valée to General Guingret, 19 May 1839.
21. Colonel de Saint-Arnaud to Leroy de Saint-Arnaud, 31 Oct 1842, in Arnaud-Jacques Leroy de Saint-Arnaud, *Lettres du maréchal de Saint-Arnaud, 1832–1854*, 2nd ed., vol. 2 (Paris: Michel Lévy frères, 1858), 436.
22. SHD 1H 76, doss. 1, Governor-General Bugeaud to General Baraguay d'Hilliers, 12 May 1841; "Arrêté qui détermine les règles suivant lesquelles doit s'opérer la répartition des prises faites sur l'ennemi," 26 Apr 1841, *Recueil des actes du gouvernement de l'Algérie, 1830–1854* (Algiers: Imprimerie du Gouvernement, 1856), 160. Quote, SHD 1H 83, doss. 3, Bugeaud, Circular, 18 Jun 1843.
23. Letter to Bernard de Montagnac, 31 Mar 1842, in Montagnac, 118.
24. Smaïl Aouli et al., *Abd el-Kader* (Paris: Fayard, 1994), 338–39.
25. Louis Blanc, *History of Ten Years, 1830–1840* (London: Chapman and Hall, 1845), 2: 482; Letter to Elizé de Montagnac, 15 Mar 1843, in Montagnac, 152–153; No. 132, Colonel Tartas, 15 Nov 1843, in *Campagnes d'Afrique, 1835–1848. Lettres adressées au Maréchal de Castellane* [...] (Paris: Plon, 1898), 332; Stephen d'Estry, *Histoire d'Alger, de son territoire et de ses habitants* [...], 2nd ed. (Tours: Ad. Mame, [1843]), 346; Archives Nationales (Paris), Papiers Chanzy, 270 AP 1^A (doss. 2), "Journal, Algérie. Octobre 1845–Février 1846," entry for 18 Nov 1845.
26. I.e. No. 234, Colonel de Mirbeck, 24 May 1847, in *Campagnes d'Afrique*, 513.
27. Kamel Kateb, *Européens, "indigènes" et juifs en Algérie (1830–1962)* (Paris: Éditions de l'Institut national d'études démographiques, 2001) 40–68.
28. SHD 1H 100, doss. 1, Soult to Bugeaud, 6 Jan 1845.
29. Letter to Bernard de Montagnac, 8 Mar 1842, in Montagnac, 208.
30. Antony Thrall Sullivan, *Thomas-Robert Bugeaud: France and Algeria, 1784–1849: Politics, Power, and the Good Society* (Hamden, CT: Archon Books, 1983), 89–91.

31. No. 105, Captain Cler, 1 Jul 1842, in *Campagnes d'Afrique*, 274–77.

32. No. 173, Chef de bataillon Canrobert, 18 Jul 1845, in *Campagnes d'Afrique*, 413.

33. SHD 1H 105, "Rapport sur l'ensemble des opérations de la colonne de Médéah pendant les mois d'Avril, Mai, Juin et Juillet 1845 chez les Ouled Naïl, au Dira et au Jurjura, Monsieur le Général Marey, commandant," [15 Jul 1845].

34. Charles-André Julien, *Histoire de l'Algérie contemporaine*, vol. 1, *La Conquête et les débuts de la colonisation (1827–1871)* (Paris: Presses Universitaires de France, 1964), 288–89, 296–97.

35. "Discours à la Chambre [...] 15 Jan 1840," in Bugeaud, 67–68.

36. Paddy Griffith, *Military Thought in the French Army, 1815–1851* (Manchester: Manchester University Press, 1989), 41. Sixty-seven of one hundred French infantry regiments saw active duty in Algeria between 1830 and 1854. Porch, "Bugeaud, Galliéni, Lyautey," 376.

37. "Circulaire no. 42 du Gouverneur-Général à MM. les Chefs des divers services pour les inviter à transmettre au Secrétarait-Général du Gouvernement, tous les faits intéressants qui parviennent à leur connaissance," (9 Mar–9 Apr 1844), in A. Franque, *Lois de l'Algérie, année 1844* (Paris: Dubos frères et Marast, [1844]), 24.

38. SHD 1H 83, doss. 3, Soult to Bugeaud, 1 Jun 1842.

39. SHD 1H 104, Soult to Bugeaud, 12 Jul 1845.

40. Laws of 13 and 21 brumaire Year 5, in André Claude Sébastien Du Mesgnil, *Dictionnaire de la justice militaire* (Paris: J. Dumaine, 1847), 40.

41. Jean-Claude Caron, *Les Feux de la discorde: conflits et incendies dans la France du XIXe siècle* (Paris: Hachette, 2006).

42. SHD 1H 101, doss. 1, Soult to General Lamorcière (interim governor-general), 16 Mar 1845; SHD 1H 85, doss. 1, Bugeaud, "Ordre général," 26 Aug 1842; B.C., "Tortures de l'Algérie," *Journal de la Société de la morale chrétienne* (Sept 1845): 44–45.

43. On Bou Maza, see Charles Richard, *Étude sur l'insurrection du Dahra (1845–1846)* (Algiers: Imprimerie de Besancenez, 1846), and Julia Clancy-Smith, *Rebel and Saint: Muslim Notables, Populist Protest, Colonial Encounters (Algeria and Tunisia, 1800–1904)* (Berkeley: University of California Press, 1994), 94–96.

44. The following is based on "Rapport du Colonel Pélissier au Maréchal Bugeaud, Gouverneur Général de l'Algérie," 22 Jun 1845, and "Lettre du colonel Pélissier au colonel de Saint-Arnaud," 20 Jun 1845, in Raoul Busquet, *L'Affaire des grottes du Dahra (19–20 Juin 1845) d'après les documents originaux* (Algiers: Adolphe Jourdan, 1908), 145–157. Details are also in the eyewitness account of a Spanish officer, first printed in the *Heraldo* (Madrid) and widely reprinted in France, i.e., *La Presse* 3367 (17 Jul 1845).

45. "Rapport du Colonel Pélissier," 147.

46. Colonel de Saint-Arnaud, letter to his brother, 15 Aug 1845, in Saint-Arnaud, 37. Early twentieth-century oral histories report that local inhabitants later rescued a few survivors. E.-F. Gautier, "Une enquête aux Grottes du Dahra en 1913," in *L'Algérie et la métropole* (Paris: Payot, 1920), 23–26.

47. Francois-Certain de Canrobert, *Le Maréchal Canrobert, souvenirs d'un siècle*, Germain Bapst, ed. (Paris: E. Plon, Nourrit, 1898), 422.

48. Centre des Archives d'Outre-Mer (Aix-en-Provence), henceforth CAOM, 2EE 11, Bugeaud to Pélissier, 23 Jun 1845.

49. SHD 1H 104, [Pélissier to Saint-Arnaud], 20 Jun 1845 (copy). Busquet reproduces the entire letter.

50. SHD 1H 104, Soult to Bugeaud, 5 Jul 1845.

51. "Massacre at Dahra," *La Réforme*, c. 12 Jul 1845, in review of "The Herald of Peace," *The Metropolitan Magazine* (London) 43 (May-Aug 1845), 523; anon to *L'Algérie*, in *La Presse* 3377, 27 Jul 1845, *La Quotidienne*, 28 Jul 1845; anon. to the *Heraldo*, *La Presse* 3367, 17 Jul 1845; officer of 36e de ligne to *Journal de Saint-Etienne*, in *La Presse* 3372, 22 Jul 1845.

52. *Moniteur universel* 193, 12 Jul 1845: 2117.

53. *La Quotidienne* 193, 12 Jul 1845; *Journal des Débats,* 11 Jul 1845; *La Presse* 3362, 12 Jul 1845; *Le Siècle* 190, 12 Jul 1845; *Le National,* 11 Jul 1845.

54. *Moniteur algérien* 695, 15 Jul 1845. See also SHD 1H 105, Bugeaud to Soult, 14 Jul 1845.

55. *La France algérienne,* 26 Jul 1845, in *Moniteur universel* 205, 24 Jul 1845: 2245.

56. *Journal des Débats* in *Times* (London), 19 Jul 1845.

57. "Chambre des Pairs. Séance du mercredi 16 juillet," *Moniteur universel* 198, 17 Jul 1845: 2165.

58. CAOM 2EE 11, Bugeaud to Pélissier, 23 Jun 1845.

59. "Essay on Algeria" (1841), in Alexis de Tocqueville, *Writings on Empire and Slavery,* Jennifer Pitts, ed. and trans. (Baltimore: Johns Hopkins University Press, 2001), 70.

60. *Moniteur algérien* 695, 15 Jul 1845.

61. I.e., *Le Courrier français,* 16 Jul 1845; *Le Siècle* 199, 22 Jul 1845; *La Presse* 3362, 12 Jul 1845. Quote, Prince de la Moskowa, "Chambre des Pairs, séance du vendredi 11 juillet," *Moniteur universel* 193, 12 Jul 1845: 2117.

62. "L'Armée anglaise et l'armée française," *Le Siècle,* 10 Aug 1845.

63. *Courrier français,* c. 12 Jul 1845, in "French Atrocities in Algeria," *Times* (London), 14 Jul 1845.

64. *Ibid.*

65. 17 Jul 1845.

66. *Le National,* 22 Jul 1845.

67. *Le National,* 12 Jul 1845.

68. "Chambre des Pairs. Séance du vendredi 11 juillet," *Moniteur universel* 193, 12 Jul 1845: 2117.

69. I.e., *Le National,* 18 Jul 1845; "De l'humanité des armées anglaises," *Le Siècle* 201, 24 Jul 1845.

70. "Humanité anglaise," *La Quotidienne* 203, 22 Jul 1845.

71. "De l'état actuel de la guerre d'Afrique et des modifications à y apporter," no. 3379, 29 Jul 1845.

72. "Appendice aux récits d'Alger," *La Quotidienne* 194, 13 Jul 1845.

73. Francisque Bouvet, *Revue de l'Ain,* cited in B.C.: 45.

74. SHD 105, Soult to Bugeaud, 12 Jul 1845.

75. SHD 105, Soult, "Note pour Bureau des Opérations militaires," 25 Aug 1845, and "Extrait du *National* du jeudi, 20 Novembre 1845." Bugeaud to Soult, 15 Aug 1845, and Bugeaud to Soult, 19 Aug 1845, in *Moniteur universel* 247, 4 Sept 1845: 2387.

76. *La Gazette des Tribunaux,* 13 Jul 1845.

77. Sullivan, 121.

4

Spaces of War: Rural France, Fears of Famine, and the Great War

Martha Hanna

The rural population of World War I France has not been well-served by literary accounts of the war. Henri Barbusse castigated the avaricious peasantry for charging the common soldier extortionate prices for even the most ordinary commodities.[1] Ernest Pérochon, author of *Les Gardiennes*, recognized that peasant women bore a heavy burden throughout the war, but he nonetheless insisted that it was their moral duty to work themselves to the bone in order to maintain the land that their husbands, brothers, and sons had left in their stewardship. This, at least, was the judgment of his young soldier, Constant, who left the farm before the war to find work on the railways: his parents, sister, and sister-in-law were, he insisted, to "work so that the soldiers have everything they need; you must work to the very exhaustion of your strength, to the death if need be."[2] This image of the sometimes angry, often alienated, and rarely indulgent front-line soldier is challenged, however, by extensive evidence from soldiers' letters, the postal control records, and other archival sources, all of which speak of peasant soldiers' reliance on, concern for, and empathy with the women and children, parents, and in-laws they had left at home.

French peasants who left their villages in August 1914 to wage war never fully abandoned their identification with the land or their interest in its productivity. Their letters are filled with concerns about the weather at home, inquiries about crop yields, and calculations about profits and losses. When Paul Pireaud had to forsake his farm in the commune of Nanteuil-de-Bourzac (located in a corner of northwest Dordogne), he corresponded regularly with his wife Marie, who kept him apprised of her efforts—and those of her parents and in-laws—to till the soil and harvest its bounty.[3] Accustomed to thinking about themselves and their land as self-sufficient, men like Pireaud, who numbered more than 3.5 million by the end of the war, took comfort in their families' reassurances that the land remained productive, relished the food packages that offered ample evidence of their continued agricultural success, and until 1916 expressed only minor concerns about the land and labors of those who worked it. In the winter of 1916–17, however, anxieties about agricultural production and fears of famine unsettled the peasant soldiers' long-established assumptions about food, farm-life, and family survival. This article will, therefore, address the theme of spaces of war by focusing on two related senses of space: first, the tangible space of rural society under duress in a time of apparently endless war; and second, the mental space of peasant soldiers' imagination and how in early 1917 their visions of, and anxieties about, home, family, and the future infused their optimism in advance of the Nivelle Offensive, contributed to their disillusionment in the aftermath of its failure, and prompted their demands for an improved leave roster that would allow them—quite literally—to bridge the spaces of home-front and military front in order to defend their families simultaneously from foreign aggression and the imagined prospect of famine.

Food, and its ready availability, mattered enormously to all combatant powers in the First World War. Indeed, those nations that failed to master the simultaneous challenge of feeding their military and civilian populations invariably found themselves on the losing side of the war. The inability of the tsarist state to provide Russia's urban workers with adequate sustenance at affordable prices contributed directly to the February revolution of 1917; the further inability of the provisional government to improve upon the tsarist record of incompetence and insufficiency did much to doom it, too, to political oblivion by October 1917. In Germany, food shortages were not as destabilizing in 1917 as those suffered in Russia; within a year, however, they were serious enough to undermine popular respect for the military high command, contributing (in Belinda Davis's judgment) to the emergence of revolutionary sentiment in 1918. In Austria-Hungary, urban hunger in 1917 and 1918 provoked strikes, exacerbated working-class resentments, and diminished regard for the empire itself. Even in Britain, where problems of food distribution were never debilitating and where, in fact, standards of living for the working poor improved during the first two years of the war, reduced food supplies caused in 1917 by the German submarine campaign and inequities in distribution roiled the working classes and exacerbated preexisting class tensions. Working-class Tommies, aware that their families were more likely than most to suffer the effects of high prices, erratic supplies, and endless queues for food, were by early 1918 disillusioned with authorities at home and increasingly demoralized.[4]

In France, where the peasantry comprised 40 percent of the population in 1914 and would account for 44 percent of all military casualties, the situation was different yet again. France depended more than either Britain or Germany on the ability of its agricultural population to feed the nation in times of peace and to fill the ranks of its army in time of war. In the years immediately preceding the war, France had needed to import only the most exotic foodstuffs; everything else—eggs, wine, butter, and cheese, grains and potatoes, meat and poultry—was produced within the hexagon.[5] And yet, as the experience of the Great War would make evident, the peasantry could maintain only with great difficulty, and considerable anguish, the nation's agricultural self-sufficiency and its military might. As more and more men were drawn into the maw of military mobilization, the land became less and less productive, and the capacity of the nation to feed itself—an important symbol of French identity, for the peasantry above all—was sorely tested. Although the first wartime harvest, reaped by the families of men recently mobilized, was reassuringly abundant, it would not be equaled thereafter. The wheat harvests of 1915 and 1916 were less than 80 percent that of 1914 and the proportionate decline in other crops even greater.[6] Near-crisis conditions emerged, however, only in the winter of 1916–17, when the cumulative effects of mobilization, crop and cattle requisitioning, and fertilizer shortages combined with bitter cold weather to cast doubt upon a reassuring truth that had held firm for the first two years of the war: that France could, by dint of serious effort and considerable sacrifice, feed its soldiers, guarantee the well-being of its women and children, and maintain its identity as a land of self-sufficiency and rich regional variety.

Food had assumed an important place in the calculations, enterprise, and affections of French families since August 1914. This was true for the urban middle-class and peasant households alike: everywhere the preparation and mailing of food packages became a national pastime. Indeed, so many packages made their way to the front that the state tried, unsuccessfully, to limit the practice by imposing high postal rates for packages, by banning the shipment of certain commodities, and by actively discouraging families intent on sending parcels to the front. All to no avail. More than 200,000 parcels made their way forward every day.[7] Most were filled with homemade delicacies, meat and fruit, and forbidden bottles of alcohol. Bourgeois soldiers recounted in their letters home how grateful they, and the men under their command, were to receive packages stocked with such necessities as tobacco and unfamiliar specialty items available for purchase only in the finest Parisian épiceries: one unsophisticated peasant soldier from Lorraine expressed something close to awe when invited to share a parcel from Paris that included canned pineapple.[8]

However interesting it might have been to savor unfamiliar delicacies nestled in Parisian packages, nothing was quite as appetizing to the peasant soldier as the familiar foods of home. Rabbits (complete with a package of lard for cooking), chickens, ham, even eggs; cherries and apples, cauliflowers and artichokes; rillettes, pâté, and brioche, chocolate and honey made their way to the front, reminders of the rich bounty of rural France and the affection of those who remained at home. Fernand Maret's parents worked a farm in the Mayenne, and from the time of his mobilization in 1915 his mother and aunts worked diligently to prepare and ship weekly food packages. Like many other soldiers, Maret both welcomed their attentions and worried about the effort exerted and expense entailed to send him provisions so regularly.[9] Indeed, he sometimes begged his womenfolk to send him packages less frequently: in January 1916, having received three packages from his parents, aunts, and grandmother, he was so overladen with the fruits of their generosity he could not carry everything in his kitbag.[10] Yet there were other times when he made special requests for fondly remembered treats. He and his comrades-in-arms especially appreciated the brandy that Maret's father made on the farm and the rum that often found its way into family packages: there was, he averred, "nothing like it for giving us a bit of courage."[11] In anticipation of Mardi Gras in 1916 he longed for crêpes rolled in sugar that would remind him of home and the feast day he and his parents would have celebrated together were it not for the war. Ten days later the crêpes arrived, delicious, well-preserved, and delightful memories of home.[12]

If many a food parcel acquired an almost Proustian potency, not every parcel was this inviting. A Parisian office clerk, in letters of acknowledgment more frank than most, thanked his mother for her food packages but lamented the damage done by rats as they gnawed away at parcels awaiting distribution to frontline troops.[13] Others wrote of meat gone off and fresh fruit destroyed in transit. Spoilage was, under the circumstances, to be expected: cherries stained Paul Pireaud's socks, apples went bad; recently slaughtered chickens and rabbits—not to mention unrefrigerated pork and lamb—were a rather risky proposition; and raw eggs rarely arrived intact.[14] Whatever could be salvaged, however, was always relished and usually shared. Germain Cuzacq reassured his wife, who prepared his packages from their farmhouse in the distant southwest department of Landes, that once he had scraped away the rotten outer surface of some recently arrived and no longer perfectly fresh meat and heated what remained there was no unpleasant taste to the meat at all.[15] Unlike raw meat and uncooked eggs, homemade brandy and liqueurs, when carefully wrapped in slices of cured ham, disguised as medicine or otherwise protected, often arrived at the front none the worse for wear.[16]

In the grimmest days of the war, when the army was hard-pressed to provide adequate food, these packages became both appreciated tokens of family affection and much needed sustenance in times of dire distress. For men serving at Verdun, parcels from home were often the only provisions they could count on. Cuzacq observed in mid-March 1916 that he and his companions were almost "at the end of their strength. I really hope that we can get out of this filthy hole soon; we're not suffering as much from the rotten weather but we are still badly fed and have no rest."[17] To prove his point he itemized the soldier's meager daily diet: "we have two quarts of soup, a quart of coffee, a quart of wine, 700 grams of bread, 150–200 grams of meat." With biting irony, he concluded, "with all that one can be strong, withstand fatigue, look healthy, and have excellent morale."[18] Conditions were not much better in Pireaud's artillery battery. "It is noon," he wrote in mid-May, "and we have not eaten our soup I have been snacking on the preserves that my mother sent me it is likely that we will not be able to heat our soup and we will have to eat it cold you can see what our situation is like and why I ask all of you for packages." The next day he repeated his plea: "I beg of you as well as my parents to continue to send me parcels because it has been several days now when we can't eat our soup until noon or one in the afternoon and if we have something to snack on we can use it there were three days when we couldn't eat at all because they couldn't heat up the soup."[19]

Conditions such as these made food packages all the more essential. The Cuzacq family was far from affluent: sharecroppers, they raised some livestock, planted oats and barley, and herded sheep. Given their straitened economic circumstances, it was no small thing for them to send packages to the front week in, week out. Yet when Germain Cuzacq was at Verdun, his wife Anna, his mother, sisters, and mother-in-law applied themselves to the task unstintingly. Between March 23 and April 26 (when his company was relieved), he received eight food parcels filled variously with ham, sausage, and uncooked (and rather gamy) lamb; brandy, jam, chocolate, and sugar.[20] At much the same time, and in comparable circumstances, Pireaud relished the provisions his wife, mother, and sister-in-law sent regularly to Verdun. Nor was he the only man in his battery to benefit from such familial attentions. In late April he recounted how the gun crew had received two packages a day for four days in a row, and everyone had "lived like princes." Still in position in early June—his unit would serve at Verdun without relief from early April through mid-June—and under heavy bombardment as the Germans advanced on Fort Vaux, he was delighted to receive two packages within two days: "I thank you my dearest this package in the situation that I find myself in is worth more to me than all the others and will do me more good."[21]

Material proof that the men who fought were remembered and loved, packages from home gave frontline troops both the capacity to fight and a reason for doing so.

The contents of the parcels were, quite clearly, essential at times to their physical survival. Cuzacq observed: "I have received the packages of food which are indispensable to me. Without that I would not be able to hold out any longer and it's the same for my comrades."[22] But parcels did more than compensate for erratic and inefficient army provisioning, and the pleasure a frontline soldier derived from them was not merely one of material satisfaction. Soldiers perceived in each parcel a token of familial affection and an expression of civilian empathy. As Stéphane Audoin-Rouzeau has argued, "the food they contained was…a breath of pre-war life, a taste or scent from home, a morsel of family life left behind."[23] For men miserably situated and in direct confrontation with death, the emotive significance of the parcel mattered as much—or more—than its contents. Cuzacq said, quite simply, that the parcel he received on Palm Sunday "comforted him." Pireaud acknowledged that those Marie and his mother prepared meant the world to him: "I thank you with all my heart my poor, beloved wife it is so pleasant to see our lot improved by those we love especially when we find ourself in need." As Fernand Maret remarked, in birthday greetings sent to his mother, however hard his conditions were at the front, he counted himself fortunate in comparison with "those who did not have a mother and father to send them good things and comforting words." Each letter and every parcel reminded him of why he was fighting, and for whom he was willing to die.[24]

Through the end of 1916, food parcels provided rural soldiers with incontrovertible, reassuring proof that the land they had left behind and the people who worked it in their name continued to thrive. But the following winter, marked by record cold temperatures and uncommonly persistent snowfall across much of the land, called into question the capacity of civilians to persevere in their rural enterprise, gave rise to the possibility that soldiers and civilians alike would suffer from scarcity, and summoned up the specter of famine. On the one hand, harsh winter conditions, which lasted well into April, made the food packages that soldiers had come to rely upon even more essential; on the other, these same conditions impeded peasant families in their efforts to till the frozen soil, plant crops, and secure the nation's food supply for the coming year. The peasantry's ability to feed themselves and, by extension, their men at the front was for the first time since mobilization uncertain.

By late 1916 the war and its insatiable appetite for men, animals, and materiel had progressively eroded the ability of peasant families to extract from the land the grains, vegetables, and meat and dairy products that constituted the foundations of their customary diet. Military mobilization had deprived villages all over France of farmers and rural artisans alike; the requisitioning of cattle had decimated the nation's dairy stock, eliminated sources of natural fertilizer, and made it ever more difficult to plough the land; and the appropriation of nitrates for armaments production had seriously reduced the availability of chemical fertilizers.[25] As a result, land that had once been cultivated fell into disuse, and land that remained under the plough was noticeably less productive. In the village of Juignac in the Charente, the harvest of 1916 had been less abundant than that of 1914 or 1915, and "a certain amount of the land usually under cultivation had been left fallow."[26] This expedient, understandable under the circumstances, nonetheless created a vicious cycle: as more and more land fell out of use, fodder crops diminished, reducing the food supply of draft animals that had not yet been requisitioned. And as animals essential to the proper maintenance of the family farm were slaughtered, sold or undernourished, women found that working the land that remained under cultivation became more arduous than they could bear.

In these circumstances, abandoning the land entirely or leaving greater portions of it fallow was many a woman's last resort: "it hurts me to see all of our land laying fallow" one woman wrote to her husband, but high prices for seed and inadequate labor left her little choice.[27] Prospects for the future were no more encouraging: "what would the harvest of 1917 be like," one farmer wondered, when so much of the arable land was left untended? "Little by little," he lamented, "everything is laid to waste."[28]

Outraged that more and more of the nation's arable land was now overrun by weeds, many urban commentators were quick to blame peasant women for the nation's looming food crisis: indolence and self-indulgence were, some claimed, now the besetting vices of rural women.[29] Peasant soldiers were much more reluctant to blame their wives or parents for failing to keep the land under full production. In the north, where manpower shortages caused by the disruptions of war and the displacement of local populations were most aggravated, men at the front urged their wives to let lapse leases on land they could no longer cultivate.[30] Elsewhere, soldiers suggested that it would be better for wives, parents, and young children to leave fields fallow than to overexert themselves to little real good: "As for obliging my wife and my young son to work even harder I don't want it. They have already worked to the limit and beyond. We will leave all of our land fallow."[31] Maret was of a similar opinion, urging his father to do only what was physically possible: why exhaust himself to plant a crop or bring in a harvest that would only provide food for shirkers and, in the final analysis, extend the war?[32] The most disaffected *poilus* even urged their loved ones to abandon their agricultural efforts entirely: peasant productivity once considered a patriotic virtue was, in their caustic judgment, now a form of betrayal that only extended the war indefinitely. Most, however, preferred a more moderate path. Loath to see their land uncultivated, they wanted to return home—for good if possible, for a week or two if not—to lend wives and parents a much-needed hand. Without the sturdy contributions of able-bodied men the next harvest would be, they feared, dangerously deficient. As one soldier put it: "If the war does not end this year, I wonder what there will be to eat next winter."[33]

The unseasonably cold winter of 1916–17 aggravated this already alarming situation. The family correspondence of frontline troops and the postal control records of early 1917 are filled with tales of soldiers waging war against the elements, of civilians beset by food shortages, and of uniform fears of what the future would bring. On the Somme, where both Maret's infantry regiment and Pireaud's artillery battery passed a wretched winter, snow, which began to fall in mid-January remained on the ground for weeks on end. Everyone grumbled about the cold and the inconveniences that accompanied it. Food was difficult to prepare, wine froze in its canteens, and morale plummeted. "The men are suffering from the cold and the snow," the postal censor reported; "they complain about it and sometimes show themselves to be a bit discouraged by the conditions, which as one of them writes makes them depressed."[34] Men in Pireaud's regiment did what they could to survive the wintry conditions. On the evening of January 17, when ten inches of snow fell, they tossed snowballs at one another and hoped that the snow would warm things up a bit. A week later, however, whatever moderating influence the snow might have temporarily had was gone: it was twelve degrees below zero (or about 10°F) and everything—bread, water, wine—was frozen solid.[35] Under such conditions, packages from home were as essential to the army's well-being as ever. One soldier, anticipating his return to the frontlines, asked his wife to send him "a few more parcels…it won't be too much because you know we are dying of hunger because it's almost impossible to feed us."[36]

Families obliged, with considerable generosity if they could. Maret's parents sent him a package every week throughout the winter: when the temperature lingered well below freezing, butter, chicken, and pork chops could travel considerable distances without risk of spoilage.[37] Indeed, he was so well provided for by his parents and two aunts who sent pastries, pâté, and flasks of rum, he had to demur: "it is not worth your bother to send me so many parcels; a package of butter once a week and another package every two weeks will more than suffice."[38] For at the same time that Maret was benefiting from his family's generosity, he was hearing—from his parents and other sources—unsettling stories of bread shortages, sugar rationing, and civilians forced by new regulations to eat only stale bread made of inferior flour.[39] Indeed, as the postal censorship records reveal, from January through April, rumors of imminent bread rationing, tales of relentless requisitioning (and its deleterious effects), and complaints about high prices and inadequate supplies became the persistent lament of one and all. Urban communities were certainly hard-hit, as the scholarship of the Great War has clearly documented,[40] but so, too, were villages and hamlets, farm wives and children.

In rural communities closest to the front, and particularly in the territory newly liberated from German occupation by the strategic withdrawal of enemy troops to the Hindenburg line, scarcity was the most aggravated and disruptions to everyday life the most pronounced. In villages from the Nord, Picardy, and Lorraine, families lamented that the essentials of everyday life were either unavailable or exorbitantly expensive. In the Artois and in the hinterland of the Somme, vegetable crops had frozen in the ground and wheat harvested in the previous summer had been left to rot, unthreshed, for lack of coal, equipment, and manpower.[41] Military requisitioning only compounded the problems wrought by bad weather. When the army demanded animals and feed for horses, farmers worried that they would soon be unable to work the land at all. One woman complained: "the state came to requisition our oats. If our horses don't have oats any more they will not be able to work. We haven't finished seeding yet and since they took everything from us we won't be able to continue."[42] The requisitioning of cattle also radically reduced the nation's milk supply. A farm woman in the Meuse, affluent enough to afford a maid or elderly enough to require one, recounted how she would send her servant to stand in line with 250 others at the local dairy farm: "Children and the sick were served first, then those with large families, and then everyone else if there was anything left. Often one could stand in line for an hour and return home empty-handed."[43] When women and children went to bed having had nothing for supper but a chunk of bread and a glass of water, hunger loomed large indeed.[44]

Hardship, scarcity, and fears for the future also afflicted agricultural regions far from the frontlines. In the southeast, rural communes in the department of the Loire suffered grain shortages from February 1917 onward, with many communities exhausting their supplies entirely by April.[45] In the wine regions of Beaujolais and Bordeaux, where all land was given over to vines, the situation was no better. Three successive years of bad weather had left small-scale proprietors in the Beaujolais in a sorry state; and in the vineyards of the Gironde, fears that the food supply would not last until the next harvest—creating the dreaded *soudure* that had proved so politically disruptive in past centuries—emerged by mid-year.[46] Even in those regions of southwest France where cash crops were the exception and peasants preferred to divide their land among crops, vines, and pasture, the new year brought bleak prospects. Soldiers returning from leave in the Gironde and Charente exchanged worrying stories about grain shortages dire enough to necessitate rationing. Indifferent to the plight of city folk, "who did damned near nothing," these peasants-under-arms advised their families to hide whatever surplus they might have or face the threat of famine. As for the people in the cities, they could "go and work the land that was lying fallow."[47] In the Dordogne, where self-sufficiency had been secure through the summer of 1916, some essential commodities were hard to come by in early 1917. Sugar (without which the summer fruit harvest could not be preserved) was scarce and bread difficult to procure in some areas. By April flour supplies were running low in several communes, forcing some bakers to close their shops at least one day a week.[48] Postal censors in Bordeaux concluded that "in cities and the countryside alike the difficulties are the same: even if one has money it is only with great difficulty that one can acquire the basic necessities of life." Villagers, they observed, were not starving, but they were anything but happy.[49]

The extended winter also delayed the planting of new crops. By mid-March and well into April, when work in the fields should have been fully underway, inclement weather continued to disrupt farm families' plans: one farmer in the north of France observed that he had been able to plant only one-third of his crop before winter set in and feared that for lack of adequate manpower he would be unable to plant the spring wheat crop at all. As late as April 12, 1917 (only four days before the spring offensive began) farmers worried that the weather would continue to disrupt their planting schedule, leaving open the possibility that vegetable and grain crops alike would be dangerously deficient.[50] Some wrote ominously of the continuation of the war through the winter of 1917–18, the prospect of which made them genuinely apprehensive: the censors noted that "famine is predicted for next winter, bad weather and a lack of manpower have prevented the sowing of spring crops and the lack of fodder has made animal husbandry difficult."[51] More and more, the censors observed, correspondents predicted that "soon there will be famine."[52] And the men at the front who were most unnerved by this grim prognosis were peasants.[53]

Anxieties about hunger and the declining well-being of farm and family prompted rural soldiers to place their confidence in the much-touted, by no means secret upcoming spring offensive to be unleashed in mid-April by British and French forces from the heights of Vimy to the chalky plains of Champagne. Only a decisive, victorious offensive that would bring a quick end to the war could save those they loved from famine. And thus there emerged amid the worry and uncertainty of early 1917 a much-repeated mantra: the spring offensive would succeed where previous campaigns had failed. In late March and early April, as civilians in the regions closest to the front observed the massing of British and French troops and materiel and concluded that something important was in the offing, French soldiers from the Aisne eastward to Lorraine revealed in their daily correspondence "an admirable enthusiasm" for the "formidable offensive" that awaited them. They seemed, in fact, almost oblivious to the danger that awaited them.[54] Not every soldier was so persuaded, of course, and as John Horne has demonstrated, the most hardened warriors were also among the most skeptical.[55] These reservations notwithstanding, the optimism was real enough and a refreshing change from the morose mood of previous months. That frontline troops, survivors of Verdun, the Somme, and the worst winter on record, could now contemplate a major offensive with genuine optimism was as noteworthy as it was unexpected. Perhaps the men who placed their confidence in Robert Nivelle's strategic genius did so because they had been convinced that their new commander-in-chief had indeed found a way to win the war; but for peasant soldiers this "admirable enthusiasm" was surely animated as well by a need to believe in imminent victory. A victorious battle would mean, in the simplest calculus, an end to the war, an end to military service, and a speedy return home. Speedy enough, in fact, to reclaim their land and rescue their families from the threat of famine.

The optimism of early April quickly turned to anguish when it became evident that the Chemin des Dames offensive had done nothing to secure victory, or even bring its date closer. And from the anguished despair of brutally disheartened men emerged indiscipline. The mutinies of 1917 that sprung quick on the heels of military disaster roiled the French army from mid-April through early June. Frontline soldiers refused to participate in ill-conceived attacks, mocked military authority, and wrought havoc on discipline. The mutineers did not demand an immediate but ignoble end to the war, for they continued to believe that French territory had to be liberated from the enemy's grasp, but they did insist that their commanding officers identify a new way of waging war, accommodate the men's material demands for better food and more generous leave, and, as Leonard Smith has argued, accept a balance of authority that was negotiated rather than simply imposed from above.[56]

The mutinies of 1917 were not driven only, or indeed predominantly, by apprehensions about hunger on the home front. Had the mutinies been spurred by agricultural anxieties alone, then rural soldiers in all sectors would have participated and the peasantry would, presumably, have been represented in disproportionate numbers. Neither was the case. Peasants constituted only one-third of all soldiers charged with or punished for participating in the mutinies.[57] And peasant soldiers serving in sectors outside the especially blighted Chemin des Dames-Champagne axis did not rise up against their military commanders. Indeed, even some regiments that did see pitched battle there (including those in which Pireaud, Maret, and his younger brother, Paul, served) did not mutiny. Thus it would be inaccurate to portray the mutinies as a predominantly rural insurrection—a Great Fear of 1917, as it were. That being said, there might still be good reason to examine how apprehensions about food supply and the deteriorating state of rural society affected military morale in the spring of 1917. Agricultural anxieties simmered not far beneath the surface of military indiscipline through the spring and summer of 1917, and were most clearly articulated in soldiers' demands for leave.

Scholars have long known that complaints about the leave roster dominated soldiers' grievances before the mutinies, and calls for immediate improvements emerged as central to the mutineers' material demands. Furthermore, insofar as General Pétain did take steps to improve soldiers' access to leave, these concessions appeared to go a long way to restoring order in the ranks.[58] Nonetheless, the particular importance of leave to rural soldiers in 1917 has not yet been examined. While it is true that every frontline soldier yearned for leave, the evidence of 1917 suggests that peasant soldiers were especially insistent that the leave roster be reformed and understandably indignant that opportunities for agricultural leave (introduced in early 1917) often passed them by. Rural soldiers demanded improvements to the leave roster not just because they, like all frontline troops, were eager for well-earned rest; they longed for leave so that they could work their land, harvest the crops, and thus secure their families' food supply for the coming winter.

"That farmers are kept under arms in spite of the fact that their presence at home is deemed indispensable for the harvest" was, the censors noted, a recurring lament.[59] One soldier complained that a delay in his leave, caused by a relocation of his company, meant that he could not be home in time to help with the potato crop.[60] Another feared that it would be "impossible to get the harvest in if I don't go home."[61] Paul Pireaud also hoped to use his next leave to help with the harvest. In regular contact with his wife, he knew how difficult circumstances were in his own village and how overburdened she was. Thus when she suggested that they would profit most from his leave (due in late July or early August) if he came home after the heavy work of the harvest was finished, he could not agree. Enticed though he was by the allure of a blissfully idle, sensually rapturous leave, he could not in fairness to her contemplate such indolence: "you told me to come home on leave after the harvest. On the contrary, if I can I would rather go now because it makes me heart-sick to see that you are obliged to do everything." It had been raining at home, and he acknowledged that "this won't make your labors any easier especially right now all of you are in the midst of reaping the countryside must be very sad right now with so much work to do and the shortage of workers. As you said, I would be very happy to spend a restful leave but I would be just as happy to help ease your burden a little with the heavy work in the fields."[62]

Some rural soldiers took comfort in the expectation that the new agricultural leave policy would allow them to return home for an extended period of time during harvest season. In practice, however, the policy, which applied in any event only to the oldest classes of conscripts, seemed rife with cronyism and corruption. Or so it seemed to those men and women who failed to benefit from it. Indeed, inefficiencies and injustice in the allocation of agricultural leave did much to aggravate peasant soldiers' misery: those who should have been able to return home were often denied the necessary leave, while those who had influential connections wangled a leave to work a nonexistent farm. Thus when one soldier complained of "guys leaving who have never worked the land…waiters in cafés and…that fat butcher, who owns land twenty kilometers from where he lives and which he rents out!", he gave voice to the complaints of many others.[63] Women also bemoaned the fact that their own husbands remained at the front, while so-called farmers took advantage of local connections to qualify for an agricultural leave: "I see that your request for leave was refused. Ah yes, if you had been a notary, a school-teacher, or something else and you had asked for agricultural leave, you wouldn't have been turned down; it's as if all those who are on agricultural leave none of them are farmers. If they had put all the farmers where they belong then we wouldn't soon be needing ration cards for bread."[64] For peasant soldiers and the families who cared for them, leave meant more than respite from frontline misery and an opportunity for welcome, albeit abbreviated, reunion. It constituted an antidote to hunger and an insurance policy against dearth.

What does this composite picture of diminished agricultural productivity, apprehensions about famine, and military indiscipline tell us about how French peasant soldiers thought about the spaces of war? First, the evidence suggests very strongly that for the duration of the war, peasant soldiers continued to inhabit, in their minds at least, both the trenches of the Western Front and the villages of home. They were neither uninterested in, ignorant of or unaffected by the mundane realities of rural life. Indeed, they were physically and psychologically sustained by its successes and disconcerted by its trials. Every parcel from home was a palpable reminder of home, the affection of one's family, and the continued viability of the family farm. Every letter of lament—about food shortages, uncultivated land, and the burdens of agricultural labor—was reason for worry. These were not the embittered, angry, and alienated frontline soldiers of collective memory and literary construction. By 1917 they were certainly angry, but their anger derived as much from concern for civilian welfare as from evidence of civilian indifference.

Having felt reasonably confident in the well-being of their families through 1916, a well-being made manifest in the abundant food packages that made their way to the front, rural soldiers watched helplessly from afar as conditions at home deteriorated in the bitter and apparently endless winter of 1916–17. Material conditions at home as much as at the front were in the months leading up to and overlapping with the mutinies sufficiently dire to provoke widespread concern within the rank and file about the welfare of their families and their capacity to withstand further deprivation. Ultimately, therefore, the evidence of 1917 prompts us to revisit and understand in a new way that well-known, putatively ironic response of the oppressed *poilu*: "pourvu que les civils tiennent." Usually interpreted as a caustic dismissal of self-satisfied civilians who were wont to complain about every minor inconvenience, this watchword of the French frontline soldier might have taken on a different tone in the spring of 1917. Far from being a scornful and ironic dismissal of the spoiled civilian, it could well have been an expression of genuine empathy for those at home struggling to keep the nation fed and the family farm operational. The war could be won "if only the civilians held out." But that could happen only if the parents and wives of peasant soldiers, confronting ever greater adversity, were reinforced in their enterprise by fighting men guaranteed leave on a timely and well-timed basis. The longing for leave that punctuated the letters and pervaded the demands of rural soldiers in 1917 was no doubt infused by an understandable war weariness; but it was also informed by the empathy and anxieties of a dominantly peasant army eager to return home and secure the material well-being of those they were fighting to defend.

Notes

1. Henri Barbusse, *Under Fire*, Robin Buss, trans. (New York: Penguin Books, 2003), 61–66.
2. Ernest Pérochon, *Les Gardiennes* (Paris: Plon, 1993), 12. Originally published in 1924.
3. I have developed this argument in much greater depth in *Your Death Would Be Mine: Paul and Marie Pireaud in the Great War* (Cambridge, MA: Harvard University Press, 2006), ch. 1, "How Sad the Countryside Is."
4. On the centrality of food shortages to revolution in Russia, see Barbara Alpern Engel, "Not by Bread Alone: Subsistence Riots in Russia during World War I," *Journal of Modern History*, 69 (1997): 696–721, and Mary McAuley, *Bread and Justice: State and Society in Petrograd 1917–1922* (Oxford: Oxford University Press, 1991). For Germany, see Belinda Davis, *Home Fires Burning: Food, Politics, and Everyday Life in World War I Berlin* (Chapel Hill, NC: University of North Carolina Press, 2000); for Austria-Hungary, Maureen Healy, *Vienna and the Fall of the Habsburg Empire: Total War and Everyday Life in World War I* (Cambridge: Cambridge University Press, 2004), and Reinhard Sieder,

"Behind the Lines: Working-class Family Life in Wartime Vienna," in *The Upheaval of War: Family, Work and Welfare in Europe, 1914–1918*, Richard Wall and Jay Winter, eds. (Cambridge: Cambridge University Press, 1988), 109–38. For Britain, see Peter Dewey, "Nutrition and Living Standards in Wartime Britain," *The Upheaval of War*, 197–220; and Bernard Waites, *A Class Society at War: England, 1914–1918* (Leamington Spa: Berg, 1987), who notes that "even more disturbing for the government was the repercussion of working-class discontent with the food situation on the temper of the country's soldiers. The extent to which anxieties and irritations amongst the civilian population affected the fighting men is powerful confirmation of the magnitude of the crisis," 230.

5. *Histoire de la France rurale*, Georges Duby and Armand Wallon, eds., vol. 4, Michel Gervais, Marcel Jollivet, and Yves Tavernier, *La Fin de la France paysanne, de 1914 à nos jours* (Paris: Seuil, 1976), 31.

6. Michel Augé-Laribé, *L'Agriculture pendant la guerre*, Histoire économique et sociale de la guerre mondiale, Publications de la Dotation Carnegie pour la paix internationale (Paris: Presses Universitaires de France, n.d.), 56.

7. Archives Départementales de la Dordogne (hereafter ADD), 1 M 89, newspaper clipping (date not included) concerning mailing of packages for Christmas and New Year. This would suggest that Stéphane Audoin-Rouzeau's judgment based on evidence gleaned from trench newspapers that parcels arrived at the front only infrequently understates the case. Audoin-Rouzeau, *Men at War: National Sentiment and Trench Journalism in France during the First World War*, Helene McPhail, trans. (Oxford: Berg Publishers, 1992), 139.

8. Pierre Maurice Masson, *Lettres de guerre* (Paris: Hachette, 1917), 30, 34, 57–58. Letters dated 27 December 1914, 3 January 1915, 5 March 1915.

9. Fernand Maret, *Lettres de la guerre 14–18* (Nantes: Siloë, 2001), letter dated 22 February 1916.

10. Maret, *Lettres de la guerre 14–18*, letters dated 21 January 1916, 27 January 1916.

11. Maret, *Lettres de la guerre 14–18*, letters dated 4 February 1916, 23 February 1916.

12. Maret, *Lettres de la guerre 14–18*, letters dated 12 March 1916, 22 March 1916.

13. *14–18 La Grande Guerre: le vécu du soldat René; lettres quotidiennes à sa mère* (Paris: G.E.P, 1990), 125. Letter dated 8 September 1916.

14. *Le Soldat de Lagraulet: lettres de Germain Cuzacq, écrites du front entre août 1914 et septembre 1916*, Pierre and Germaine Leshauris, eds. (Toulouse: Eché, 1984), 52. Letter dated 8 May 1915.

15. *Le Soldat de Lagraulet*, 47, 53. Letters dated 24 March 1915, 19 May 1915.

16. *Le Soldat de Lagraulet*, 62, 65. Letters dated 24 June 1915 and 7 July 1915.

17. *Le Soldat de Lagraulet*, 104. Letter dated 17 March 1916.

18. *Le Soldat de Lagraulet*, 105. Letter dated 19 March 1916.

19. Section Historique de l'Armée de Terre (hereafter SHAT). 1Kt T458. Correspondance entre le soldat Paul Pireaud et son épouse, 10 janvier 1910–1927. Letter from Paul Pireaud to Marie Pireaud, 12 April 1916, 23 May 1916, 24 May 1916. All subsequent references to the Pireaud correspondence will be to this collection.

20. Cuzacq acknowledged receipt of one parcel on 23 March, two on the 29th, and one each on 31 March, 5, 9, 16, and 24 April 1916. *Le Soldat de Lagraulet*, 105, 106, 107, 109, 110, 113.

21. Paul Pireaud to Marie Pireaud, 25 April 1916, 26 May 1916, 31 May 1916, 3 June 1916.

22. *Le Soldat de Lagraulet*, 108. Letter dated 6 April 1916, to his sister.

23. Audoin-Rouzeau, *Men at War*, 140.

24. *Le Soldat de Lagraulet*, 110. Letter dated 17 April 1916. Paul Pireaud to Marie Pireaud, 31 May 1916. Maret, *Lettres de la guerre 14–18*, letter dated 12 August 1917.

25. On the importance of rural artisans to the well-being of villagers and farm workers, Yves Pourcher observes that villagers in Brittany in 1917 petitioned the local authorities to secure the reassignment from the front of the village's only *sabotier* because without their

wooden shoes, adults and children could not work the fields: *Les Jours de guerre: la vie des Français au jour le jour, 1914–1918* (Paris: Plon, 1994), 139–41. Henri Gerest notes the importance of skilled artisans who supervised the operation of village threshing machines in the rural communes of the department of the Loire: Henri Gerest, *Les Populations rurales du Montbrisonnais et la Grande Guerre* (St. Étienne: Centre d'études foreziennes, 1975), 126–29.

26. Archives Départementales de la Charente (hereafter ADC), J 84, Histoire de la Commune Juignac (Charente) pendant la guerre, renseignements fournis par Mmes Pinassaud et Gerbaud, institutrices.

27. SHAT, 16N 1449, Commission de contrôle postal de Bar le Duc. Report of 29 March–4 April 1917.

28. SHAT, 16N 1448, Grand Quartier Général [GQG], 2ème Bureau, Contrôle postal créé de Abbeville, Amiens, 1916–1918. Report dated 15 March 1917.

29. Margaret Darrow, *French Women and the First World War: War Stories of the Home Front* (New York: Berg, 2000), 182; Françoise Thébaud, *La Femme au temps de la guerre de 14* (Paris: Stock, 1986), 155.

30. SHAT, 16N 1440; Commission de contrôle postal d'Amiens. Xe Armée. Report of 6 January 1917.

31. SHAT, 16N 1448, GQG, 2ème Bureau, Contrôle postal créé de Abbeville, Amiens, 1916–1918. Report dated 16 August 1917.

32. Maret, *Lettres de la guerre 14–18*, letters dated 16 February 1916, 26 April 1916, 18 May 1916.

33. SHAT, 16N 1441. Report of the 30 Régiment d'artillerie, 11 May 1917.

34. SHAT, 16N 1440, Commission de contrôle postal, X Armée. Report of 10 January 1917, 33 Régiment Artillérie. 16N 1452, Commission de contrôle postal, Boulogne-sur-mer. Report of 12–18 February 1917. 16N 1440, Commission de contrôle postal, X Armée. Report of 21 January 1917, 108 *R.A.L* et 4 Génie; Report of 8 January 1917, 141 Régiment Territorial.

35. SHAT, 16N 1440, Commission de contrôle postal, X Armée. Report of 26 January 1917, 112 *R.A.L.* Maret, *Lettres de la guerre 14–18*, letter dated 31 January 1917; Paul Pireaud to Marie Pireaud, 24 January 1917.

36. SHAT, 16N 1440, Commission de contrôle postal, X Armée. Report of the 19 January 1917, 144 Régiment Infanterie.

37. Maret, *Lettres de la guerre 14–18*, letters dated 1 January 1917, 6 January 1917, 13 January 1917, 19 January 1917, 20 January 1917, 27 January 1917, 29 January 1917, 1 February 1917, 3 February 1917, 5 February 1917, 7 February 1917, 13 February 1917, 16 February 1917, 20 February 1917, 21 February 1917.

38. Maret, *Lettres de la guerre 14–18*, letter dated 3 February 1917.

39. Maret, *Lettres de la guerre 14–18*, letters dated 5, 7, and 11 February 1917.

40. Jean-Jacques Becker, *The Great War and the French People*, Arnold Pomerans, trans. (New York: Berg, 1985), 206–07.

41. SHAT, 16N 1448, GQG. 2ème Bureau, Contrôle postal créé de Abbeville, Amiens, 1916–1918. Reports dated 22 January 1917, 15 February 1917.

42. SHAT, 16N 1449, Commission de contrôle postal de Bar le Duc. Report of 12–18 April 1917.

43. References to milk shortages appear in the postal control records from February 1917 onward. SHAT, 16N 1448, GQG, 2ème Bureau, Contrôle postal créé de Abbeville, Amiens, 1916–1918. Reports dated 15 February 1917, 24 May 1917. 16N 1449, Commission de contrôle postal de Bar le Duc. Reports of 29 March-4 April 1917, 26 April–2 May 1917.

44. SHAT, 16N 1449, Commission de contrôle postal de Bar le Duc. Report of 26 April–2 May 1917.

45. Henri Gerest, *Les Populations rurales du Montbrisonnais et la Grande Guerre* (St. Etienne: Centre d'études foreziennes, 1975), 129.

46. Archives Départementales du Rhône (hereafter ADR), 4 M 234, Etat d'esprit de la population. Report dated 1 May 1916. Archives Départementales de la Gironde, 1 M 414, Etat moral de la population, 1916–1917.

47. SHAT, 16N 1452, Commission de contrôle postal, Boulogne-sur-mer. Report dated 12–18 March 1917.

48. ADR, 4 M 234, Etat d'esprit de la population. Report of the Prefect of the Rhône to the Minister of the Interior, 1 March 1916. Marie Pireaud to Paul Pireaud, 11 March 1917. ADC, J 94. "Histoire des communes par les instituteurs (1914–1918) Salles La Vallette"; J 84. "Histoire de la Commune Juignac (Charente) pendant la guerre"; J 87. "Historique de la guerre: Montignac-le-coq." ADD, 8 R 51, Correspondance générale sur la situation du ravitaillement en céréales. Arrondissement de Ribérac. Decree of the Mayor of Montpon, 22 May 1917. ADD, 1 M 82, Report of the Prefect of the Dordogne, 17 June 1917.

49. SHAT, 7N 985, Commission Militaire de contrôle postal de Bordeaux, 1915–1918. Report for April 1917.

50. SHAT, 16N 1448, GQG, 2ème Bureau, Contrôle postal créé de Abbeville, Amiens, 1916–1918. Report dated 12 April 1917.

51. SHAT, 16N 1448, GQG, 2ème Bureau, Contrôle postal créé de Abbeville, Amiens, 1916–1918. Report dated 15 March 1917; 7N 985, Commission Militaire de contrôle postal de Bordeaux, 1915–1918. Report for March 1917.

52. SHAT, 16 N 1448. GQG. 2ème Bureau, Contrôle postal créé de Abbeville, Amiens, 1916–1918. Reports dated 22 January 1917, 15 February 1917.

53. Lionel Lemarchand, *Lettres censurées des tranchées 1917: une place dans la littérature et l'histoire* (Paris: L'Harmattan, 2001), 97.

54. SHAT, 16N 1452, Commission de contrôle postal, Boulogne-sur-mer. Reports dated 2–8 April and 9–15 April 1917.

55. John Horne, "Soldiers, Civilians, and the Warfare of Attrition: Representations of Combat in France, 1914–1918," *Authority, Identity, and the Social History of the Great War*, Frans Coetzee and Marilyn Shevin-Coetzee, eds. (Providence, RI: Berghahn, 1995), 237–38.

56. Leonard V. Smith, *Between Mutiny and Obedience: The Case of the French Fifth Infantry Division During World War I* (Princeton: Princeton University Press, 1994), 179.

57. Guy Pedroncini, *Les Mutineries de 1917* (Paris: Presses Universitaires de France, 1967), 222.

58. Pedroncini, 233.

59. SHAT, 16N 1452, Commission de contrôle postal, Boulogne-sur-mer. Report for the week of 9–15 July 1917. 7N 985, Commission Militaire de contrôle postal de Bordeaux, 1915–1918, Report for the period 15 May–15 June 1917.

60. SHAT, 16N 1452, Commission de contrôle postal, Boulogne-sur-mer. Reports dated 11–17 June 1917, 2–8 July 1917.

61. SHAT, 16N 1448, GQG, 2ème Bureau, Contrôle postal créé de Abbeville, Amiens, 1916–1918. Report dated 16 August 1917.

62. Paul Pireaud to Marie Pireaud, 8 July 1917, and undated letter [summer 1917].

63. SHAT, 16N 1448, GQG, 2ème Bureau, Contrôle postal créé de Abbeville, Amiens, 1916–1918. Report dated 23 August 1917.

64. SHAT, 16N 1452, Commission de contrôle postal, Boulogne-sur-mer. Report for the week of 2–8 July 1917; report for the week of 16–22 July 1917.

5

The Search for Civilian Safe Spaces: Re-evacuating Nord and Pas de Calais in Response to British Bombing, September 1940–March 1941

Nicole Dombrowski Risser

A ir warfare dramatically reconfigured traditional relationships between the geographical and psychological spaces of war during the twentieth century. Most crucially, this change expanded civilian noncombatants' vulnerability to wartime violence.[1] Urban-based families especially suffered as victims of the collateral damage accompanying the bombing of legally "sanctioned" strategic military targets near industrial centers. With an evolution in states' understanding of collective psychology, bombing campaigns have also intentionally targeted civilian centers as a strategy of psychological warfare. Many historians tracing the history of warfare in the twentieth century have inaccurately partitioned the study of war into two distinct spaces—home front and battlefront.[2]

The April 11, 1918, German bombing of a Paris maternity ward, Clinique Baudelocque, exemplified the redefinition and expansion of war's combat spaces.[3] While, no soldiers died from this bomb, a midwife, who had just delivered a newborn baby to its mother, joined the ranks of the fallen. Air power thus ruptured any remaining nineteenth-century bourgeois conceptual view of the home as a protected and private sphere.[4]

Sven Lindquist's chronology of bombing marks one of the first attempts to record and unmask nation states' military and political choices to collapse the distinction between home front and battlefront.[5] Domestic civil defense programs play an important role in this boundary erasure. Civil defense programs have attempted to defend the distinction between civilian safe spaces and combat zones, yet these same programs aid in military efforts, thus eliminating this spatial distinction.[6] We see a conflicting process at work at the level of international institutions too, whereby the Hague Conference of 1907 prohibited wanton destruction of cities, villages, and towns, yet failed to limit the proliferation of air power.[7]

John Williams first showed how the Great War's "wholesale dedication of the belligerent powers to the waging of war…brought changes to the civilian scene unprecedented in any previous conflict." Mass mobilization for war removed the "man of the house," weakening the perceived protection of home.[8] The concept of a *home front* as a distinct space emerged over the course of the war. From 1914 to 1918 the function of the home front was not to protect civilians, but to lend industrial and psychological support to the combat zone.[9] The lexicological pairing of "home" with front unites the geographic and conceptual realms of home and combat in a manner that renders the concept of home meaningless as a realm of safety. The new word identifies the home as a front enmeshed within the complex psychological, economical, and geographical web of militarization.

Interwar civil defense planning also expanded the state's role as organizer of the home front. However, in France, protective services for civilians achieved neither priority status in military preparedness campaigns, nor uniform access. In northeastern departments, civilian survivors of Germany's World War I occupation lobbied French municipal governments for protection from war's future encroachment into civilian spaces. Germany's invasion of the Rhineland in March 1936 motivated these civilians to advance a set of comprehensive claims against all levels of government for protection from war's future physical, economic, and emotional devastation.[10] A politics of civilian protection spread beyond northern border departments to Alsace and Lorraine and then to Paris after Germany's annexation of Austria in 1938.

Britain's bombing of northern France from September 1940 to April 1941 exposed the limits to interwar achievements in civil defense policies and procedures. British bombing of German-occupied, French coastal towns provided historians with an opportunity to examine the coordination difficulties encountered by French and German authorities, charged by international legal conventions to ensure protection of civilian noncombatants exposed to hostile fire. In particular, French officials in the occupied zone confronted newly vexing negotiations with civilians, the Vichy government, and the Germans as British destruction of coastal towns intensified local residents' pressure on German and French authorities to identify or create civilian safe spaces.[11]

In Le Havre, Calais, Dieppe, Dunkerque, Boulogne-sur-Mer, and Rosendael, town mayors, and department prefects appealed for aid to national, Paris-based, civilian security planners at the Service of Refugees and to the regional German military command. Mayors requested designs and finances for new, occupation-time, civilian protection programs. Because coastal cities fell within territory defined by the German Armistice Commission as a Forbidden Zone, Paris-based officials faced many difficulties in providing support to threatened localities.[12]

German partitioning of France into militarily fortified, Forbidden Zones diced the country geographically and added a new layer of bureaucracy to the coordination and delivery of regional civilian protection programs (figure 5.1). Despite the partition's inconveniences for French civilians and administrators, General Falkenhausen, who extended the German Military Administration in Belgium's authority (MBH) over Pas-de-Calais and Nord, defended the practical and ideological benefits of French territorial sub-partitioning. Falkenhausen advocated partition as a tool for annexing strategically "key" and "ethnically desirable" departments such as Alsace and Lorraine.[13]

In addition to Pas-de-Calais and Nord, the Forbidden Zone included Ardennes, Aisne, Haute-Normandie, and Basse-Normandie. Civilian security in these departments

Figure 5.1   Partition of France by the Line of Demarcation: Forbidden Zone Sub-partitioning, July 1940. Courtesy of the University of Texas Libraries, University of Texas at Austin.

had previously collapsed in May 1940. In August, a month prior to British attacks, the Germans refused to readmit residents who had fled the region during the May invasion. Louis Marlier, honorary prefect and special representative for refugees, acted on Vichy's agenda to accelerate the reconstruction of the coastal regions. In July, Marlier pressured the Germans to permit Forbidden Zone residents, now refugees in the Free Zone, to return to their homes.[14] Anticipating future combat along the coast, the Germans resisted Vichy's efforts to repatriate these refugees.

National partition and conflicting interests among different parties generated confusion, thus jeopardizing civilian security interests. The case of Mme. Marie Choin illustrates the partition's impact on individual lives. A 67-year-old widow, Mme. Choin had lived in Carignan, Ardennes. She had spent nine months in a refugee camp in Charente, near France's Atlantic coast. Mme. Choin welcomed Vichy's refugee repatriation initiatives. Given the choice of remaining in the Free Zone or returning to her home in German-occupied Ardennes, Mme. Choin chose home. Nearing home, Mme. Choin encountered an obstacle in Reims (Marne): German authorities refused her readmission to the Forbidden Zone. Unaware of impending combat or geopolitics, Mme. Choin requested that the mayor of Reims at least authorize her to remain in Reims: "I am a 67 year-old widow without any assistance. Having had to travel all over France, it [France] cannot even give an old woman 10 francs a day so she can live or at least return me to my home."[15] The Returnee Relief Allocation Commission of Reims refused Mme. Choin's request on the grounds that the only acceptable motive for her to move from Charente to Reims would be to join underage children. The mayor of Reims suggested that the 67-year-old widow try returning to Charente.[16] Elderly refugees especially faced the compounded problems of physical deterioration and the precipitation into poverty, which accompanied old age.

Prior to the British attack, Vichy had pushed the Germans to admit refugees to the Forbidden Zone for three main reasons: to boost economic reconstruction by repatriating vital personnel; to reclaim the Occupied Territory symbolically by repopulating the area with French citizens; to respond to the demands of internally displaced refugees living in the south or in refugee relocation centers along the Line of Demarcation who longed to return home.[17] Historical hindsight reveals that by repatriating civilians to the Forbidden Zone, French authorities exposed repatriated civilians to British bombing campaigns. Those civilians who managed to repatriate now required re-evacuation. But to where?

By September 1940, residents of Pas-de-Calais and Nord who had successfully repatriated from the Free Zone experienced more pressing economic constraints due to their initial displacement and the destruction visited upon their home communities by German invasion and occupation. By the autumn of 1940, Lynne Taylor finds that "[t]he breakdown in the regional economy and the resulting massive unemployment, shortages of every kind and galloping inflation meant that daily life had devolved into a struggle to subsist."[18] As in Reims, Forbidden Zone mayors could not dispense state refugee relief to most coastal returnees who had managed to cross the Line of Demarcation without a German-approved laissez-passer. For many coastal returnees, their homecoming further soured upon the discovery that bombs had demolished their homes or that Germans troops had occupied them.[19] Population reintegration statistics show that significant numbers of coastal residents did return to their homes, perhaps believing repatriation safer and more affordable than life in refugee camps (figure 5.2).

In Le Havre, reintegrated civilians and local authorities managed the daily messiness of bombing, attempting to carve out new civilian safe spaces within the British target zones. From September 16 to 28, the British attacked Le Havre nightly.[20] Ninety-four civilians died, and explosions injured 179 others. Unable to manage the ground disaster by relying uniquely on local forces, Le Havre's police commissioner urgently wired Paris, describing the damage: "The two hospitals are completely overflowing with the unending flow of the wounded. The hospitals have no security. One has been hit by a bomb, and the staff has moved surgical operations to a more secure spot."[21] Over a period of 16 days, bombs completely demolished 165 buildings including the stock exchange, city hall, the post office, and the armories of Eblé and Kléber.[22] The commissioner petitioned Marlier, at the Service of Refugees in Paris, to assist with the re-evacuation of urban children.

City	Pop. Pre-May 1940	Repatriated Pop. Returned March 1941	Long-term Displaced Not Returned	% Long-term Displacement % Not Returned
Dunkerque	31,000	12,000	19,000	61
Rosendael	17,000	16,000	1,000	6
Malo-les-Bains	11,000	7,200	3,800	35
Couderkerque Branche	13,600	10,000	3,600	26
Petite Synthe	7,100	Unknown	Unknown	Unknown
Saint-Pol-sur-Mer	12,700	Unknown	Unknown	Unknown
Cappelle-la-Grande	22,300	15,000	7,300	33
Calais	67,500	49,050	18,450	27
Boulogne-sur-Mer	52,000	30,000	22,000	42

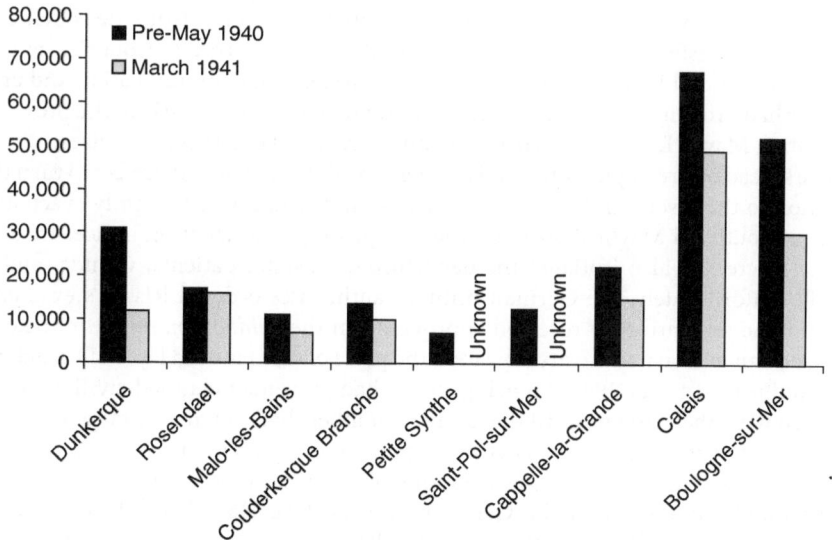

Figure 5.2   Population Figures for Pas-de-Calais and Nord from May 1940 to March 1941.

Isolated from Vichy, and even working at odds against Vichy's repatriation policy and German strategic interests, local officials tripped over a myriad of obstacles in their attempts to implement new rescue and relocation strategies. The commissioner emphasized: "The people of Le Havre, who...had previously evacuated the city in June 1940 are shaken with panic....A new exodus has begun, lamentably, we have no means for transportation," he continued.[23] Families described as "stricken by the liveliest terror" began to take refuge in nearby communes. At dusk each evening a flood of refugees evacuated the city by foot, bicycle, or wagon to find shelter in fields or woods.

"The fear of bombings from ships at sea accentuates even more the hysteria that has overtaken the entire population," wrote the commissioner.[24] In the margins of the received status report, Prefect Marlier jotted down his own spin on the locals' analyses: "Emotion but not panic."[25]

News of the fires consuming Le Havre reached Parisians not through newspapers, but via mobilization and transfer of Parisian firefighters. Injuries had so depleted Le Havre's volunteer rescue reserves that, the Service of Refugees, the mayor's office in Le Havre, and the Délégation Générale du Gouvernement Français pour les Territoires Occupées (DGGFTO), each petitioned the Germans to authorize a regiment of Parisian firefighters to travel to the coast. Fearing interzonal travel, the Germans limited the transfer of emergency workers to those Parisians who could show "evidence of family ties" in Le Havre.[26]

Once in Le Havre, rescue workers faced an uphill battle to contain the urban blazes. RAF bombings had completely paralyzed the city's water systems. Residents, medical workers, and firefighters competed for dwindling supplies of water. The fire chief ordered a mobile water pump from the Latil Company in Paris, but the company insisted upon a payment of 175,000 francs before delivery. Mayor Risson lacked the funds and urged Marlier to press General La Laurencie for funds from the DGGFTO.[27] Approval to transfer DGGFTO funds to Mayor Risson had to travel through several layers of German and French bureaucracy. French systems of communication and credit transfer had broken down, delaying relays of financial aid from Paris to the provinces, frustrating Mayor Risson's efforts to coordinate regional emergency services.

The spread of fires, epidemics, and hunger forced the mayor and prefect to turn their attention to the rescue and relocation of children. Unsupervised, nightly evacuations of urban youth led Mayor Risson to order the posting of evacuation notices throughout Le Havre. He also initiated the departure of hospital patients, women, and the elderly. Unfortunately, the German military authorities ordered Risson's evacuation orders removed until they received approval from the *Feldkommandantur* of Rouen. The German military authority now took the place of the removed layer of French military authority. In May 1940, French generals had prevented planned civilian evacuations from northern towns until the German bombers had flown over targeted cities.[28] The Germans supported, indeed insisted, upon removing civilians from coastal towns, but they refused to finance new evacuations. The Service of Refugees backed Risson's petition and requested that the Germans transport Le Havre's residents to Rouen. German military officials rightly feared that Rouen remained well within the British fly zone. Yet the German Armistice Commission forbade evacuees from crossing the Line of Demarcation into Vichy's Free Zone. On this point the Germans and Vichy actually agreed. Vichy had, after all, directed Marlier to limit new relocation schemes to Rouen at German expense.[29]

While the Germans delayed authorizing the relocation of Le Havre's population, a sharp division of opinion emerged among civilian families. One group claimed the right to remain in Le Havre, while the other demanded the right to evacuate. Informal debates over the choice to evacuate or remain in British bombarded cities unfolded outside the prewar democratic framework of municipal councils. Opponents of re-evacuation surfaced predefeat arguments defining departure as a form of capitulation. Military defeat and economic hardship now weighted the discussion with a sharper and more broadly diffused moral and political language than that of May 1940.

Civilian complaints about new evacuation plans reflected a sense of previous government betrayal and communicated civilians' raised expectations regarding receipt of government cash subsidies to offset their evacuation expenses.[30] Still, despite the daily deprivations accompanying life in a bombing zone, many civilian veterans of Germany's spring offensive questioned the merits of evacuation. In Dunkerque and Le Havre, civilians knew that evacuation programs could distance them from British bombing but could also exacerbate other miseries of war, including dependency upon unreliable government assistance.

Confronted with persistent civilian endangerment and civilian skepticism toward evacuations, French security planners explored local security options.[31] German destruction of regional rails from the May–June campaign complicated plans for transferring coastal populations to the French interior.[32] Broken regional rails interrupted delivery of even the most vital supplies and foodstuffs "both from outside and within the region."[33]

In Le Havre and Dieppe, the accumulation of damage undermined the logic of those civilians who wished to stay. By October 1, Royal Air Force bombers had sacked 950 to 1,000 buildings in Le Havre alone, leaving an additional 12,000 Havrais residents homeless.[34] Mayor Risson reported that destruction of factories added 1,500 civilians to his unemployment list, which now totaled 17,000 for the entire city. The only fortunate residents of Le Havre were the 15,000 refugees who had never repatriated from the May–June exodus. Having failed to return for various reasons, those 15,000 residents probably faced hunger and unemployment in refugee centers in the Free Zone, but did not face nightly exposure to British air assaults.

In this messiness of war and occupation, civilians also contributed to merging the spaces of war. In their effort to minimize war's intrusion into their daily lives, they improvised the dubious performance of short-term, short-distance evacuations. Of Le Havre's 150,000 residents who had managed to repatriate to their homes since June 1940, each night 15,000 to 20,000 of them fled to the forests of the surrounding countryside. Interruptions in communication between Paris and Le Havre forced Marlier to send a fact-finding team from Paris to survey the region's damage and to estimate the dimensions of future long-term evacuations. The Service of Refugees' field staff actually encountered 500–600 people retreating by foot and riding bikes, in an effort to escape to Rouen. Children comprised the majority of these self-guided evacuees, as some parents in Le Havre and Dieppe encouraged their children to leave on their own. However, adults did seem to accompany the very youngest evacuees.

The impact of these nightly retreats on the health of civilian youth has been insufficiently studied. Lynne Taylor's research on the region attributes enduring shortages of foodstuffs and inflation to the "inexorable physical wasting of the population through chronic malnutrition."[35] Studies of the regions' children found a marked wartime underdevelopment compared to other departments in the Occupied Zones, particularly among children aged 6 to 14, precisely the group most frequently found in nightly evacuations. These children gained only 9 kg between 1938 and 1944, as opposed to the norm of 18 kg. Taylor found reports that children from the region suffered, "weak muscular development, abnormal fatigue, rickets, serious dental problems, retarded puberty and a serious vitamin deficiency."[36] However, these distinctions among the region's children were common to the working class before the war. French public health researchers attributed the regional difference to bad diet, lack of exercise, and a lack of interest in hygiene.[37] But historians of children, public health, and war must reconsider this evidence recognizing that the region suffered occupation under two wars. In World War II, the combination of military occupation, repeated evacuation by foot, bombing and destruction of the health, agricultural, educational, and financial infrastructure undoubtedly dramatically increased children's illness over two generations. Records indicate that the inability to evacuate Le Havre's 800 elderly people who resided in the town's two hospitals meant that bombing hit the elderly in disproportionately higher numbers too. Local doctors estimated that 500 elderly could survive transport to other locations, but that 300 could not be evacuated. Medical reports described how failures in the supply of water and electricity contributed to growing cases of typhoid, which health workers feared could infect the remaining civilian population.[38] Indeed, by December a full-blown typhoid and paratyphoid epidemic hit Pas-de-Calais and Nord, only to be temporarily contained until a new outbreak struck in 1943.[39]

As an alternative to evacuations, civilian security planners advocated an unusual solution for safeguarding the population of Dieppe. Here, nature offered protective promise. The prefect of Haute-Normandie and the Service of Refugees identified caves embedded in cliffs along Channel coast beaches.[40] The cavities reached four to five meters deep, providing an earth-covered refuge with rock walls that measured 50 meters in thickness. Marlier's field scouts lamented that shallow water covered the cave floors, requiring drainage. The team recommended that the Service of Refugees seek German approval to "modernize" the caves for civilian habitation. Engineers estimated that with a budget of 200,000 francs they could make the caves livable for approximately 800 evacuees. Civilians would move into these lodgings once a "modernization plan" could be completed. The engineers outlined eight steps to prepare the caves for human habitation: improve aeration; install light; install bunk beds; install a system of toilets; install rock-moving equipment inside the caves (in the event of collapsed walls); install telephone service; wire warning lights; create partitions to accommodate separately women, men, children, and families.[41]

The barbarism of modern warfare forced modern civilian security planners to return women and children to the dwellings of primitive man. The inability to secure civilians within modern urban centers or to move them across occupation boundaries limited the French to the planned relocation of 800 women and children to coastal caves. In World War I, civilians in the Champagne region used cellars as bomb shelters. The Dieppe project capitalized on the region's technical expertise gained in mining. It also signaled the advent of a postwar trend in civil defense to expand the use of subterranean nuclear fallout shelters.[42] The return to caves, arguably man's most primitive dwelling, to seek sanctuary from man's most advanced systems of weaponry—aerial launched bombs—underscored the de-civilizing projects engaged in by all warring powers during World War II. What seems perhaps more remarkable in this case is the fact that the Service of Refugees prioritized a plan for modernizing cave dwellings over a plan for long-term population relocation to the Free Zone or to refugee reception centers established along the Line of Demarcation. In the fall of 1940, the refusal by both Vichy and the Germans to absorb the costs of evacuating imperiled civilians forced French civil defense planners to consider "modernizing" caves for human habitation.

It must be emphasized that the vagueness of prewar international law about the requirements of an occupying power to protect occupied-zone civilians allowed the

Germans to resist funding and administering civilian evacuations. As a point of comparison, it matters that in France, unlike in the Eastern occupied territories, the Germans actually attempted to cooperate with local government authorities to provide civilian security, if only to facilitate their military operations.

French working people had lost their jobs, savings, property, and loved ones during the exodus of May-June 1940. Dockworkers in Dunkerque, for example, unconvinced of the benefits of re-evacuation, rearticulated the narrative that began to develop in May, which ascribed treason with retreat and identified retreat as a bourgeois dereliction.

In Boulogne-sur-Mer located in the department of Pas-de-Calais, the mayor also confirmed that socioeconomic background determined civilian access to safe zones. Thirty thousand residents of the original 52,000 had returned after the May evacuation. The mayor reported that all the middle-class families had evacuated of their own free will. Only families of workers remained behind. Of the remaining 30,000, 10,000 children under age 14 awoke to nightly air raids. Housing in Boulogne-sur-Mer offered another problem. Of the 9,600 houses constructed in the town, bombing had demolished 720; 2,200 houses had no water or electricity, but residents made do. German authorities had requisitioned approximately 2,200 livable housing units to quarter their troops (see figure 5.3).[43]

Sustained destruction to Dunkerque, Calais, and cities along the coast only increased throughout the spring of 1941, confusing civilians and clarifying for French security officials a need to impose mass evacuations. In February, Dunkerque's mayor reported that the city withstood bombings on an average of two to three times per day.[44] Unlike the population of Le Havre that attempted to flee the city every evening, residents of Dunkerque and its surrounding areas insisted on staying home.

Finally between February and March 1941, the Service of Refugees and the prefect of Nord coordinated a relocation plan for the women and children of the coastal cities. They petitioned the Germans to transfer 21,000 children and women to the manufacturing center of Lille, located marginally deeper within the department.[45] In March 1941, after six months of intensive British bombing, the Service of Refugees produced the

City	Buildings Destroyed	Buildings Damaged	Evacuated	Wounded	Dead
Dunkerque	3,250	Unknown	19,000	Unknown	1,000 / 200* (6/40) 191 (9/40–3/41)
Rosendael	800	1,500 ! 1,700 @	1,000	270	520
Boulogne-sur-mer	720	2,200 ! 2,200 #	Unknown	350	220

Figure 5.3   Destruction to Pas de Calais and Nord May 1940–March 1941.

Notes: *Missing, presumed dead
!Seriously damaged
@Pillaged and temporarily uninhabitable
#Occupied by German troops

Source: A.N. F/60/1507, Service of Refugees, Marlier to Ambassador of DGGFTO, March 2, 1941, 2–11.

results of its survey of the conditions of civilians living in the Forbidden Zone subjected to sustained bombardment. The report compelled Marlier finally to instruct the prefects of Nord, Pas-de-Calais, and Haute-Normandie to coordinate with Secretary General of Physical Education and Sports, Jean Borotha, planning the relocation of 61,000 coastal residents to inland areas within each department. This plan avoided both German and Vichy objections of moving evacuees deeper within the Occupied or Free Zones.[46] In concluding his recommendation, Marlier conceded, "[t]he recommendation to move these children as far away as possible from the hostilities constitutes a minimal solution."[47] The goal for the future, he added, would be to enlarge the evacuations to include all women, children, and the elderly from all towns endangered by bombings, water restrictions, and interruptions to electrical service.

Finally, Marlier joined the ranks of the few visionary French military planners such as Lieutenant-Colonel Arsène Vauthier, who in 1930 wrote in his pamphlet, *The Aviation Danger and the Future of the Country*, that "the entire civil population will be placed abruptly on the warfront, despite the efforts of ground army troops, despite all obstacles and all fortifications."[48] Lille and even Paris certainly fell within the British fly zone, and not infrequently the British bombed them too. Moving populations deep into France's interior to the Corrèze and the southwest had succeeded in 1939 in keeping many civilians out of bombs' way during the initial invasion.[49] In April 1940, moving populations a distance of 80 kilometers from Verdun to Reims, for example, had gained security for no one (figure 5.4).[50]

Louis Marlier had fought forcefully with the German Armistice Commission in July and August 1940 to gain readmission of Forbidden Zone refugees to their homes and jobs located along the Channel coast. It must have been a bitter pill to swallow now to issue new evacuation orders to this distressed and continually displaced population.[51] A native of Aisne, Marlier hoped that the return of French residents to Forbidden Zone departments would launch economic recovery and reunite families. Ironically, British resistance to German military aggression scuttled his plans and re-exposed civilians to renewed bombing.

The Service of Refugees' recognition, in April 1941, that socioeconomic conditions determined civilians' ability to access safe spaces and improve their chances of

City	Children (7–14)	Relocation Destiny
Dunkerque	12,000	Lille
Bray Dunes	5,800	Lille
Loon Plage	11,900	Lille
Grande Synthe	3,500	Lille
Calais	7,300	Unknown
Boulogne-sur-Mer	10,000	Unknown
St. Léonard and St. Martin	2,500	Unknown
Total	53,000*	

Figure 5.4   Evacuation Plans for Nord and Pas de Calais March 1941.

Note: * Request for transfer of 61,000 differs by 8,000, suggesting adults formed part of request.

Source: A.N. F/60/1507, Marlier to DGGFTO, March 2, 1941, 2–11.

surviving aerial bombing marked a major victory for wartime civilians. Marlier reaffirmed this understanding again in April 1942: "Only those [children] of well-off families can assure their departure at their own expense, as they have done, leaving at their own convenience once they have received permission from the proper authorities."[52] Socioeconomic position played a major role in producing parents' fear of relocation. When the British began bombing Paris later in the war in 1943, Marlier extended the principle to the Paris Red Belt: "We have confirmed that the last bombings have above all fallen around the families who live closest to the factories already attacked by bombs that risk being targeted again. Obviously, it is working-class families by and large who live in these densely populated neighborhoods."[53] The complication, as Marlier formulated it, was that "[t]hese families do not desire to entrust their children to the official administrative services or its workers. Instead they prefer that their children be accompanied by a relative such as a mother, a grandmother, an older sister or a stepparent."[54]

The lessons learned on the ground by local mayors, prefects, and the Service of Refugees emerged from the dialogue between French government officials and French civilians living under German occupation and British bombardment. Throughout the period of hostilities, working-class women had argued, strongly and in class-based terms, that without state aid they could not take adequate advantage of civilian protection schemes. These dialogues began to produce a body of principles to govern the obligations of protective powers, such as the German occupying authorities and the Service of Refugees, with regard to ensuring civilian security during periods of bombing.

Once Marlier accepted the reality of the socioeconomic constraints of evacuation, he suggested that the public welfare agency, *Assistance Sociale*, issue support to re-evacuated refugees. With this decision, Marlier affirmed the need for the state to subsidize the evacuation of low-income and working-class civilians. Marlier's ultimate acceptance of the idea, that only with receipt of state-paid refugee assistance could low-income civilians participate in state-sponsored evacuations, marked an important evolution in his own convictions. More importantly, it committed the French state to an important financial principle with regard to ensuring civilian protection, a principle it had tried to avoid institutionalizing from the moment of the war's declaration on September 3, 1939.

Conclusion

Civilians demanding protection from belligerent powers' "wanton destruction of cities, villages and towns" won a Pyrrhic victory in 1941. The Service of Refugees and the DGGFTO agreed to subsidize low-income families' relocation costs, expanding popular accessibility to safe spaces.

The 1907 Hague Conference had too vaguely articulated the financial responsibilities of occupying powers and their obligations to cooperate with local authorities to deliver civilian protection programs. The on-the-ground activism of bombing victims, and the unprecedented death toll of civilians between 1939 and 1945, thus mandated improvements to international standards for civilian protection, generating the category of crimes against humanity in the postwar Geneva Conventions.

Like the Channel coast caves, tougher protections for civilian noncombatants offer questionable shelter from the evolving practices and technologies of twenty-first-century warfare. In 1914 the term *home front* masked combat's encroachment into home's protective space. During and after the war, civilians needed international law and states to provide greater financial and physical protection. However, in the post-atomic era, these laws and programs perpetuate an illusion of security. Only a program of global peace can offer the necessary protection to civilians searching for safe space. This history shows that fortification against bombing proved temporary. The security agenda of states and engaged civilians must now be the fortification of the peace.

Notes

1. Thirty-seven million civilians died worldwide during World War II. Approximately 25 million soldiers died. John Dower uses the figure of 55 million total casualties. John Dower, "Race, Language, and War in Two Cultures: World War II in Asia," in *Japan in War and Peace: Selected Essays* (New York: New Press, 1995), reprinted in Gordon Martel, ed., *The World War II Reader* (New York: Routledge, 2004), 226. Tony Judt reports that "[t]he number of civilian dead exceeded military losses in the USSR, Hungary, Poland, Yugoslavia, Greece, France, the Netherlands, Belgium and Norway." Tony Judt, *Postwar: A History of Europe Since 1945* (New York: Penguin Press, 2005), 18.
2. Karma Nabulsi, *Traditions of War: Occupation, Resistance and the Law* (New York: Oxford University Press, 1999). Nabulsi's chapter on the Paris Commune offers a nineteenth-century example of how civilian spaces and combat spaces frequently collapse in the context of civil unrest. Nabulsi cites 1871 as the modern starting point for the process of diminishing the distinction between civilian noncombatant and legal combatant.
3. Mindy Jane Roseman, "The Great War and Modern Motherhood: La Maternité and the Bombing of Paris," *Women and War in the 20th Century: Enlisted With or Without Consent*, Nicole Dombrowski, ed. (New York: Garland Publishing, 1999; New York: Routledge, 2004), 56.
4. Roseman, 57. Also Michelle Perrot, ed., *A History of Private Life: From the Fires of Revolution to the Great War* (Cambridge, MA: Belknap Press of Harvard University, 1990).
5. See Sven Lindqvist, *A History of Bombing*, Linda Haverty Rugg, trans. (New York: New Press, 2000), 1–2. Lindqvist identifies the beginning of bombing as November 1, 1911.
6. For a history of attempts to maintain a distinction between civilian and combatant, see Helen M. Kinsella, "The Image before the Weapon: A Genealogy of the 'Civilian' in International Law and Politics," Ph.D. Dissertation, University of Minnesota, July 2004.
7. Work on wartime commemoration reveals that the spaces and traces of war continued to bleed into postwar culture and landscapes, yet in ways that mimicked the attempted compartmentalization of war during its combat stages. See Jay Winter, *Sites of Memory, Sites of Mourning: The Great War in European Cultural History* (Cambridge: Cambridge University Press, 1998); Annette Becker, *War and Faith: The Religious Imagination in France, 1914–1939* (New York: Berg Publishers, 1998).
8. Roseman, 58.
9. John Williams, *The Other Battleground: The Home Fronts in Britain, France and Germany, 1914–1918* (Chicago: Henry Regnery, 1972), 1, quoted by Mindy Jane Roseman, 57. The militarization of the home front is first and best analyzed by Jurgen Kocka, *Facing Total War: German Society, 1914–1918*, Barbara Weinberger, trans. (Cambridge, Mass: Harvard University Press, 1985). For a discussion of the Great War's impact upon gender identities, see Susan Grayzel, *Women's Identities at War: Gender, Motherhood, and Politics in Britain and France During the First World War* (Chapel Hill: University of North Carolina Press, 1999), 87. Grayzel shows that during World War I, "mothering came to represent for women what soldiering did for men, a gender specific experience, meant to provide social unity and stability during a time of unprecedented upheaval," 87.
10. See Nicole Dombrowski, "Beyond the Battlefield: The French Civilian Exodus of May–June 1940," (Ann Arbor, MI: University of Michigan, 1995), 514–31.

11. For initial inquiries into state evacuation policies in France during World War II, see Jean Vidalenc, *L'Exode de mai-juin, 1940* (Paris: Presses Universitaires de France, 1957); Dombrowski, "Beyond the Battlefield; Julie Suzanne Torrie, "For Their Own Good: Civilian Evacuations in France and Germany, 1939–1945," unpublished thesis, Harvard University, 2002. On Franco-Prussian war evacuations, see Rachel Chrastil, "Rebuilding France: Rights and Responsibilities in the Provinces following the Franco-Prussian War, 1871–1892," Ph.D. Thesis, Yale University, Department of History, New Haven, CT, 2005.

12. A.N. F/1a/3660, Minutes for the Service of Refugees' Meeting at the Hôtel Majestic, Paris, July 5, 1940. Nord and Pas-de-Calais fell more precisely under the Oberfeldkommandantur 670 (OFK 67), headquartered in Lille. See Lynne Taylor, *Between Resistance and Collaboration: Popular Protest in Northern France, 1940–45* (New York: St. Martin's Press, 2000), 11–12. On partition, see Eric Alary, *La Ligne de Démarcation 1940–1944* (Perrin: Paris, 2003), 27; Philippe Souleau, *La Ligne de démarcation en Gironde: occupation, résistance et société 1940–1944* (Périgueux: Fanlac, 1998), 330.

13. Alary, 27–28.

14. Dombrowski, 327. A.N. F/1a/3660, German Armistice Commission and the Service of Refugees, "Minutes," Paris, July 5, 1940.

15. A.M.V.R. 3514, March 24, 1941, Mayor of Reims to Mme. Choin.

16. Ibid.

17. For a full discussion of the repatriation negotiations, see Dombrowski, "Beyond the Battlefield," 317–86.

18. In Nord, unemployment registered at 18,090 in May 1940 and increased to 248,048 in August before the first round of refugees gained readmission. Taylor, 23.

19. For a discussion of repatriation, see Dombrowski, "Beyond the Battlefield," 330–55.

20. The British Royal Air Force (RAF) cycled in a series of bombers beginning in December 1937. The early model Vickers Wellington (B Mark 10) carried two tons of bombs and had a range of 2,200 miles and a flight radius of 1,100 miles. The Halifax 1 was a four-engine bomber that entered into service by November 1940 and claimed responsibility for the bombing of Hamburg. By July 1941 the British had approximately 732 bombers in operation with 353 Vickers Wellingtons, 40 Halifax, and 24 Sterling bombers. On January 9, 1941, the British tested the Lancaster Mark 1, which emerged as the British heavy bomber.

21. A.N. F/60/1507, Report, "Secours aux populations civiles, victimes de bombardements," written to Louis Marlier from the Commissioner of Police of Le Havre, September 1940.

22. Ibid.

23. Ibid.

24. Ibid.

25. Ibid.

26. Ibid.

27. Ibid.

28. Dombrowski, "Beyond the Battlefield," 140–42.

29. In Tokyo in 1934 the International Committee of the Red Cross (ICRC) drafted a document titled "International Convention on the Condition and Protection of Civilians of Enemy Nationality who are on Territory Belonging to or Occupied by a Belligerent." Article 9 vaguely stated that "[e]nemy civilians shall be protected against measures of violence, insults and public curiosity." Article 10 targeted belligerent powers and suggested that "Measures of reprisal directed against them [civilians] are prohibited." See http://www.icrc.org/ihl.nsf (accessed January 28, 2008).

Vichy finally acquiesced agreeing to relocation in the Free Zone, "*only* [my emphasis] if the Germans had absolutely no other means of accommodating them in the Occupied Zone." A.N. F/60/1507, Louis Marlier to Mayor Risson, September 28, 1940.

30. On resistance and collaboration in northern France, see Lynne Taylor, 61–72.
31. For complaints by northern mayors and civilians about evacuation relocation distances, see Dombrowski, "Beyond the Battlefield,"42. A.N. Series F/23/229. "Un rapport suisse sur la défense passive de Paris et le problème de l'évacuation." Prepared for the Prefect of Paris and dated March 12, 1940. Also Archives Départementales de la Marne, Annex Reims, Series 6W/R/2747.
32. The approximate number of refugees in both zones in August 1940 totaled one million. 540,000 resided in the Free Zone, and 460,000 were displaced in the Occupied Zone. Of these, 700,000 were lodged in camps or private homes and 40,000 were homeless in the Forbidden and Open zones of the Occupied Zone. See A.N. Series F1 a 3660, Report of the Service of Refugees, 1941, 1. No specific date indicated.
33. Taylor, 20.
34. A.N. F/60/1507, Report, "Secours aux populations civiles, victimes de bombardements," to Louis Marlier from the Commissioner of Police of Le Havre, September 1940.
35. Taylor, 57–58.
36. 44W 3980 9/3, Services de la Santé du Pas-de-Calais, to M. le Docteur Vielledent, Directeur Régional de la Santé, no date. Cited in Taylor, 58.
37. Ibid.
38. A.N. F1 a 3660, Report of the Service of Refugees, 1941, 1. No specific date indicated.
39. Taylor, 56.
40. Ibid.
41. A.N. F/60/1507, Letter, Louis Marlier to Mayor Risson, September 28, 1940.
42. On antinuclear civil defense in postwar United States, see Laura McEnaney, *Civil Defense Begins at Home: Militarization Meets Everyday Life in the Fifties* (Princeton, N.J.: Princeton University Press, 2000).
43. A.N. F/60/1507, Service of Refugees, Marlier to Ambassador of DGGFTO, March 2, 1941, 2–11. The week of February 22, British attacks claimed 15 lives. Six hundred residents left the town by foot.
44. A.N. F/60/1507, Service of Refugees, Marlier to Ambassador of DGGFTO, March 2, 1941, 2–11.
45. Ibid.
46. Ibid. In Saxenhausen prison, Borotha had a cell near Paul Reynaud. Paul Reynaud, *In the Thick of the Fight 1930–45*, James D. Lambert, trans. (New York: Simon and Schuster, 1955), 652.
47. A.N. F/60/1507, Service of Refugees, Marlier to Ambassador of DGGFTO, Report of March 2, 1941, 2–11.
48. Lieutenant-Colonel Arsène Vauthier, *Le Danger aérien et l'avenir du pays* (Paris: Berger-Levrault, 1930), 72, in Roxanne Panchasi, "Future Tense: The Culture of Anticipation in France between the World Wars," Ph.D. Dissertation, Graduate School, New Brunswick Rutgers, The State University of New Jersey, 2002, 138.
49. Dombrowski, "Beyond the Battlefield," 32–37. For an account of daily life of these Alsatian and Lorraine evacuees, see Laird Boswell, "Franco-Alsatian Conflict and the Crisis of National Sentiment during the Phoney War," *Journal of Modern History*, 71 (1999): 552–84.
50. During 1940, the RAF used Wellingtons and Sterlings, which had a fly range extending to France's interior and back to airfields in southern England.
51. A.N. F/1a/3660, "Minutes for the Service of Refugees' Meeting," Hôtel Majestic, Paris, July 5, 1940.
52. A.N. F/60/1507, "Notes," Dr. Kübler and Prefect Marlier, April 17–18, 1942, April 22, 1942.
53. Ibid.
54. Ibid., 3.

6

The Anglo-American Troops as Seen by French Labor Conscripts: Forms of Ambivalent Critical Support

Patrice Arnaud

The question of which side to support in the war was a simple one for the French prisoners conscripted by the Reich to perform forced labor. A German victory would likely extend their stay in Germany, while a rapid victory by the Allies would result in their quick return home. Few conscripts adopted a position contrary to their interests, yet they were sometimes critical of the Allied military strategy, especially concerning the use of bombing raids. Once they were liberated, how did these French soldiers view Allied soldiers? To what extent is this image clouded by cultural stereotypes formed during the interwar period and later reinforced by U.S. foreign policy after 1945?

Conscripts in the Allied Camp: Impatient Support

Postal inspection statistics from January 1944 indicate that only 6 letters out of 10,000 mention a German victory, down from 8 letters the previous month. In contrast, some 366 letters express hope and certainty concerning an Allied victory, as opposed to 292 a month earlier. Germany's proponents number 2.66 percent in December 1943 and only 1.6 percent in January 1944, figures that illustrate a natural support for the Allies' war objectives that corresponds to anti-German sentiment and the harsh experience of expatriation.[1]

Impatient to return home, however, the French conscripts were critical of the Allies' all too slow advances. Each bit of good news, such as the landing in Italy or Normandy, rekindled their hopes for a military reversal in two or three months. A *Sicherheitsdienst* (SD) report from August 2, 1943, mentions the "great joy" caused by news from Italy, leading several French work details to decide not to report at the Hermann Goering factories June 26 and 27, 1943.[2] The Berlin state prosecutor noted that in September 1943, workers at Finsterwalde appeared for work wearing ties, and that some of them even sported tricolor cockades.[3] From fall 1943 onward, French conscripts lived in hope of a landing in France. As summer approached, their impatience became tangible, sometimes turning to exasperation: "By God, won't they ever land? What the hell are they doing?"[4] The teacher Paul Fourtier Berger sums up the ambiguous nature of the critical support for the Allies: "Each extra minute prolonged terribly the massacre…for all those who suffered under the Nazis….Thus we were unkind in judging the higher authorities who held our fate in their hands, but we had great affinity for those who paid with their life to save us."[5] The majority of the French were excited by the landing, as the last letters intercepted by the censor reveal: "I believe you'll be better off with the liberators than with the Germans, whom unfortunately I've gotten to know all too well." "I hope you're in good hands with the Allies. Food supply conditions will also get better, even though the papers say the opposite."[6]

On June 6, almost everyone anticipated being home for Christmas, a symbolic date on which expectations focused.[7] The following days were more trying since propaganda mentioned heavy Allied losses. Until August the wait was "endless."[8] The news of the liberation of Paris intensified hopes for a quick outcome, so that, according to Paul Cèze, "anyone predicting spending yet another winter would have been met with hoots of contempt."[9] In September, the Allies' entry into Belgium led Pierre Destenay to think that the Americans would force Germany to capitulate within a month. "The American soldier was seen as a friendly and powerful spirit who had come from another land to uphold the law and defend violated freedom….In general, the French thought that the war would be over in October and that the American army would break through the Westwall in two weeks."[10] The counter-offensive in the Ardennes plunged the conscripts into depression, and the increasing number of bombings made them more impatient. In Leipzig in April 1945, Robert Lasaffre's wait took on a critical tone: "Patton's first division took up their position in the western suburb of Leipzig. The Hasag factory was in the north, and all the workers were in a fit. What are the Yanks waiting for since they're no longer meeting any resistance?"[11]

To get news about the advancing front, some conscripts tried to listen to the Allied radio. Listening to the BBC or Swiss radio stations represents one-eighth of our sample of judicial documents concerning political offenses. Arnaud Boulligny estimates that this is why 12 percent of his sample of 663 French workers in Sachsenhausen were imprisoned for "civil resistance."[12] The French often organized ways to spread news in the camps. A recurrent theme in the conscripts' stories is listening to the BBC, which they did most often by making a galena crystal radio set.[13] After the summer of 1944, the police forbid the French from having radios in the camps where sets had previously been set up, for fear that French national radio would be listened to.[14]

The French in Germany got their orders from the Allies by way of tracts. Reading these tracts was difficult since it was often necessary to translate from a text originally written in German. Tracts established solidarity among foreign workers, as was the case with two French electricians who distributed General Eisenhower's orders to foreign workers in September 1944.[15]

Support for the Allied side was also expressed in several songs. Out of 17 known songs, only one mentions the United States and another one alludes to the USSR, whereas 6 explicitly praise England's combat efforts. This preference can be explained by the date when these countries entered the war. The author of "La Chanson des requis" (The Conscripts' Song) states that Hitler deported the French when he saw their support for the English, and he expresses his hopes for an English landing.[16] In "Vive de Gaulle" (Long Live de Gaulle) another Frenchman mentions "glorious England" and her "heroic soldiers."[17] In "Viens mon petit fridolin" (Come Here My Little Kraut), the English give the Germans swimming lessons in the Channel, and the "return of the English forever" announces the return of happier days.[18] The author of "Heil Hitler" praises Churchill and de Gaulle.[19] A song titled "La Marche des requis" (The March of the Conscripts) surprisingly praises the bombings.

> When night falls, the Tommies show up…
> They have no fear of the DCA…
> They know a thing or two, these English and Americans,
> And from the darkness of the rubble, all the Krauts run off to hide…
>
> They wanted a total war, and they got a total war.
> We poor exiles in this god-forsaken place,
> Let's cry out to our Allied comrades, "Come one, come all."
> Everyone's needed to get the job done
> To exterminate this race of good-for-nothing hitlerians.[20]

The Ambiguous Attitude of the French Concerning the Bombings

The behavior of the French during the bombings was described in various ways. On May 28, 1943, the *Sicherheitsdienst* expressed its pleasure in the attitude of the majority of foreign workers. The French in particular supposedly participated actively in struggling to put out fires, with some of them crying out, "This isn't war, it's murder." However, a few rejoiced, saying when the air raid warnings sounded, "our friends are here." The report claimed that Frenchmen in Bremen and Koblentz lit torches to guide the planes, a surprising gesture since bombs don't choose their victims. In fact, according to the *Militärbefehlshaber in Frankreich* archives, the bombings caused 26.32 percent of the French deaths occurring between January 1943 and July 1944 (see figure 6.1).[21]

In Berlin, the proportion over the entire war comes to 45.6 percent.[22] However, as Regina Wallet notes, following the bombing of Leipzig on December 4, 1943, the French expressed great pleasure over what the English had done.[23] In Munich, Louis Girerd discovered the bloody head of a pilot who had been shot down, and he praises the sacrifice of this combatant who was killed far from home. "This was the first time we had seen the bloody face of one of our liberators. He was black and probably an American. 'My thanks, unknown friend, who sacrificed your life for our freedom when yours was not threatened.'"[24]

At times the French admired the bombings, such as the one of Linz on July 25, 1944, which during the 35 minutes it lasted struck only the Hermann Goering factories. "Perfect work," observed Fernandré Jules Viannec.[25] A few conscripts gradually changed their mind. Gabriel Vasseux, for instance, doesn't mince words in criticizing the military: "We said, 'the damn fools up there who were paid to toss bombs on us weren't any better than the bunch of idiots, including us, who got them for free.' I was a big fan of airplanes, and that made me a little more anti-military."[26] When he saw 21 comrades get killed, Michel Caignard couldn't understand "why the Allies go after the

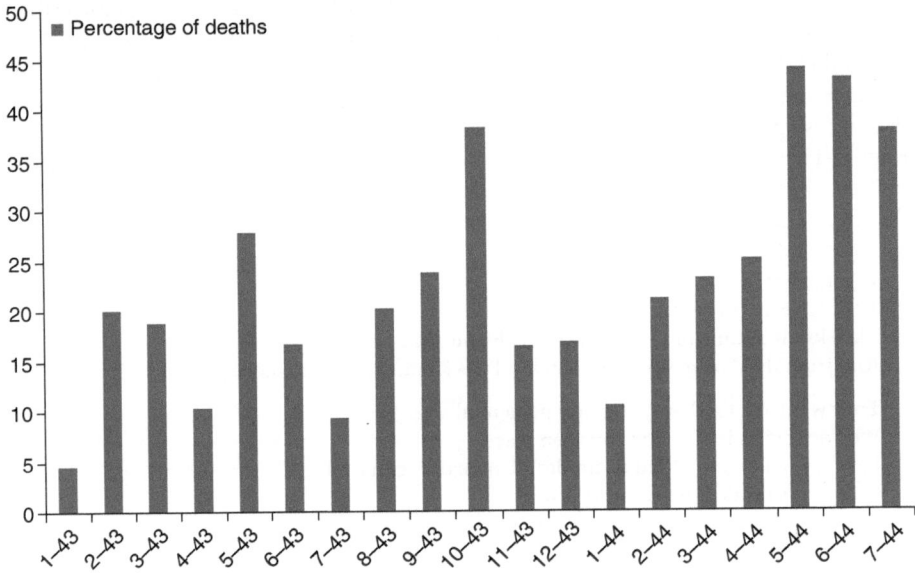

Figure 6.1   Mortality Table for Bombing Deaths of French Conscripts.

camp relentlessly, while the factory remains unscathed!"[27] Paul Fourtier-Berger notes that this observation led to political discussions in the barracks. "Since there was no longer any question about the Allies' supremacy, I suspected they were keeping the war going for a lot of reasons that my buddies slowly came around to.... [The brass— always in cahoots with Big Business, they know how things were done in previous war.... Not enough damage has been done yet."[28] Cavanna goes so far as to imagine that the American military-industrial complex influenced the decision to wage all-out air war in order to profit from bomb production.

> These damn Yankees are all over us at noon sharp.... They do it the American way— the whole load. Why aim for anything? All you have to do is toss out all the bombs you want, and the whole countryside gets wiped out for a couple of kilometers around the target.... Not a square centimeter is left untouched.... Sure it costs a lot, but aren't there wars so that factories are kept busy making bombs?[29]

Many STO's are torn between their hatred of Germany and their incomprehension of the Allied strategy of targeting civilians first. In Hamburg, André C. complained in July 1944, "After a few days of calm, the air raid alerts started up again more frequently. No mistrake about it—we all were asking when they would leave us alone!"[30] As in France, the English (unlike the Americans) were admired for the relative precision of their air raids.

> FB: We knew when it was the English. The English came in very low and aimed a lot better. There was no escaping the Americans because the Americans dropped their bombs from 3,000 meters up.
> PM: They started five kilometers before and the finished five kilometers after.[31]

Charles-Henri-Guy Bazin invokes the Wars of Religion to condemn "the cowardice of the pilots" who every night committed one or more Saint Bartholomew's Day massacres.[32] Alain Robbe-Grillet accuses the English and the Americans of cultural crimes in Nuremberg for having reduced to ashes Baroque churches and medieval wooden houses.[33] Yves Clavel uses the same terms as German propaganda when he describes the Allies as "gangsters in the air relentlessly pursuing their goal of massive destruction."[34] Surprisingly, the parallel with atrocities committed by the Nazis frequently reappears. Paul Fourtier-Berger accepts the parallelism between the genocide of the Jews and the gypsies and the use, late in the war, of time bombs on Munich. "Who could imagine a second attack right after the first one...that would blow up ambulances and kill the wounded?...Well, that's what happened on the evening of January 7 and the night of April 20, 1945. ...Who on the Allied side could have given orders like that?"[35]

Albert Kindig uses the term "holocaust" to characterize the sacrifice of the entire city of Königsberg to "the war gods," comparing the streets and cellars to "crematorium ovens."[36] Seeing the two sides in the same light raises the question of who the enemy is. Even the Allies' supporters had their convictions shaken, as is suggested by the journal of José Cabanis who, noting that Allied planes cut down children walking along a path, concludes with disillusion that "it was the war of Law against barbarism."[37] The Nuremberg tribunal has sometimes been accused of imposing the justice of the victors while remaining silent about Allied crimes. Henri Baudon claims that "all the war criminals haven't been judged."[38] Robert G. Riffé goes a step further, holding that Arthur Thomas Harris "and a lot of others, including Churchill," should have been on trial, concluding "apparently the massacre of civilian populations isn't a crime against humanity unless committed by soldiers or people of the conquered country."[39] Going beyond World War II, André Gabs-Mallut denounces the discrepancy between the discourse promoting human rights and the military strategy adopted to restore them.

> Phosphorus....On what grounds? What have we done?...On the TV, when they want you to applaud that, I want to throw up...Serbia, Kosovo, Iraq...Do Human Rights kill while poll numbers keep climbing?...February 14 and 15, 1945. The prisoners had been withdrawn to Dresden....Two nights of butchery....British cooking....More civilians were killed in these two nights than American soldiers were during the entire war....Crimes that went unpunished![40]

These examples show that in stories about World War II, reality is perceived through the lens of the author's political opinions, the actions of the United States being judged after 1945 in a way that plays a fundamental role in how memories are related. Not surprisingly, the conscripts who are farthest to the political left, such as Cavanna, or to the extreme right and leaning toward the Front National, such as Gabs-Mallet or Robert Riffé, criticize the Allies most harshly.

The clumsy tracts the Allied dropped weren't well received, and the psychological war backfired. The French in Berlin were just as fearful as the Germans when they read the tract that declared "Cologne was a test—Hamburg is a success—Berlin will be a triumph."[41] Pierre Destenay didn't enjoy the irony of the Allies as they boasted of destroying Hanover: "'Umsonst gratis,' the tract said....Were the Allies hoping that this would incite a popular revolution in Germany? The method was dubious."[42] The worker Maurice Georges expresses his bitterness over the Allied tracts that were addressed directly to them and clumsily wished them "good luck." "The Allied...tracts dropped from the sky repeatedly told us to remain idle and wished us good luck...in the face of bombs that couldn't choose their victims."[43]

The French Conscripts' Ambiguous View of the Allied Liberators

Except for the collaborators, none of the French feared the arrival of the Americans, who were awaited impatiently and with a certain curiosity. The French were surprised by the jeep, which they called a "strange little car with a flat hood"[44] or "a four-wheeled matchbox,"[45] and they wondered skeptically whether "this is what the Americans would win the war with."[46] They were also surprised by the uniforms. Cavanna mentions "odd-looking soldiers wearing little shirts that were too short and pants that were too tight," and Victor Dufaut describes "round, globe-like helmets that were almost medieval."[47] The Americans are rarely described physically, as if they were no different from the Europeans. A few conscripts stress how tall they were. Victor Dufaut uses some of the few things he knows about the United States to describe an American captain who is "as tall as a building, thin and slender. An arm of the Mississippi."[48] François Cavanna is more disparaging, equating the Americans with cow boys and stressing that "they all have big butts, even the thin ones, and they're hunchbacked, too."[49] Once again, cultural elements form a grid through which memories are reinterpreted according to current French stereotypes about America.

The French can quickly identify an American above all by his attitude. Unlike a Germans' military way of doing things, the Americans seem surprisingly relaxed in the way they fight the war. Bazin finds them "phlegmatic."[50] Their salute, according to Cavanna, is "limp."[51] Even in combat, they never abandon their "rubbery gait," and when they push back the civilians they do it "with a lackadaisical gesture"[52] The Polytechnique student Paul Lemoine is surprised by how "easy-going these soldiers are" and by the guards he sees "slumped in an armchair."[53] Others are surprised to see Americans with "their feet on the door, peacefully chewing"[54] or "their feet on the hood, an unusual position for German soldiers.[55]

Chewing gum, which beginning with the Liberation symbolizes the soldier from across the Atlantic, is connected to this attitude. Chewing gum isn't always referred to as a likeable trait,[56] and it sometimes reinforces a disparaging description. The rewriting of idiomatic expressions ("*les noirs s'en fichent comme de leur premier chewing-gum*," the Blacks don't give a stick of gum about it) stresses the link with a way of being. For Jean-Louis Foncine, sweets reinforce the stereotype in the image of "the gum-chewing guy from Kansas"[57] For André Castex, the expression "somebody who eats chewing gum"[58] reflects a critical distance, with chewing gum used symbolically in a disparaging way.

Despite the frequency of condescending descriptions, a few conscripts describe the enthusiastic reception the Americans received. Georges Caussé mentions an "indescribable state of excitation" and "unforgettable moments."[59] Pierre Destenay remembers "a delirious crowd" that accompanied "the Americans triumphant entry." Jean Dupin preserved the same memories of men "who marched to the applause of all the foreigners."[60] In the Stadt-des-KdF-Wafens, everyone is yelling wildly, with shouts of "Come on guys, On to Berlin, Done for, Long live America."[61] Victor Dufaut even compares the Americans to gods. After describing the "shrill cries" and the "hysterical laughter" of the "crazed crowd," he deifies an American soldier who can make war as well as cook crêpes.

> Stupidly we watched him, a kind of demigod on his mythological chariot. Pierre made a gesture like a beggar asking for food. "OK," the god answered. He opened a huge tin can, poured a thick liquid into the pan that was placed on a portable stove. He rolled two crêpes and offered them to us.... Just then we heard a roaring in the air.... Our demigod just turned from the stove to a machine-gun.... The gunfire didn't last long.... [He] laughed and handed us a second crêpe.[62]

All the liberations weren't the same. The moment of freedom was sometimes deceptive because of the combat situation. Bazin and his buddies were met with "shouted insults and vehement reproaches from officers"[63] because their vehicle was blocking traffic. Paul Fourtier-Berger's liberation expresses the civilians' deception in the face of the logic of military efficiency.

> Jean raced over to them, said something in English, touched a shoulder and tried to embrace them. No way! These guys were on a battle mission. Their look was impassive. With their free hand they pushed back my friend.... Bursting with joy, we all watched the passage of liberating army that we had placed our hopes in for so long.[64]

The French were frustrated that they couldn't celebrate and had to watch the passage of an army on the march. Paul Winkler agrees.

> We had been liberated but we weren't free yet.... There was no demonstration of joy. The English didn't give us the chance for it since they were bent on neutralizing the Germans.... We were happy, but with an unexpressed joy we kept inside. Who could we share it with anyway? Not with the Tommies for whom the war wasn't over, even less with the Germans.[65]

The French were even more irritated when they were subjected to the same harassment as the Germans. André Castex admits "having had his pride hurt" when he heard the GI's insults ("that's the enemy"), and he wouldn't excuse "their stupidity for having spoiled the moment of liberation that had been so long awaited." He bitterly concludes that "all the exhilaration this event could have held for him seems to belong only to his dreams."[66] This moment was also spoiled by a period of waiting that had gone on too long. In April 1945, Robert Lesaffre felt that "because spirits had been stretched too tight they lost their resilience" and that "the explosion of joy was ambiguous."[67] Similarly, Georges Caussé was surprised not to feel as joyful as he had thought he would: "Our happiness didn't reach the heights of our previous hopes. ... We thought we would die for joy, that we would be ecstatic.... And there we were, happy, delighted, but that was it!... No trumpets, no flowers or exhilaration, no shouts, displays or dancing."[68] André Gabs-Mallut's story from Watenstedt explains why the Liberation left the French less than fully ecstatic:

> April 11...The Americans!...Crowds were alongside the road...a hundred times more people than those who were arriving. People were on the look-out...anxious even. The

crowd...was thoughtful, all in a line, unbelieving, flabbergasted. We didn't wait any more. No spectacles, no shouting: no strength for it, not used to it. Liberated. Maybe.[69]

Beginning in summer 1944, many people saw the Liberation as inevitable, yet slow in coming. Thus the Americans' time frame is sometimes seen as a point against the Allies.[70]

Linguistic misunderstandings didn't make relations any easier. When the Americans arrived in Hanover, the only word the workers knew was "French," which produced this strange trilingual expression of gratitude on a streamer: "French *Arbeiter merci*."[71] Paul Winkler, a *polytechnicien*, can't find in his "poor high-school English the words or sentences with which to begin communicating."[72] Jean Edmond "mixes together German, French, and a few words of English that surface in [his] memory" to communicate with a Briton.[73] Describing the moment of Liberation allowed a few conscripts to express their opposition to the foreign policy of the United States. In Pomerania, François Cavanna claims to have waited for the Red Army because of the Americans distance and the unfavorable image he had of them. "The Americans don't hit strategic installations; they prefer cities full of people....I don't like these city-smashing jerks. I can see too many bodies in the rubble.[74] The U.S. strategy of protecting their soldiers' lives first was sometimes incomprehensible to the French who, like Robert Lesaffre, were quick to accuse them of cowardice because of their crushing material superiority.

> The center of town had hardly been taken when suddenly things got worse. A handful of young Hitler fanatics had gathered together, and in a last-ditch attempt these kids began to pierce tanks almost at point-blank range....With their pride wounded, the gum-chewers pulled back a few kilometers to bombard the sector day and night....When death finally calmed things, the soldiers just came back again and took up position on the main intersections.[75]

American war objectives were also virulently denounced. Because of American postwar superiority, authors are more critical in their memories than they were in 1945. André Michel indignantly sees American throwing "cigarettes and tin cans of food...[and] tossing milk chocolate," and he imagines that "the war was won in order to maintain the economic superiority of the US."[76] In Eisenach, Charles-Henri-Guy Bazin states that the "Amerlos" (Yanks) who "waited five years to help us came first to flood Europe with their junk: cans of inedible beans, the music of degenerates."[77] Luxury goods are repulsive for Cavanna, who remains indifferent to "American lavishness, American cars [and] American cigarettes."[78] Some conscripts share Lamothe's point of view: "the Yanks think about themselves first, then about others."[79] Following a plane accident, the French unanimously denounce the Americans' carelessness and their "hurry up and go" attitude.[80] Crossing a train on a boat bridge unleashes accusations against "the guys from the Far West" who "weren't bothered by the fine points of the art of river engineering."[81]

The "immorality" of the soldiers is another recurrent theme. When Jean Edmond asked for a smoke from a Briton with a trunk full of cigarettes, he was astonished to be asked for his watch in exchange.[82] André Deutsch encountered a black American driver who showed him a collection of stolen watches,[83] while an American sentinel on the Weser demanded a watch before he would let the French cross the river.[84] Paul

Lemoine had his "fantastic lueger" spirited away in the street by a GI.[85] These examples are few in number, however, compared with the thefts committed by Soviet soldiers. What the French condemn particularly is associating with German women. The civilians' resentment is accompanied by a kind of inferiority complex on the part of these former prisoners toward soldiers crowned in glory. In his story, Victor Dufaut makes a connection between sexual activity, masculinity, and heroism. By possessing a woman, he tries to compensate for his nonparticipation in battle, but he admits that this pleasure belongs above all to the winners.

> My heart and my body wanted love....I made more of myself than just a man: a superman, a hero....I was as good as any of them, these soldiers from America. We have to get over this hang-up....They say the Yankees make it with "Ruskas" and Polish women. A real orgy!...We didn't do everything we did just for the fun of watching an ex-gangster from Chicago panting in the arms of a Ukrainian peasant girl.[86]

The term "ex-gangster from Chicago" shows how much stereotypes shaped the mental images of the French. Reflecting on what was happening in France, Charles-Henri-Guy Bazin commented to his friend that "for the Americans and the Germans, France is Paris and girls."[87] André Michel was outraged that the Americans "are even making off with the French women from our side."[88] Catholic priests are the most severe toward sexual permissiveness. Jean Damascène's account shows that anti-Americanism was shared by a clerical right wing.

> Never did I encounter immorality equal to that of some members of the American army....In the first hours after the city had been taken, when the Germans had nothing left to eat, I saw an American proudly...go off with a pitiful German woman who, for a piece of bread, was going to give herself to him....The victor's exhibitionism filled us with shame and anger....I passed by a brigade of guardsmen...and I saw on the sidewalk...a dozen rubbers—used. I can't help but think...about the irony of the three little words people used to describe the Germans: they are proper. Maybe their successors will be too.[89]

The last comment reproaches the United States for occupying a country without understanding anything about its history and culture. This criticism pertains to the German example, but certainly it is reinforced by the American presence in the world after 1945. Prevented from looting a cellar by American soldiers, Jean Dupin writes, "as soon as they arrived, we began to curse them."[90] Bazin too was surprised that an officer made him give back a cart stolen from a German woman.[91] On May 13, André Michel observed the establishment of an occupation police force and the arrest of "everyone who tried to get away on their own."[92] François Cavanna could bear being locked up, which he equated with a harassment that reversed the roles of victor and vanquished.

> And what's the first thing these Yanks do? They shut us up behind barbed wire! Sentinels, MPs, no exit. "To avoid incidents!" Otherwise we might provoke the peaceful German citizens. We saw these peaceful German citizens...in their Sunday best. Young girls were on the arms of American officers, all proud. The ugly ones had to make do with just a soldier...I didn't see why I had to be locked up. If these people were innocent, what about me?[93]

For Robert Lesaffre, the Americans didn't see any difference between torturer and victim, and they wanted them to wash their dirty laundry in public.[94] Others attack the concessions made to the USSR. Jean-Louis Foncine criticizes the "wall of refusal" of the MPs who prevented thousands of refugees from rejoining the side of "democracy." Listening to a former soldier play the "Horst Wessel Lied" and the "International" on his harmonica, he wondered what "a gum-chewer from Kansas can understand about the subtleties of such provocations."[95] Several criticize the Americans' indifference to problems of repatriation[96] and their incomprehension of human problems. "Men from the other side of the Atlantic were camped out in unfamiliar surroundings, seeing nothing and understanding nothing. Their goal was to occupy a bit of land: they weren't interested in the humans who were there and who, in fact, got in their way."[97]

Some of the French went so far as to claim that the Americans didn't share the values of European civilization. Victor Dufaut describes a commander who was "dazed to find himself in a foreign country" and who, unlike European generals, was a leader "just as easily as he could be a machine-gunner."[98] Father Jean Damascène notes that "we saw reactions in them that were allogeneous with respect to the standards of propriety we called for."[99] In the closing of his *L'Europe buissonnière*, Antoine Blondin evokes the fall of the house of Europe and its replacement by American domination, comparing it to a barbarous foreign civilization:

> It was the rush to the West.... Superniel [a character in Blondin's story] smiled. He knew that the house would finally collapse. He thought about the searchers who would find in the rubble the young girl's body and the two faces joined on her chest, that of the Frenchman and the American, as if they were placed beneath the pyramid of a dead civilization.... They walked along for quite a while on the same path, hanging on to life. They walked until the day they met a black. It was then they realized they had made it home.[100]

Antoine Blondin's racism is an isolated example. In general, the French attitude toward blacks was rather positive. One conscript describes black GM drivers as "real likeable dare-devils."[101] Since they were used to colonial troops, the French sometimes established close relations with black Americans, either drivers or cooks, who gave them food and offered them advice on how to avoid MP patrols.[102]

Despite the extent of these criticisms, the attraction exerted by the United States is undeniable. Highly nutritional food provisions, although sometimes unfamiliar, along with tobacco reconciled the French with the occupying troops, especially after a long period of doing without. The conscripts were amazed by the lightness of American cigarettes. Victor Dufaut evokes the American dream "contained in long blond cigarettes" and the "thousand lips of the crowd... that quiveringly, avidly, seized this manna from heaven."[103] Constant Guimault remembers having received cigarettes, chocolate, tin cans of margarine and corned beef,[104] René Oradou a carton of cigarettes,[105] and André Deutsch some chocolate, sugar, coffee, concentrated milk, cigarettes and chewing gum.[106] America became synonymous with abundance and generosity: "our liberators finally entered, their arms filled with provisions."[107] In displaced persons' camps, people were astonished to be able to choose from different menus.[108] If Bazin from Bordeaux was "disappointed" by the gastronomic nature of canned food, most of the conscripts were disillusioned when they returned to France and encountered rationing.[109] American military power also commanded respect. Jean-Charles mentions an "avalanche of materiel" in Watenstadt,[110] while in Munich Joseph Lépinasse recalls a "very long line of men and materiel that stretched on for hours," so that those "who had just been liberated realized once again how terribly powerful the United States was."[111] The conscripts were surprised by technological advancements such as DDT, "a Yankee thing" that François Cavanna calls "a magic product."[112]

In conclusion, one cannot help being surprised by how the conscripts' memories of the English and the Americans differ. This difference stems more from the perpetuation of anti-Americanism, which became stronger during the postwar period, than from the situation in 1945, when a strong Anglo-phobia existed and gained expression especially after Mers-el-Kebir. If former labor conscripts focused their criticism on the Americans and largely spared the English, with the exception of Sir Thomas Harris, this difference in treatment was based on cultural stereotypes that were inveterate in the case of the United States, while biases against the neighbor across the Channel were diminished by the construction of Europe. What legitimately can be seen as ingratitude also stems from the difficulties the American military had in capitalizing on the favorable disposition on the part of liberated civil populations.

Notes

Chapter translated by Daniel Brewer

1. Statistical report on French workers in Germany, January 12, 1944, AN 3W118.
2. *Aufnahme und stimmungsmäßige Auswirkung des Regierungswechsels in Italien bei den ausländischen besonders italienischen, Arbeitskräfte und Kriegsgefangene im Reich*, SD report of August 2, 1943, BA, 58R167.
3. Berlin court state prosecutor, monthly report, September 9, 1943, BA, 22R3356.
4. Paul Fourtier Berger, *Nuits bavaroises ou les désarrois d'un STO* (Reims: P. Fourtier, 1999), 141, 188.
5. Fourtier Berger, 186.
6. Translated from German,"Auswirkung der militärischer Ereignisse im Westen auf die französischen Zivilarbeiter und Kriegsgefangenen," BA, 55R1233.
7. Maurice Georges, *Le Temps des armes sans arme: une tranche d'histoire à Berlin* (Beaugency: Elvire, 1990), 181.
8. Georges-Charles Demay, *Le Travailleur sans histoires—Paris-Berlin-Paris (1943–1948): chronique historique* (Yerres: Georges-Charles Demay, 1996), 158.
9. Paul Cèze, *Chronique des années noires* (Digne-les-Bains: Édition de Provence, 1994), 55.
10. Pierre Destenay, *Babel germanique* (Nancy: Berger-Levrault, 1948), 168–70.
11. Robert Lesaffre, *Des bruyères d'Auvergne aux ronces du STO* (Paris: Les Lettres Libres, 1986), 183.
12. Arnaud Boulligny, *Les Déportés de France arrêtés en Europe nazie (hors la France de 1939)*, DEA thesis, directed by Jean Quellien (Université de Caen, 2004), 127.
13. Interview with Georges Gandon, conscript at LMT, March 10, 1998.
14. Georges, *Le Temps*, 182.
15. Berlin KG prosecutor's indictment against Marcel E and Cyrille E, December 12, 1944, AN 40AJ1520.
16. Sentence of M [1923] by the Hamburg OLG, November 28, 1944, BAMC—Caen, Hamburg court.
17. Sentence of Victor R by the Hamm OLG, August 3, 1944, AN 40AJ1512.

18. Sentence of Raymond C by the Jena OLG, June 14, 1944, AN 40AJ1505.
19. Song created by L, Sta Dü, Rep 112–18619.
20. Victor Savary, *Journal de Victor Savary STO Berlin 1943–1944—les tribulations d'un Chanzéen dans le grand Reich* (Nantes: 1988), 127.
21. Death figures for French civilian workers in Germany, AN 40AJ863.
22. Cause of death for French war victims in Berlin, 1939–45, AN 9F3690.
23. Régina Wallet, *J'aimais un prisonnier* (Paris: André Bonne, 1953), 214–28.
24. Louis Girerd, *1920–1945: un passé toujours présent*, printed memoirs communicated by the author, 92.
25. Fernandré Jules Viannenc, *Le Danube était gris: histoire de la déportation des travailleurs français en Allemagne* (Nouméa: F.-J. Viannenc, 1981), 109–10.
26. Gabriel Vasseux's response to our questionnaire of August 27, 1998, 25.
27. Michel Caignard, *Les Sacrifiés: un ancien du STO raconte* (M. Caignard: Périgueux, 1985), 81.
28. Fourtier-Berger, 184–85.
29. François Cavanna, *Les Russkoffs* (Paris: Livre de poche, 1979), 312–13.
30. Postcard from André C. to his wife, July 9, 1944, correspondence generously offered by Mme Darracq.
31. Interview with Pierre Martin and François Bertrand, former *cadres des Chantiers*, spring 1998.
32. Charles-Henri-Guy Bazin, *"Déporté du travail" à la BMW-Eisenach 1943–1945* (Cubnezais: C. H. G. Bazin, 1986), 154.
33. Alain Robbe-Grillet, *Le Miroir qui revient* (Paris: Editions de Minuit, 1984), 124–25, 150.
34. Yves Clavel, *Les Amants de guerre: odyssée d'un Français à Berlin* (Paris: Promotion et Edition, 1967), 14–27.
35. Fourtier-Berger, 259.
36. Albert Kindig, *L'Odyssée d'un STO à travers l'Europe* (Paris: La Pensée Universelle, 1991), 120–21.
37. José Cabanis, *Les Profondes Années: Journal 1940–1945* (Paris: Gallimard, 1976), 202.
38. Henri Baudon, *J'étais le déporté du travail n° 6219*, typescript, n.d. (1980s), 75.
39. Robert G. Riffé, *STO 1943/1945* (Nice: Alandis Editions, 2000), 223.
40. André Gabs-Mallut, *Entre vivre et mourir* (Paris: Edition Godefroy de Bouillon, 2000), 45–48.
41. Jean Satanil, *René Marie: journal d'un déporté* (Paris: Calmann-Lévy, 1946), 80.
42. Destenay, 97.
43. Georges, 131.
44. Georges Moullet-Echarlod, *La Faim au ventre: STO* (Paris: La Pensée Universelle, 1978), 497.
45. Jean Raibaud, *Témoins de la fin du IIIe Reich: des polytechniciens racontent* (Paris: L'Harmattan, 2004), 353.
46. Raibaud, 353.
47. Victor Dufaut, *La Vie vient de l'Ouest* (Paris: Promotion et Édition, 1969), 197–98.
48. Victor Dufaut, 205.
49. Cavanna, 401.
50. Bazin, 332.
51. Cavanna, 401.
52. Fourtier-Berger, 335–36.
53. Raibaud, *Témoins*, 99.
54. Lesaffre, 183–84.
55. Michel, 114.
56. Bazin, 341.
57. Jean-Louis Foncine, *Un Si Long Orage*, vol. 2, *Les Eaux vertes de la Flöha* (Pouilly-sur-Loire: Héron, 1995), 273–76.

58. Lesaffre, 183.
59. Caussé, 159.
60. Jean Dupin, *Nous avions 20 ans: des monts du Lyonnais aux usines de l'Allemagne nazie en passant par les Chantiers de Jeunesse* (Saint-Laurent-de-Chamousset: Centre social et culturel, 1994), 61.
61. Jean-Charles, *La Vie des Français en Allemagne: Ceux du Tac (Stadt des KDF-Wagens 43–45)* (Bordeaux: Bière, 1945), 94.
62. Dufaut, 199.
63. Bazin, 332.
64. Fourtier-Berger, 335–36.
65. Raibaud, 329 (Paul Winkler's account).
66. André Castex, *Au-delà du Rhin 1943–1945* (Pau: n.d.), 95.
67. Lesaffre, 183–84.
68. Georges Caussé, 159.
69. Gabs-Mallut, 101.
70. Raibaud, 318–19.
71. Moullet-Echarlod, 497.
72. Raibaud, 329.
73. Jean Edmond, *La Vie en Allemagne nazie 1943–1945: c'est là que j'étais*, account written in 1996, put on the Web in 2000 and placed in the IHTP [ARC 067], 88.
74. Cavanna, 334.
75. Lesaffre, 183.
76. Michel, 122.
77. Bazin, 351.
78. Cavanna, 334.
79. Quereillahc, 291–92.
80. Bazin, 352.
81. Lesaffre, 190.
82. Edmond, 88.
83. André Deutsch, *Zalika: sous les bombes a germé une rose* (Paris: Osmondes, n.d [2001]), 89.
84. Jean-Charles, 105–07.
85. Raibaud, 99.
86. Dufaut, 203–04.
87. Bazin, 1986, 341.
88. Michel, 113.
89. Father Jean Damascène de la Javie, *Prêtre-ouvrier clandestin* (Paris: France Empire, 1967), 173–74.
90. Dupin, 62.
91. Bazin, 340.
92. Michel, 118.
93. Cavanna, 403.
94. Lesaffre, 184.
95. Foncine, 273–76.
96. Raibaud, 353 (Robert Clausse's account).
97. Lesaffre, 184.
98. Dufaut, 205.
99. Father Jean Damascène de la Javie, 192.
100. Antoine Blondin, *L'Europe buissonnière* (Paris: 1949; reed. La Table ronde, 1953), 529.
101. Daures, 238.
102. Raibaud, 319.
103. Dufaut, 197–98.

104. Guimault, 183.
105. Account of René Oradou, teacher, class 1920, http://perso.wanadoo.fr/ror/allemagne1.htm (accessed January 14, 2008).
106. Deutsch, 80.
107. Odette Chambroux, *L'Exilée* (Les Monédières: Lucien Souny, 1990), 108–09.
108. Bazin, 341.
109. Caussé, 164–65.
110. Jean-Charles, 94.
111. Joseph Lépinasse, *Les Bradés de l'an 40* (Saint-Just-la-pendue: J. Lépinasse, 1984), 172.
112. François Cavanna, *Les Russkoffs* (Paris: Livre de poche, 1979), 404.

7

De-Prussianizing the Saar: Lessons from the Second World War and the French Path to European Integration

Bronson Long

In 1919 the Versailles Treaty gave France control of the German province of the Saar and its valuable coal mines. French rule under the auspices of a League of Nations' mandate was highly unpopular in the province. After 1933, however, the possibility of its return to Germany also meant incorporation into the Third Reich. After a heated campaign in late 1934 and early 1935, Saarlanders voted overwhelmingly in a plebiscite held on January 13, 1935, to rejoin Germany. During the Second World War, the Saar's strategic location along France's eastern border as part of the *Westwall* and its coal and steel industries made it a target for allied bombing. By May of 1945 much of the province, especially its capital Saarbrücken, lay in ruins. Following a brief American occupation at the end of the war, elements of the French First Infantry were left in sole control of the Saar on July 10, 1945.[1] With the arrival of French troops, the Saar became part of the French zone of the occupation in Germany under the command of General Pierre Koenig. This zone also included Rhineland-Pfalz, the southwestern parts of Baden and Württemburg, and the northwest section of Berlin. The French presence in the Saar, however, was quiet different than in the rest of its zone of occupation, and the policies implemented there had a much more prolonged impact on postwar Europe.

Curiously enough, the French occupation of the Saar after the Second World War has largely been forgotten. Historians, however, often overlook or underestimate the importance of military occupations. For example, Tony Judt's excellent work, *Postwar: A History of Europe since 1945*, gives little attention to the actions on the ground taken by British, American, Soviet, and French military officials to alter German society. Judt discusses the failure of re-education efforts, but does not adequately address the broader cultural, social, economic, and political ramifications of the occupation.[2] Governments have used wars as instruments of policy, and the violent, unpredictable nature of warfare has greatly influenced the past. Nevertheless, military occupations have also had an important role in reshaping nations, societies, and international relations.

Military forces often spend as much time, and sometimes more, occupying conquered territory as they do engaged in warfare itself. In explaining the importance that militaries have had upon the past, historians need to account for the impact of military occupations. This article suggests that historians should seek to broaden their examination of the impact of war to include a central place for military occupations. It argues that France's occupation of the Saar in 1945 and the High Commission that followed it were of great significance to postwar Europe because they complicated Franco-German relations and the early stages of European integration in the early 1950s. The Saar dispute, therefore, is an important example of how military occupations have influenced past events.

In many respects, France's occupation of the Saar was similar to many other military occupations. The basic duties of French troops included maintaining order, seizing, and destroying *Wehrmacht* munitions and equipment, and arresting high profile Nazis and war criminals.[3] On the whole, French troops performed their tasks admirably, and relations between them and the population improved over time to such a degree that their commander remarked in 1949 that Saarlanders "no longer consider French soldiers as occupiers, but as friends."[4] The French occupation of the Saar, however, was far from an ordinary military operation. During the occupation and in the years immediately after its conclusion the French attempted to separate the Saar from Germany. France's representative in the Saar, Gilbert Grandval, wanted to transform this province into a new nation or a "second Luxembourg" as he came to describe it on numerous occasions.[5] In some sense, Grandval's views on the Saar after the Second World War fell within the general scope of French thinking about Germany and French concerns about German militarism, nationalism, and anti-democratic tendencies. Yet his long-term plans for the Saar were quite radical, extending well beyond decentralizing Germany or pursuing the French government's economic-minded goal of creating an autonomous Saar economically united to France. Moreover, Grandval's agenda often clashed with those of important figures in Fourth Republic governments in the late 1940s and 1950s, especially in the Quai d'Orsay. Grandval and his allies in the Saar based their idea of Saar nationhood on two essential elements, Saar "particularism" and European identity. First, after a complete "de-Prussianization," a period of French guidance, French *pénétration culturelle* and the establishment of strong cultural and personal ties with France, the Saar's own unique culture would emerge. This culture was to be Francophile yet Germanic, though distinct from German culture. Finally, by the late 1940s, Grandval and his allies in the Saar believed that the cosmopolitan mixture of French and German culture coupled with the Saar's geographic position would lead to strong popular support for European unity. Grandval's conception of a united Europe, however, was one under French leadership. The attempt to foster a "particular" and European Saar meant that the French occupation of the Saar was part of Grandval's exercise in nation building.

The literature on the French occupation of the Saar and French relations with this province in the late 1940s and in the 1950s has underestimated the degree of radical change that Grandval had in mind and the ensuing impact these plans had on efforts to foster Franco-German reconciliation and the early stages of European integration. Grandval genuinely believed that the Saar was a distinct nation that had long suffered under the boot of Prussian oppression. His idea to use Luxembourg as the model for postwar Saar nationhood found much resonance among the Saar's political elite. Both they and Grandval were very early supporters of the movement to unite Europe and saw Saar independence as unfolding within the context of this new Europe. Much of the recent literature has either neglected the full scope of Grandval's plans or has portrayed him as merely seeking to strengthen Saar autonomy against French planners such as Michel Debré who advocated strong French control of the province.[6] Although Grandval's designs developed and became more precise over time, he remarked as early as February 1946 that French goals for the province should go beyond economics and should "constitute the first stone laid down in the construction of a solid and durable European edifice."[7] European integration, therefore, was central to Grandval's thinking on the Saar.

The Growth of Francophile Forces in the Saar

France's Saar policy developed gradually after the end of the Second World War. Gilbert Grandval was the most important French official in the shaping and implementation of this policy between 1945 and 1955. Grandval arrived in the province in early September 1945. He served as France's highest ranking official in the Saar, first as military governor from 1945 to 1947, then as High Commissioner from 1948 to 1952, and finally as ambassador from 1952 to 1955. Grandval was a Gaullist and later went on to head the colonial administration in Morocco and to serve as labor minister under Prime Minister Georges Pompidou in the 1960s. For Saarlanders he was France's direct representative in one capacity or the other for nearly ten years. Grandval himself came from an old Alsatian Jewish family that lived in Paris and was well connected in Parisian society.[8] During the Second World War, Grandval served in the French resistance. He rose to the rank of colonel, and General Koenig, his immediate commander in London, put him in charge of resistance forces in eastern France, including almost all of Alsace and Lorraine.[9] Grandval's experiences in the resistance had a deep influence on his views of Germany.

Political life in the Saar began anew after General Koenig issued an order for the entire French zone on December 13, 1945, that allowed political parties to form again. Grandval found many important allies for his plans for the province among its postwar political leaders. Many of these leaders had fled the Saar in 1935 and went into exile in France, and those returning from exile or imprisonment played a greater role in Saar politics than in the politics of any other German province after 1945.[10] During their ten years of exile, these leaders had developed markedly Francophile tendencies as well as a suspicion of German nationalism. In addition, their experiences in exile led them to consider the traditional nation-state as incapable of coping with the problems of the postwar world, and they saw Saar independence from Germany and European integration as a necessity. During the French occupation, the Saar's two dominant parties, the *Christliche Volkspartei* (CVP) and *Sozialdemokratische Partei Saar* (SPS), came out in favor of an autonomous political status for the Saar and economic union with France. In their earliest resolutions in April 1946, these two parties also came out strongly in support of European unity.[11] The most radical group in the Saar after the war was the *Mouvement pour la Libération de la Sarre* (MLS), which was created in Paris on March 25, 1945. The MLS successful united most exiled Saarlanders and at the end of the war moved to Saarbrücken, changing its name to the *Mouvement pour le Rattachement de la Sarre* (MRS) in February 1946.[12] The MRS advocated a gradual French annexation of the Saar and wrote to Bidault in February 1947 that "the majority of the Saar population will furnish the French state with good and loyal citizens."[13] The MRS grew in the immediate years after the French occupation and by February 1947 claimed 150,000 members.[14] As the MRS came to include members in several different Saar political parties, it became a powerful force in the province in the years immediately following the war.

In the spring of 1946, the MRS began a campaign for closer relations with France. This culminated with the "French days" of May 18 and 19. These pro-French celebrations were the largest public manifestation put on by the MRS. They included speeches, fireworks, singing, a procession of people dressed in French revolutionary costumes, a French military parade, and a parade of around 5,000 children in Saarbrucken carrying French flags.[15] In the town of Saarlouis the MRS and a like-minded group called the *Souvenir du Maréchal Ney* unveiled a statue dedicated to the memory of Marshal Ney, one of Napoleon's generals and a favorite son of the town.[16] At the unveiling ceremony, the *Souvenir du Maréchal Ney*'s president Josef Maria Felleten proclaimed that Prussia had lied when it said Saarlanders were German. He implored the people to "consult your genealogical trees and you will see that the roots are French!"[17] With Governor Grandval and General Koenig in attendance, the mayor of Saarlouis, Pierre Bloch, also spoke and expressed the view that "the majority of the Saar population approves the policies of the French government and is only waiting for it to take the decisive step to create a territorial union between Saar and the great French nation."[18] Louis XIV's military architect Vauban had founded Saarlouis as a French border fortress, and the Saar was a French department during the French Revolution and under Napoleon. The MRS and their associates attempted to use this past to their advantage and to claim that the true cultural and national identity of the Saar was with France, that the Prussians had imposed their own identity upon the Saar, and that France should restore the borders of 1814.

The "French days" in the Saar of May 18 and 19, 1946, were curious in many respects. They were described as the largest public celebration in the Saar's history, with 120,000 people out of a population of approximately 850,000 turning out for the various festivities.[19] Yet did this indicate that many in the Saar saw themselves as exclusively Saarlanders or even French? Were Saarlanders abandoning any sense of German identity scarcely a year after the fall of the extremely nationalistic Third Reich of which they were a part? Or was participation in the French days a convenient way of forgetting, ignoring or even hiding the recent Nazi past? While the outward manifestations of the French days might have suggested that Saarlanders had begun to embrace a separate, Francophile Saar identity that French authorities alleged years of Prussian domination had suppressed, the reality was more complicated.

That Saarlanders came to have a positive view of the French by 1946 was hardly surprising. At the end of the war, poor material conditions in the Saar, especially the need for housing, food, and the revival of its coal and steel industries, made economic cooperation with France particularly advantageous. Concerns for the present and the future, especially economic ones, thus fueled Francophile sentiment in the Saar. As Peter Sahlins has argued with the case of the Basques and the Pyrenees, people living along borderlands often tend to identify with the state that provides the most material benefits irrespective of their linguistic or cultural identities.[20] This was certainly the case in the Saar in the years immediately after the Second World War when France and the French state had far more to offer Saarlanders economically than Germany, which no longer even existed. It is therefore difficult to assert that most Saarlanders participating in the French days and other MRS meetings were expressing a newfound Saar or French identity. Instead, as Armin Heinen and Armin Flender have argued, many may have simply taken part in the French days because it was an event that included access to more food as well as to entertainment.[21] Yet economic benefits were not the only reasons for the turn toward France and pro-French expressions in the Saar. The results of the war and the false promises of Nazism were plain for all to see, leaving many Saarlanders disillusioned with extreme German nationalism. As the end of the war had meant the demise of both Germany and Prussia as political entities, many found some sort of close association with a victorious France an inviting prospect.

The postwar economic and political situations alone did not explain the Saar's attraction to France. Saarlanders also wanted to escape or forget the recent Nazi past. The Saar, however, had been a part of the Third Reich for ten years. Nazi propaganda, pressure, and intimidation as well as a genuine desire to return to Germany were factors in the outcome of the 1935 plebiscite. The League of Nations administered this plebiscite in which over 90 percent of Saarlanders voted in favor of uniting the Saar not simply with Germany, but with Nazi Germany. Support for the Nazis and the fervor of German nationalism in the Saar thus played important roles as well. Yet this plebiscite took place nearly two years after Hitler's seizure of power. In this manner Saarlanders had a better idea of exactly what Nazi rule would mean than did other Germans in 1933, and despite the appeals of an anti-fascist coalition of socialists, communists, and Catholics chose it over union with France or a benevolent international rule under the League of Nations.[22]

Nazi control meant a total loss of civil liberties in the Saar, the rule of a totalitarian state, the persecution of the Saar's Jewish community, and ultimately war. During the war Saarlanders fought in the *Wehrmacht*, and they participated in the harsh occupation of neighboring French Lorraine. Foreign workers were forced to work in the Saar during the war and endured cruel treatment and even executions. Saar coal production reached its highest point ever of 16,000,000 tons a year in 1943 under the Nazis, and while Saar coal helped fuel the Third Reich, its steel industry was active in the construction of tanks and submarines.[23] The Saar also played a role in the Holocaust. A Gestapo camp called Neue Bremm was located in the Saar on the outskirts of Saarbücken. Neue Bremm was not officially a concentration camp and was quite small in comparison to Dachau, Bergen-Belsen, and especially Auschwitz. Yet many were deported from Neue Bremm to such concentration camps, and prisoners at Neue Bremm lived in poor conditions and were subjected to some of the same brutalities as in other concentration camps. Some of its inmates who were shipped to concentration camps later testified that Neue Bremm was even worse than Buchenwald or Dachau.[24] Although it was certainly dangerous for Saarlanders to speak of Neue Bremm during its operation, its location just outside the city of Saarbrücken meant that many residents of the city knew that it was more than a work camp.

Francophilia in the Saar after 1945 blossomed in large part because it served to mask a dark past and to paint the Saarlanders as victims of Prussia instead of participants in actions of the Third Reich. The trial of Neue Bremm's personnel in 1946, which did include testimony from victims, received coverage in the Saar press.[25] In addition, in March 1946, the French military government opened a large exhibit at the Saarland-Museum in Saarbrücken whose aim was to show Nazism's crimes to the public. Governor Grandval and the leaders of the CVP, SPS, and Communist Party all attended the grand opening of this exhibit, which included photos of what had transpired at Neue Bremm, Dachau, Bergen-Belsen, and Mauthausen.[26] In order to allow as many people seeing these photos as possible, the Saarland-Museum was open daily and admission was free.[27] The Saar press also covered the proceedings of the Nurenburg trials, and the convictions and executions of high-ranking Nazi officials in the fall of 1946 made front page news.[28] While certainly the awareness of Nazi atrocities was hardly complete in 1946, it was not coincidental that 1946 saw both the first major revelations of the horrors of Nazism and the greatest outpouring of Francophile sentiment in the Saar. The motivations behind Francophilia, even as expressed by the MRS, were frequently to escape identification with and thus responsibility for Nazism. Yet the widespread embrace of Francophilia in the Saar immediately after the war also indicated that the trauma of the war and the defeat of German had caused a crisis of German national and cultural identities, which were thus being drastically reshaped.

French "pénétration culturelle" and the Construction of a Saar National Identity

By the beginning of 1948 the Saar had its constitution and a democratically elected government. The French military government in the Saar officially came to an end with the establishment of this Saar state and the creation of a customs and monetary union between France and the Saar. Grandval, however, remained in Saarbrücken as France's High Commissioner, and in this capacity he had veto powers over the Saar's parliament. French troops also continued to be garrisoned in the Saar, although their presence was more discreet. The formal end of the occupation thus hardly led to a strong reduction of French authority in the province. Although Saar Prime Minister Johannes Hoffmann had merely called for Saar autonomy before taking power, his actions afterward clearly indicated that his ultimate goal was to make the Saar into a new nation independent from Germany. Both Grandval and the Saar government used ceremonies and political symbolism to reinforce the legitimacy of the Saar's new order and to create a new sense of identity in the Saar, or Saar "particularism" as Grandval called it. One of first symbolic changes made in the Saar was to rename many streets. Grandval wrote Laffon on February 11, 1946, that Saar municipalities had already begun to change the names of streets that honored Nazism, Nazi leaders, and the *Wehrmacht*. In Grandval's view, the purging of such street names needed to take place not only in the Saar but in the entire French zone of occupation, eliminating street names that "glorify German militarism or German nationalism."[29] Some new street names were quiet ordinary, for example changing "Adolf-Hitler Straße" in Saarbrücken to "Bahnhofstraße." Others, however, reflected the new political reality, as with the renaming of a street in Saarbrücken after exiled Social Democratic leader Max Braun, who had become a kind of martyr figure for Social Democrats in the Saar after 1945. Overall, 127 streets and squares were renamed in the Saar between 1945 and 1947.[30] Giving streets new names was both part of de-Nazification and the construction of a new political order.

Even more important than renaming streets was establishing new public holidays and celebrations in the Saar in 1947 and 1948. The most striking holiday adopted in the Saar after 1945 was France's national holiday, Bastille Day. Widely celebrated in the Saar for the first time in 1947, Bastille Day became an official holiday the following year. July 14 celebrations featured large displays of French flags and banners, French military parades, speeches, fireworks, and other festivities.[31] In this manner, they resembled the French days of 1946. Yet Grandval and Saar political leaders tried to give July 14 a different meaning from that given to the French days. Instead of reminding Saarlanders of their alleged French roots, July 14 was presented to Saarlanders as a celebration of the universal principals of liberty, brotherhood, and equality.

Bastille Day was of particular interest to French authorities. In a front-page article in the *Saarbrücker Zeitung* on July 14, 1947, Grandval described the day as an international one that concerned the rights of man and the victory of democracy over tyranny. Moreover, he noted that such Germans as Kant and Goethe were part of revolutionary times, and that the heritage of the French revolution was German as well.[32] For their part, the SPS stressed the connection between socialism and the French revolution, with the party's newspaper the *Volksstimme* arguing in 1946 that "we socialists have always felt that July 14 was our holiday, as in the 150 years following it the ideas of socialism and freedom developed."[33] Christian Democrats on the other hand emphasized the connection between human rights and July 14.[34] The celebrations of July 14 in the Saar thus came to serve the interests of those in power in the Saar and to validate the new political order.

What did Saarlanders think of July 14 celebrations? Was there a similar dynamic at work as with the "French Days"? According to Armin Heinen and Armin Flender it was an important celebration from 1946 to 1950, but was not celebrated widely in the Saar after 1950. For Heinen, July 14 celebrations simply did not take hold in the Saar, and ultimately "Saarlanders celebrated in other ways than did the French."[35] Enthusiasm about July 14 may well have decreased after 1950. Yet French reports after 1950 described large crowds of Saarlanders attending July 14 celebrations as well as large numbers of French and Saarlanders who attended Grandval's customary July 14 garden party.[36] Even as late as the summer of 1955, the French government's report on the festivities indicated that crowds were visibly relaxed and as large as they had been in previous years.[37] Celebrations of July 14 thus appear more important after 1950 than described in the accounts of Heinen and Flender. Nevertheless, what meaning the Saar population took from them is unclear, and July 14 celebrations and especially the speeches of both Saar and French political leaders were always occasions in upon which the meaning of the holiday was associated with their policy goals.

Celebrations on July 14 were not the only ones that French and Saar authorities utilized to legitimate the new order in the Saar. The Saar government celebrated each anniversary of the proclamation of the Saar Constitution on December 15, 1947, with great fanfare. Events often included music and speeches, and the Saar government took great care to use these occasions to remind Saarlanders that the Constitution established a democratic order in the Saar that brought prosperity through the economic union with France.[38] The most prominent use of December 15 to foster Saar "particularism" occurred in schools. December 15 was a school holiday after 1949, and schools had celebrations leading up to this day that always concluded with teachers reading the Constitution's preamble proclaiming the Saar's separation from Germany.[39]

Hoffmann was a strong proponent of Saar "particularism" and used December 15 as a day to reinforce the legitimacy of his regime and to teach Saar children about their national identity.

Education was important to Grandval and Hoffmann in their attempts to construct a new Saar. In addition to "de-Prussianizing" the Saar, Grandval thought that "French success in the Saar will largely be determined by the results of our cultural policies, which themselves are dependent on using everything at our disposal to improve education."[40] During the occupation, Grandval had taken several measures to spread French culture in the Saar. Despite material difficulties, French became obligatory in all schools in the Saar beginning in the second year of primary school at the start of the 1946–47 school year. The stress on early French language acquisition was markedly different from the rest of the French zone of occupation where French instruction was only required beginning in the sixth year of school.[41] While the French military government emphasized the instruction of children, they also encouraged adults to learn French. The French created a program in 1947 that showed basic French lessons in the Saar's movie theaters in the 20 minutes before the beginning of all films.[42] In 1946, the military government started French night classes for adults. Over 7,000 Saarlanders took these classes, and the French had to import 10,000 textbooks for them.[43] Although the French brought in artists, musicians, and theater groups and sponsored a number of exchanges between France and the Saar, schooling and language instruction were by far the most important aspect of their cultural policies in the Saar.

The large number of French military and mining officials in the Saar after the war required the French to establish schools for their children, which Saar children were also allowed to attend. By December 1945, the French had opened nine primary schools of their own. On January 3, 1946, the first French secondary school in the Saar, the Collège Maréchal Ney, opened with 155 students.[44] This school in turn was transformed into the Lycée Maréchal Ney on September 15, 1947, and Grandval argued that it should become a "Franco-Saar lycée."[45] Over the course of the next several years the Lycée Maréchal Ney became exactly what Grandval hoped it would, namely, a Franco-Saar school in which both French and Saar students received a French education. Indeed, the school began with 510 French and 92 Saar students. By 1950, Saar students outnumbered the French 645 to 634. Enrollment continued to increase over the next five years. By the fall of 1955 it peaked with a total of 1,700 students, 900 of whom were Saarlanders.[46] In the end, Grandval achieved his goals for the school. His cultural director Pierre Woelfflin recalled in 1980 that "it is not unusual today to meet alumni from Lycée Maréchal Ney in the highest positions of the Saar's government administration, professions, and businesses."[47] The Lycée Maréchal Ney was thus a success for the French in their efforts to utilize education to construct a new Francophone and Francophile elite.

The Franco-Saar Cultural Accord of 1948

With the end of the occupation, France and the Saar sought to regularize their cultural relations. Talks for a cultural accord between France and the Saar began in earnest in May 1948 and concluded in early December. This accord was a broad and far-reaching one, and its 33 articles covered everything from the new university, the teaching of French in Saar schools, to sports and academic and artistic exchanges. A major component of the cultural agreement was the establishment of the new university, the *Universität des Saarlandes*, which had moved from Homburg and opened as a full university in Saarbrücken on the site of former army barracks under rector Jean Barriol on November 1, 1948.[48] In describing the university, the Franco-Saar cultural accord stated that it was through "the creation of a Saar university that the spirit that enlivens Franco-Saar cultural life expresses itself with the greatest originality and audacity." Furthermore, the university's mission was to become an "institution with bilingual instruction that draws professors and students from all nations, proclaiming its ambition to become the symbol of European hope."[49] The cultural accord heavily committed France to the university and stipulated that the French pay half of its budget and work out a system to recognize exams passed and degrees awarded in the Saar and France. In addition, France agreed to pay the salaries of titled professors, lecturers, assistants, and other personnel from France in such disciplines as French, history, philosophy, law, and economics. The new university was governed by a rector, who was at first to be French and was responsible to an administrative council made up of an equal number of Saar and French representatives with a French president. Finally, France pledged to send French artists, musicians, and theater productions to the Saar as well as to support various student exchanges and scholarships for Saar students to French universities, especially the prestigious *École Nationale d'Administration* where many of France's elite government administrators were educated.[50] The cultural agreement thus gave higher education in the Saar a strong foundation and a European scope, especially with support from France.

Cultural relations between France and the Saar were not limited to higher education. Indeed, France's initial concern after the war was primary and secondary education in the Saar, and the cultural accord addressed these areas as well. Articles 20, 21, and 22 required that Saar children have access to French schools in the Saar, mandated the teaching of French in Saar schools starting in the second year, gave French the preponderant position among foreign languages taught in the Saar, and allowed French officials to visit Saar classrooms in order to inspect French courses.[51] The guarantee of French *pénétration culturelle* that the agreement brought, however, was what both French and Saar officials wanted in order ultimately to foster Saar national sentiment. Johannes Hoffmann was a central figure in the signing of the cultural accord, and the celebrations surrounding the official signing of the agreement on December 15, 1948, were the most lavish of his entire administration. French foreign minister Robert Schuman and French education minister Yvon Delbos came to Saarbrücken to sign the agreement, with Hoffmann signing for the Saar and Schuman for France.[52] In his speech as part of the ceremonies Grandval placed the importance of cultural understanding in European terms, stating that France and the Saar were signing a cultural agreement in which they were "certain that we have created something durable that will greatly serve France, the Saar, Europe, and world peace."[53] In many respects, the cultural accord, with everything that it placed on firm legal ground after the conclusion of the French occupation, from the university to bilingual schooling in primary and secondary education, was Hoffmann's most important accomplishment as head of the Saar government. The university was by far his most enduring one.

Conclusion: The Saar and the Importance of Military Occupations

French officials in the Saar supported the Saar's new leadership and pursued a policy of "particularism" in the Saar after 1945 in which they sought to strip away Prussianism's grip on the Saar and use French cultural influence to give birth to a Saar with its own specific political and cultural characteristics. While France succeeded in creating a Saar that was economically joined to France and autonomous from Germany, Grandval admitted in April 1948 that "in contrast to the Rhineland, this new and industrial country which was settled by Prussian immigrants over the course of the nineteenth century and held under Prussia's firm tutelage now shows only a few signs of the particularism that we hope to awaken." Grandval, however, remained hopeful, stating that France still might succeed in its goals for the Saar "by developing and assembling a certain number of elements that are still fragmentary or embryonic through patient and methodic effort."[54] Nevertheless, Grandval's chief of cabinet Henri Vivard's lengthy report for Grandval in March 1949 on the Saar's status was less optimistic, arguing that "we should realize that since 1945 we have directed a policy of particularism and autonomy in the Saar, which is the only province behind the Oder that does not possess a past particularism or the least anti-Prussian sentiment."[55] For Vivard, Saarlanders were fundamentally a Germanic people and would in the long run choose Germany over the Saar's current situation if the economic situation improved there. Saar political leaders were struggling not to appear as French puppets and sought to reinforce Saar particularism and the authority of the Saar state. Overall the trend was beginning to move toward "a progressive weakening of the French position in the Saar despite the construction of a constitutional, state, legal, political, economic, and administrative edifice."[56] According to Vivard, European unity held much potential for French and Saar officials since Saarlanders "as good Germans, are idealists...they believe in the United States of Europe and the building of peace...they are waiting on us to offer them a new Gospel."[57] While Grandval and the Saar government were genuine supporters of European unity, their problems in instilling a deeply-felt particuarlism in the Saar made their increased focus after 1949 on the second pillar of Saar nationhood, namely, support for European unity and eventually a special role for the Saar in the new Europe, even easier.

By the spring of 1949 the policies of Grandval and his Saar allies to make the Saar into a new nation had not quite succeeded. Nonetheless, they greatly altered the province's political and cultural life. They also did much to separate the Saar from Germany. Grandval and his allies in the Saar such as Johannes Hoffmann not only fostered an independent Saar state and encouraged the development of a unique Saar culture and identity, but by 1949 they were also actively looking toward the prospect of utilizing the movement for a united Europe and future European organizations as a means of keeping the province from returning to Germany. This in turn set the stage for the protracted dispute between France and West Germany over the Saar.

The Saar dispute significantly dampened Franco-German relations and hindered efforts to create a united Europe between 1949 and 1957. The establishment of a constitution and the election of a Saar government in 1947 that led to economic union with France and political autonomy from Germany was a major achievement for French and Saar officials. Nevertheless, this left much uncertain. Internationally, the Saar's legal standing, along with that of Germany itself, was not settled. The growth of a movement for European unity and the establishment of the Council of Europe in 1949 offered a possible direction for the Saar, as did the revival of a German state in the west that same year. It was these two different directions that after 1949 caused much tension in efforts to create European institutions and strengthen Franco-German relations that were at the heart of the new Europe. Indeed, the Saar dispute of the 1950s between France and Germany created complications in the Council of Europe, as well as in the negotiations for the European Coal and Steel Community and the European Defense Community. The French occupation of the Saar thus strongly shaped the early course of European integration until Saarlanders rejected a European statute in 1955 that would have made their province a European territory and the headquarters of all European organizations. The results of the 1955 referendum finally led the French government to return the Saar to West Germany in 1957.

Problems over the Saar in the 1950s arose out of the French occupation of the Saar and Gilbert Grandval's attempts to construct a new nation out of the Saar. After 1949, Grandval's Saar project only reinforced the division of Germany, and the West German government took a strong stand against Saar independence. For his part, the lessons that Grandval drew from his experiences in the resistance during the Second World War included the need to "de-Prussianize" Germany and allow repressed provinces such as the Saar to emerge as their own entities under French cultural influence. In this manner, Grandval simply applied the ideology of national resurgence that dominated much of the French resistance's thinking on France to Germany and particularly to the Saar. His experience in the resistance was thus central to his outlook. The French resistance had an important influence on postwar France, and in the case of Grandval it also shaped France's occupation of the Saar.

The far-reaching occupation policies that Grandval implemented in the Saar caused enormous problems between France and West Germany in the 1950s. This in turn shows that the actions of those in charge of military occupations often have an impact on the course of postwar societies and international affairs, and that the importance of occupations is often underestimated. In explaining postwar eras, therefore, historians should reconsider military occupations and the long-term ramifications of the policies of those in charge of occupied territories. In the end, military leaders have dramatically influenced past events. Yet their influence has extended well beyond the battlefield and, as the Saar dispute shows, has been particularly important as occupiers.

Notes

1. All translations are my own. Service Historique de l'Armée de la Terre, Paris (hereafter SHAT), 10 P 193, "1ère Armée Française, Etat Major, 3ème Bureau, Télégramme Officiel (en clair) P.C." July 3, 1945, 2.
2. Tony Judt, *Postwar: A History of Europe since 1945* (New York: Penguin Press, 2005), 53–60.
3. For detailed descriptions of these activities, see the numerous reports found in Ministère des Affaires Etrangères, Archives Diplomatiques, Paris (hereafter MAE), série EU Europe 1949–1955, Sarre 196.
4. MAE, série EU Europe 1949–1955, Sarre 196, "Gilbert Grandval, Haut Commissaire de la République Française en Sarre à Son Excellence, le Ministre des Affaires Etrangères." August 4, 1949, 2.

5. See, for example, Grandval's speeches and correspondence with French foreign minister Robert Schuman. Historisches Institut der Universität des Saarlandes, Saarbrücken, Germany (hereafter HIUS), Privatarchiv Gilbert Grandval, film 5, "Allocution improvisée par M. le Haut-Commissaire à l'occasion de la fête jubilaire de l'Aurbed." July 29, 1951, 2, Archives Nationales, Paris (hereafter AN), 81 AJ 143, "Allocution prononcée lors de la cérémoine jubilaire de l'usine de Burbach." July 31, 1951, 1 and MAE, série EU Europe 1949–1955, Sarre 181, "L'Ambassadeur de France Haut-Commissaire de la République en Allemagne à Son Excellence Monsieur Robert Schuman Ministre des Affaires Etrangères-Direction Europe, très secret." April 18, 1952, 2.

6. See for example Rainer Hudemann, "Konflikt und Kooperation: Zu Frankreichs Saarpolitik nach Kriegsende," in Von der 'Stunde 0' zum 'Tag X.': Das Saarland 1945–1959; Katalog zur Ausstellung des Regionalgeschichtlichen Museums im Saarbrücker Schloß Saarbrücken 1990 (Merzig: Merzig Druckerei und Verlag GmbH, 1990), 97–120; Marlis Steinart, "Die Europäiserung der Saar: eine echte Alternative?," in Grenz-Fall: Das Saarland zwischen Frankreich und Deutschland 1945–1960, Rainer Hudemann, Burkhard Jellonnek, Bernd Rauls, eds. (St. Ingbert: Röhig Universitätsverlag, 1997), 63–95; and Raymond Poidevin, "Robert Schuman et la Sarre (1948–1952)," in Rainer Hudemann and Raymond Poidevin, eds., Die Saar 1945–1955: Ein Problem der europäischen Geschicht / La Sarre 1945–1955: un problème de l'histoire européenne (München: R. Oldenbourg Verlag, 1992), 35–48.

7. MAE, série Z Europe, 1944–1949, Sarre 18, "Le Colonel Grandval Gouverneur de la Sarre à Monsieur l'Administrateur Général Adjoint pour le Gouvernement Militaire de la Zone Française d'Occupation." February 14, 1946, 2.

8. Dieter Marc Schneider, "Gilbert Grandval: Frankreichs Prokonsul an der Saar 1945–1955," in Vom 'Erbfeind' zum 'Erneuerer': Aspekte und Motive der französischen Deutschlandpolitik nach dem Zweiten Weltkrieg, Stefan Martens, ed. (Sigmaringen: Jan Thorbecke Verlag Sigmaringen, 1993), 202.

9. Gilbert Grandval, Libération de l'Est de la France (Paris: Libraire Hachette, 1974), 15.

10. Gerhard Paul, "Die Saarlander fühlten sich durch solche Leute an Frankreich verkauft: die saarlandischen Remigranten und ihr gescheiterter Staat," in Grenz-Fall, 139–40.

11. "Résolution du parti socialdémocrate sarrois, 10 avril 1946" and "Le Parti chrétien social sarrois, 16 avril 1946" in La Documentation française: notes documentaires et études, 506 (January 8, 1947) (Paris: Direction de la Documentation, 1947), 13.

12. Armin Heinen, Saarjahre: Politik und Wirtschaft im Saarland 1945–1955 (Stuttgart: Franz Steiner Verlag, 1996), 69.

13. AN, papiers Georges Bidault, 457 AP 71, "Memoire du Mouvement pour le Rattachement de la Saare à la France concernant le règlement du problème de la Sarre." February 14, 1947, 5.

14. The MRS stated that 110,000 Saarlanders requested a subscription to their newspaper Die Neue Saar to support their claim to have such a large membership. Yet even French commandant Lanzerac, who was highly sympathetic to the MRS' goals, viewed their membership numbers as exaggerated. AN, papiers Georges Bidault, 457 AP 71, "Réunions délégats du MRS Sender, Diwo, Arend, Levy et Becker avec Georges Bidault." February 19, 1947, 1, HIUS, Privatarchiv Edgar Hector, film 21, carton 49, "Commandant Lanrezac. Note sur l'état d'esprit des Sarrois."April 1946.

15. For an itinerary and description of these events, see Landesarchiv Saarbrücken, Saarbrücken, Germany (hereafter LA Saarbrücken), Nachlass Heinreich Schneider NL. Schneider H 278, "Wort und Bild zu den französischen Festtagen am. 18 und 19. Mai in Saarlouis." May 18–19, 1946. MAE, série Z Europe 1944–1949, Sarre 570, "Le Général de Corps d'Armée Koening Commandant en Chef Français en Allemagne à Monsieur le Commissaire Général aux Affaires Allemandes et Autrichiennes, Secret, Baden-Baden." May 22, 1946, 2.

16. The Souvenir du Maréchal Ney was a very small group that was associated with the MRS. Its membership roster listed only 23 members from the entire Saar. LA Saarbrücken, Nachlass

Heinreich Schneider NL. Schneider H 278, "Marshall-Ney Verein in Saarlouis." No date listed, probably 1946, 1.

17. "Le Statut de la Sarre depuis le Traité de Versailles," in *La Documentation française: notes documentaires et études*, 506 (January 8, 1947) (Paris: Direction de la Documentation, 1947), 14.

18. Ibid.

19. Armin Flender, *Öffentliche Errinerungskultur im Saarland nach dem Zweiten Weltkrieg: Untersuchungen über den Zusammenhang von Geschichte und Identität* (Baden-Baden: Nomos Verlagsgellschaft, 1998), 37.

20. Sahlins's case study on the Basques and the Pyrenees border is an important one for the field of borderland studies. See Peter Sahlins, *Boundaries: The Making of France and Spain in the Pyrenees* (Berkeley: University of California Press, 1989).

21. Armin Heinen, *Saarjahre*, 69, and Armin Flender, *Öffentliche Errinerungskultur im Saarland nach dem Zweiten Weltkrieg*, 38.

22. The opposition to the Nazis in the Saar in 1934–35 was in many respects weakened by its own divisions. There was much mistrust between socialists and communists. Journalist Johannes Hoffmann led Catholic opposition to the Nazis, though his movement was a splinter group from the main body of Catholics and went against bishop Bornewasser of Trier, whose diocese included most of the Saar, as well as the leader of the Saar *Zentrum* party, Batholomämus Koßmann. Both Bornewasser and Koßmann supported the return of the Saar to Germany. Ludwig Linsmayer, "Die Macht der Erinnerung," in *Der 13. Januar: Die Saar im Brennpunkt der Geschichte*, Ludwig Linsmayer, ed. (Merzig: Merzieger Druckerei und Verlag, 2005), 45–47, and Heinrich Küppers, "Zwischen Vaterland und Hitler: Johannes Hoffmann bis zum Jahre 1945," in *Johannes Hoffmann: eine erste Bilanz*, Markus Geister, ed. (Saarbrücken: Gollenstein Verlag, 2004), 28–29.

23. *Renaissance de la Sarre* (Saarbrücken: Saarlandische Verlangsanstalt und Druckerei, 1947), 7.

24. Thalhofer argued that while Neue Bremm was not classified as a concentration camp, its inmates experienced it as one. Elisabeth Thalhofer, *Neue Bremm: Terrorstätte der Gestapo und seine Täter* (St. Ingebert: Auflage Röhig Universitätsverlag, 2004), 132, 136.

25. For specific articles, see "Sadisten oder vorsätzliche Morder?", *Neue Saarbrücker Zeitung*, 23 May 1946, 2; "Die Kriegsverbrecher der 'Neue Bremm' gerichtet," *Volkstimme*, 5 August 1946, 1; and "Sühne für 'Neue Bremm'", *Saarländische Volkszeitung*, 3 August 1945, 1.

26. "Hitlers Verbrechen: Eroffnung der Ausstellung im Saarland-Museum," *Neue Saarbrücker Zeitung*, 19 March 1946, 3.

27. This is according to the advertisement for the Saarland-Museum in the *Neue Saarbrücker Zeitung*, 21 March 1946, 3.

28. See the front pages of the *Neue Saarbrücker Zeitung* for coverage of the opening of the trials on November 21, 1945. For its conclusion, see the front pages of the *Saarbrücker Zeitung* on October 3 and October 7 of 1946, and the front pages of *Volksstimme* and *Saarländische Volkszeitung* on October 5, 1946.

29. AN, papiers Georges Bidault 457 AP 66, "Colonel Grandval, Gouverneur de la Sarre à Monsieur l'Administrateur Général Adjoint pour le Gouvernement Militaire de la Zone Française d'Occupation." February 11, 1946, 1.

30. This is Armin Flender's count. Armin Flender, *Öffentliche Errinerungskultur im Saarland*, 81–82.

31. Ibid., 44–45.

32. Gilbert Grandval, "Gedenktag der Verbrüderung," *Saarbrücker Zeitung*, 14 July 1947, 1.

33. The article also notes that the *International* was founded in Paris. "14. Juli Französischer Nationalfeiertag Fest der Freiheit-Gedenktag der Sozialistischen Internationale," *Volksstimme*, 13 July 1946, 1.

34. Albert Dorscheid, "Der 14. Juli an der Saar," *Saarländische Volkszeitung*, 13 July 1948, 1.

35. Armin Heinen, *Saarjahre*, 240, and Armin Flender, *Öffentliche Errinerungskultur im Saarland nach dem Zweiten Weltkrieg*, 45.

36. See reports concerning July 14 celebrations from 1951–1955 found in MAE, EU Europe 1949–1955, Sarre 221 and MAE, EU Europe 1949–1955, Sarre 222.

37. MAE, EU Europe 1949–1955, Sarre 222, "M. Eric de Carbonnel, Ambassadeur de la France, Chef de la Mission diplomatique française en Sarre à Son Excellence M. le Ministre des Affaires Etrangères." July 22, 1955, 2.

38. Until the first education minister Emil Straus left office in 1951, the anniversary of the Saar Constitution was also celebrated in Saar schools. Armin Flender, *Öffentliche Errinerungskultur im Saarland nach dem Zweiten Weltkrieg*, 53.

39. Armin Flender, *Öffentliche Errinerungskultur im Saarland nach dem Zweiten Weltkrieg*, 53, and Rolf Wittenbrock, "'Du Heiliges Land am Saaresstrand': Konfessionsschule und Identitätssuche," in *Von der 'Stunde 0' zum 'Tag X.'*, 267.

40. MAE, série Z Europe, 1944–1949, Sarre 34, "Gilbert Grandval Haut-Commissaire de la République Française en Sarre à Monsieur le Ministre des Finances et des Affaires Etrangères." June 7, 1948, 2.

41. Richard Gilmore, *France's Postwar Cultural Policies and Activities in Germany: 1945–1956*, thesis (Geneva: Université de Genève, 1973), 95.

42. "Les Œuvres culturelles française en Sarre," *Association Française de la Sarre, Bulletin*, 3 (1 April 1947), 2–3.

43. Ministère des Affaires Étrangères, Archives Diplomatiques, Archives de l'Occupation Française en l'Allemagne et en Autriche, Colmar (hereafter AOFAA), Sarre, CAB 94, "Gouvernement Militaire de la Sarre, Direction des Affaires Administratives, Section Education Publique. Étude Générale sur l'Administration Française en Sarre depuis juillet 1945 (Enseignement)." June 1946, 7.

44. AOFAA, Sarre, Cabinet, CAB 94, "Gouvernement Militaire de la Sarre, Cabinet, Rapport Mensuel, Décembre 1945, Education Publique, Secret." December 1945, 3–4 and Note Explicative du sujet du rapport Mensuel de décembre 1945." December 1945, 2.

45. AOFAA, Sarre, Cabinet, CAB 94, "Gilbert Grandval Gouverneur de la Sarre à Monsieur l'Administrateur Général Adjoint pour le Gouvernement Militaire de la Zone Française d'Occupation."August 6, 1947, 1.

46. MAE, série EU Europe 1949–1955, Sarre 405, "*Lycée Maréchal Ney Sarrebruck*, Mission Diplomatique Française en Sarre, June 30, 1953, 17," and HIUS, Privatarchiv Gilbert Grandval, Film 3, "Note sur l'enseignement français en Sarre par Pierre Woelfflin." December 1980, 2.

47. HIUS, Privatarchiv Gilbert Grandval, Film 3, "Note sur l'enseignement français en Sarre par Pierre Woelfflin." December 1980, 3.

48. Armin Heinen, "Sachwänge, politisches Kalkül, konkurrierende Bildungstraditionen," in *Universität des Saarlandes 1948–1988*, Armin Heinen and Rainer Hudemann, eds., 37–38.

49. MAE, série Z Europe 1949–1955, Sarre 34, "L'Accord Culturel Franco-Sarrois." December 15, 1948, 2.

50. See especially articles 1–4, 11–19 and 26–30 of the cultural agreement. MAE, série Z Europe 1949–1955, Sarre 34, "Accord Culturel du 15.12.48 entre le gouvernement de la République Française et le gouvernement de la Sarre." December 15, 1948, 1–2, 4–6, 8–9.

51. Ibid.

52. Grandval gives the exact itinerary for Schuman's visit in a letter to him. MAE, série Z Europe, 1944–1949, Sarre 34, "Le Haut Commissaire de la République Française en Sarre à Monsieur le Président Robert Schumann Ministre des Affaires Etrangères." October 18, 1948, 2.

53. AOFAA, Sarre, CAB 93, "Gilbert Grandval. Discours." December 15, 1948, 9.

54. MAE, série Z Europe, 1944–1949, Sarre 20, "Gilbert Grandval Haut-Commissaire de la République Française en Sarre à S.E. Monsieur le Ministre des Affaires Etrangères." April 31, 1948, 2–3 and 15–16.
55. AOFAA, PA-AP 363, 26, "Memoire: l'intention de Monsieur le Haut-Commissaire de la République Française sur l'évolution de la situation psychopolitique en Saare 1945–1949, Henri Viard." March 11, 1949, 63.
56. Ibid., 9.
57. Ibid., 4.

Part II

War Remembered or Forgotten

8

Pain and Memory: The War Wounds of Blaise de Monluc

Michael Wolfe

A body long in motion, a body now in pain, a life's experiences etched into the skin and bones of an old man, half his face blown away from a musket blast. This old man, Blaise de Monluc, sits down a year or so before his death in 1577 to dictate his memoirs, all with an eye—for only one eye remained—toward his future reputation. Born in 1500 to a minor Gascon noble family, Monluc became a renowned commander of French infantry in a career that lasted over 50 years. His *Commentaires* constructs a self-portrait framed by explicit reference to the movements and agonies of his body as it hurtled through the dynastic and sectarian wars of the sixteenth century. In dictating his life story, Monluc summoned forth memories by looking upon, touching, and feeling the pain of old wounds. The memoir's composition began in fact as a result of the shot in the face he received at the 1570 siege of Rabastens, which ended his career. Wounds prompted his memories and framed the ensuing narrative, which began as a story inscribed on his body before it made its way to the text.

A considerable literature exists on the subjects of pain and suffering, representing a variety of disciplinary perspectives. Some preliminary discussion of these studies will assist in determining how best to approach the story of Monluc's wounds. Much of this work grows out of the considerable interest taken in the human body over the past 25 years. Historians who broach this larger subject must consider the findings provided by literary criticism, feminist theory, comparative religion, and the social and behavioral sciences.[1] All of these various approaches to the body in pain should be kept in mind when studying Monluc, for the stages of his life became marked by a succession of wounds and scars, each acquiring different kinds of meaning. His scars and pain served as reminders, even proof, of past experiences and as badges of honor to show others, including us, his readers.

In *Discipline and Punish*, Michel Foucault identified the ideal soldier before the modern age as someone in possession of his own body: "[T]he soldier was someone who could be recognized from afar; he bore certain signs: the natural signs of his strength and his courage, the marks, too, of his pride; the body was the blazon of his strength and valour"[2] Monluc certainly fits this ideal, as he readily offered up his body to his king. However, Foucault goes on to argue that during the eighteenth century the soldier's body is something made, shaped by the disciplinary techniques of the state, in effect, a machine. Premonitions of this later change can again be discerned in Monluc, for the crown assiduously neglected to pay him the honor he believed was his due for the pains he endured. In the wake of Foucault's work, a chorus of critical commentary has averred that the body is "an entirely problematical notion," especially when considered in terms of gender. Judith Butler's seminal work on the "constructed" nature of the gendered body, together with her call for scholars to reconfigure the notion of individual identity, focuses on bodies in pain, and abject bodies force us to rethink physicality itself.[3] Monluc's sense of physicality was quite acute, as he very much defined his self-worth and identity by the wounds that marked both his life and his body. Some historians have drawn on Butler's insight; the pioneering work of Carolyn Walker Bynum on medieval female mystics stands out in particular.[4] For Monluc, the register needs to shift to his own complex sense of masculinity, a subject of rising interest.[5] Recent attempts, primarily in sociology, psychology, and medical anthropology, to collapse the old dichotomies between language and pain emphasize its interpersonal dimensions, so that suffering comes to be seen as a social experience shaped by culture, politics, and other exogenous factors.[6] In the case of Blaise de Monluc, his milieu as a minor provincial noble, as fervent in his Catholic faith as he was fierce on the field of battle, informed his experiences of pain and suffering, of joy and relief. Ideas about both masculinity and personal honor deeply informed Monluc's meditations on his injuries, physical as well as emotional. Finally, the new field of disability studies offers suggestive ways of approaching Monluc's text.[7] For example, as Monluc's bodily integrity became compromised over his career, his sense of moral integrity, if not righteousness, grew correspondingly. Monluc's readiness to commit atrocities against Huguenots in the 1560s during his campaigns in Guyenne cannot simply be attributed to bigotry, but resulted perhaps from a toxic mixture of hot anger and deep aching pain that inured him to the suffering of others.[8] Much work remains to be done in the history of disability, in all its permutations.[9]

Any exploration of the relationships between pain, memory, and the body must confront the issue of the ineffable divide between the reality of pain and its expression—in the form of suffering—to others. The central problem of the incommunicability of pain across social networks of language using words, gestures, and even touch has bedeviled scholars. In his *Philosophical Investigations*, Ludwig Wittgenstein argued that the pain of another finds you, just as your pain finds another, largely through the emotions of pity, empathy, and sympathy.[10] In her landmark study, *The Body in Pain*, Elaine Scarry offers a deep meditation on the wide-ranging phenomenological consequences of the fact that we possess bodies that feel pain. She concentrates on the massive involuntary pain inflicted by war and torture as expressed through discourse. Where she sees human degradation, Monluc finds self-affirmation in the painful wounds he received in war. For Scarry, the sensation of pain marks a literal "unmaking" of the world we normally inhabit, whereas for Monluc it perhaps provided its key constitutive element.[11] Scarry's abiding focus on linguistic attempts to represent pain largely eschews social and cultural factors, however. Joseph Amato, by contrast, presents a more historical approach to the study of pain and suffering based on Christian theology and intellectual commentaries. He explores their moral dimension, a key factor certainly for Monluc, who emphasized the element of sacrifice in both the chivalric and Christian senses.[12] Amato, however, privileges ideas about pain and suffering rather than their lived experience. Monluc attempts to bridge this divide in his memoir. Thomas J. Csordas's "experimental argument" about the paradigmatic scope of embodiment offers some analytic utility in tackling pain and suffering at the vital level of lived reality. Yet his efforts, and those of like-minded scholars, to arrive at an "objective" understanding of the experience of pain again diminishes the importance of how different cultural settings shape the process and perception of embodiment.[13]

For Monluc, the high possibility of losing an arm or leg as a result of war wounds left him with a deep sense of dread. Such impairment likely risked ending his military career, and with it his identity. Yet, as he often mentions in his memoirs, medical advances in his day offered some hope of ameliorating such a condition as surgeons made important strides in developing prosthetics and plastic surgery. Jean Tagault, a prominent medical expert who taught at the University of Paris, published a treatise on surgery in Latin in 1543 that contained a number of woodcuts illustrating the latest in artificial limbs.[14] Prosthetic design was based upon physiological and anatomical study of joints and musculature to replicate mechanically specific bodily functions, in this case walking. Yet Monluc's stark fear of amputation likely reflected his lack of faith in such solutions.

The only body part Monluc lost during his long career turned out to be his nose. Perhaps because of his advanced age, he opted for only a crude leather patch; a younger man could have considered plastic surgery, early modern style. Plastic surgery, in particular rhinoplasty, went back to ancient India; there is evidence of its practice in Europe during the thirteenth century.[15] A papal physician named Guy de Chauliac became well known for his surgical abilities; his *Chirurgia magna*, written in 1363, circulated widely and inspired further such work in subsequent generations of medical practitioners.[16] In the fifteenth century, a Sicilian barber-surgeon named Antonio Branca used a skin flap from the upper arm to reconstruct noses. Soon the Branca family became renowned for plastic surgery, attracting interest from doctors in the university medical faculties. The University of Bologna, long a bastion of traditional Galenic medicine, eventually became an important center of surgical theory and practice in the Renaissance. Members of its medical faculty experimented with innovative techniques of skin-grafting and reconstructive surgery. Leonardo Fioravanti, for example, a Bolognese doctor and contemporary of Monluc, reputedly visited Sicily to learn the Branca techniques firsthand.[17] He in turn inspired Gaspare Tagliacozzi, another member of the University of Bologna's medical faculty and the foremost pioneer of plastic surgery in the early modern period. In his 1597 textbook published in Venice, Tagliacozzi carefully demonstrated in words and illustrations his methods of nasal reconstruction, especially for noses ravaged by syphilis.[18] He also provided information on restoring lips and ears. These methods set the surgical standard for rhinoplasty and facial plastic surgery into the nineteenth century.

Healing wounds was also important, both in terms of developments in medicine at the time, such as the innovative surgical techniques of Ambroise Paré, and in a metaphorical sense, for a close analogy was often drawn between Monluc's private debilities and those of the body politic. Paré developed his new, more effective methods for treating gunshot wounds during the very wars in which Monluc campaigned, though there no evidence exists that the two men ever met. His *Méthode de traicter les playes*, first published in 1545 and subsequently revised in 1550, 1561, and 1572, shows the progressive development of his techniques, which owned their success as much to the use of antiseptics as to the rejection of traditional therapies, such as cauterization.[19]

Today, of course, there exists a much fuller scientific understanding of the dynamics of pain and suffering. Psychogenic trauma combines the neurological mechanisms of the sensation of pain with the psychological and existential aspects of its manifestation in consciousness and subsequent expression through words (including moans) and gestures (even involuntary ones such as shuddering and shaking). The earlier sense of the word *trauma* was taken in fact from its Greek root meaning "belonging to wounds or the cure of wounds" (*OED* 1656). While evident in Monluc's memoirs, the extension of trauma from body to mind comes in the concept of traumatic memory, which began to be noted by scientists and scholars in the eighteenth century.[20] Shock is another phenomenon that combines bodily and mental features, manifested in nervous, circulatory, and respiratory symptoms and psychological conditions of fear, anger, or catatonia. Pathogenic memory occurs when such behavior and symptoms become evident post-trauma; physiological effects induced by endorphins, for example, may even encourage addictive behavior, literal cravings to relive the pain and its subsequent relief.[21]

The *Commentaires* must obviously be considered as a literary text in its own right. Indeed, a recent study by Dalia Judovitz argues that representations of the body in French literature provide markers in the genesis of modernity.[22] Monluc's memoir should therefore be considered alongside Michel de Montaigne's frequent rumination about his body in the celebrated *Essais*.[23] Literature plays a key role in orchestrating the language within a moral community of users that validates or invalidates certain experiences as suffering in the mode of the hero, martyr, or victim. Aristotle considered suffering (pathos), usually emplotted on a catastrophic scale, as indispensable to tragedy, while medieval doctors often prescribed comic mirth as a therapy for pain.[24] The novelistic techniques for achieving the mixture of distinct styles, voices, and discourses that Mikhail Bakhtin calls "heteroglossia" sets up a dynamic clash of differing dialects and diction that subverts establishes hierarchies. It is inherently comic. Shakespeare's *King Lear*, for example, explored the origins and meaning of suffering through several characters. The treacherous, self-serving Edmund attributes suffering both to the influence of malign gods and to brutal human appetites, while the blind, foolish Gloucester blames his suffering on fortune and the pious Kent invokes the mystery of an all-seeing divine providence. Lear, in moods ranging from compassion to madness, offers still more variations on the question of what it is to suffer. And all in one play!

Monluc's autobiography was by no means unique. Indeed, memoir writing became a well-established genre during the sixteenth century, though scholars disagree on both its originality and import. Some scholars have argued that writing memoirs emerged in France during the fifteenth century, with Philippe de Commynes usually credited as the genre's inventor,[25] while others maintain that it only began as a result of the upheavals that rocked France after 1550.[26] Pierre Nora's claim that Renaissance memoir writing started as a uniquely French phenomenon has also been challenged, as has the contention that the genre revolves primarily around the question of self, identity, and life story.[27] Renaissance military memoirs like Monluc's *Commentaires* form an important subset of this type of writing, one that, according to Yuval Noah Harari, opens up the world and worldview of the Renaissance warrior nobility.[28]

Warrior nobles formed a distinct minority among both noblemen and soldiers. Self-control of body, mind, and emotions distinguished noble warriors from commoners according to many sixteenth-century commentators. David Rivault de Fleurance, for example, wrote that it was not possible to be a perfect noble warrior without displaying such qualities as endurance, decisiveness, steady nerves in the face of battle, and disdain for death.[29] Strength of both body and mind formed the crux of the noble sentiment of *générosité*, a quality the noble warrior showed in battle with elation (*allégresse*) and forced his adversaries to acknowledge.[30] Men such as Blaise de Monluc consciously cultivated a manly ethos that set them apart from the sophisticated courtiers and pen-pushing officials who increasingly dominated the royal government. Monluc's background as a *hobereau* (country squire) had also long nourished a sense of inferiority.

Monluc dictated his memoirs in the last years of his life, though they appeared in print only in 1592 thanks to the efforts of Florimond de Raemond, a hardline Catholic jurist and publicist from Bordeaux. Monluc possessed a phenomenal memory, if occasionally a selective one, as he recalled the sights, sounds, smells, and other sensations of combats, conversations, and encounters he experienced during his long, eventful career. Internal evidence suggests that the manuscript underwent two major revisions during Monluc's lifetime, and then further changes once in Raemond's hands. The memoir combined history, using multiple viewpoints, with a personal life story, both of which come through in two distinct voices. The first is a guidebook for captains in which Monluc uses the voice of a hard-bitten Gascon soldier whose guile is reserved for the field of battle, not backroom intrigues. The other narrative seeks to exculpate Monluc from charges of disloyalty leveled against him by the maréchal Damville in November 1570 by offering a complete account of all his services and sacrifices, most especially bodily ones, for the monarchy.

Monluc makes a point of showing his adeptness with vernacular languages, including Italian, English, and Spanish, but he does not show any knowledge of Latin despite his obvious homage to Caesar's Gallic conquests in the title.[31] Jacques Amyot's translation of Plutarch's *Lives* into French in 1559 nourished a desire among French nobles to emulate the virtuous exemplars from antiquity, a sentiment likely already quite alive in Monluc, who named his eldest son Marc-Antoine.[32] In his life of Coriolanus, for example, Plutarch pointed out how in the early Republic it was the custom for candidates for the consulate to go to the Forum in a cloak, without a tunic, which allowed them to show everyone they met (and solicited a vote from) the proofs of their valor seen in the battle scars on their bodies. Latin authors called such marks *cicatrices ostendere*.[33] The dead appear in the *Aeneid*, in dreams and the Underworld, riddled with wounds; bodily resurrection for Christians presumably promised the same honored marks of mutilation. The body of the Homeric hero, such as Achilles and Ajax, also served as a canvas upon which acts of valor were sketched. Indeed, for the ancients, ritual sacrifice using a bloody cut opened communication to the gods.

While Monluc ridiculed noblemen who avidly read tales about chivalrous knights, he clearly drew on tropes from medieval and Renaissance romances to construct his life story. His errant treks across France, Italy, and Flanders echoed motifs from the tales of Lancelot, Percival and Tristan, in whose works the theme of wounds figured prominently. Knightly literary motifs abound in Monluc's narrative, such as courage, panache and temerity, liberality, generosity, and an abiding faith in God, all traits found in the medieval chivalric tradition of the warrior hero. Monluc's expressed desire to retire to a monastery invoked the soldier-turned-hermit as a result of physical or emotional injury found in Ariosto and the real life by Charles V of Spain in 1555.[34] Lastly, the *Commentaires* contain numerous picaresque elements common to vernacular fiction at the time.

While prolix, Monluc's account was by no means unique among Renaissance warriors cum writers. Pierre Terrail, sieur de Bayard, for example, exemplified the noble martyr to the king in the hagiographies after his death in combat in 1524.[35] Just like the brave Bayard, Monluc emphasized how much he suffered as a result of court jealousies and intrigues, his career stymied as he routinely faced death proudly yet in poverty.[36] Like Bayard, Monluc also combined chivalric traits with a ready willingness to engage in subterfuge to outwit his enemies. A number of other warrior writers penned their life stories while convalescing. Agrippa d'Aubigné composed the opening lines of his *Tragiques* in June 1577 as he lay in bed recovering from grave wounds received in battle.[37] Wracked by his own war wounds, Ignatius Loyola composed his *Spiritual Exercises* in Paris in the 1530s to formulate a new spiritual discipline and militant religious order predicated on the virtues of self-control and obedience. The great Huguenot captain, François de la Noue, became known by the sobriquet "Bras de fer" after losing his arm in battle, while Henri, duc de Guise, came to be known as "le Balafré" for the deep scar from a sword wound that marked his face. Robert de la Marck, who styled himself the "Jeune Adventureux," describes rather perfunctorily the 47 (!) wounds, some quite serious, he received at the horrific battle of Novara in 1513. He was hauled, half dead, out from a pile of corpses. During his nearly two months of recovery, his damaged body never prompted any introspection or self-pity even though he faced the prospect of an end to his promising military career and permanent disability.[38] Götz von Berlichingen was 23 years old when he lost his hand in battle. It was replaced by an iron glove. He did not dwell on the loss, though it lent him the sobriquet "Iron Hand."[39] Charles V was unusual because he did complain often about the pain caused not by battle wounds but rather by attacks of gout. The Danish astronomer Tycho Brahe was perhaps better known during his lifetime for his golden nose than his astral observations. He lost his nose in 1566 in a swordfight while a student at the University of Rostock.

Monluc's swaggering bravura as a valiant Gascon military captain sprang from his own wartime experiences, but it was also modeled after the careers of other Gascon warriors, as celebrated in other personal memoirs as well as the portraits penned by Brantôme.[40] Brantôme describes with admiration, for example, the death of the comte de Buré who, bed-ridden with a wasting disease, insisted on greeting Death placed upright in a chair, dressed in his armor and helmet, sword in hand.[41] Ronsard's hymn to Death, which implored the one he called "Déesse" to spare him a quiet, bed-bound passing, echoed a common noble conceit at the time.[42] Brantôme mentions a Spanish common soldier who showed him half a dozen scars he had, explaining at which battle he got each. The scars thus served him as a bodily memoir that could be immediately transformed into an oral life story.[43]

Let us now take an inventory of Monluc's wounds. He suffered his first major wound in northern Italy in June 1525 when a musket ball shattered his right leg. It left him lame for a very long time and prevented him from taking part, much to his regret, in the battle of Pavia. Three years later, during an assault on Forcha di Penne, another musket shot pierced the *rondelle* on his forearm, while another struck him in the upper shoulder socket. He fell unconscious into a ditch. As his men tried to drag him out by his legs, he tumbled back into the ditch, breaking his wounded arm in two places.[44] In November 1537, during a skirmish in Piedmont, Monluc suffered another gunshot wound, this time to his left arm. Luckily, it did not reach the bone, so he recovered fairly quickly.[45] In October 1544, several English arrows pierced his body mail at the siege of Boulogne, resulting in only minor scratches. Seven years later, back in Italy, Monluc slipped and fell during an assault on the walls of Chieri, near Turin. He crashed on his left side and dislocated his hip, an injury that ever after caused him pain. He seriously aggravated his leg later in 1568, which forced him to take to his bed for three months.[46] And finally, of course, there's the fateful face wound he sustained during the assault on Rabastens on July 23, 1570. He received it as he began to climb up a ladder to take the wall. His first sensation was that of blood pouring from his mouth, nose, and eyes. No doubt in shock, he kept up the attack despite the horrified looks on his own men's faces as they stared at his. Eventually, Monluc grew faint and had to be carried away. It took nearly three days before he stopped bleeding. His wife then came to fetch him home, where he remained to rest, reminisce, and dictate his memoirs.

Besides war wounds, Monluc also succumbed to a variety of fevers and other illnesses, all of which he again described in great details, including his delirious visions, parched lips, and crippling fatigue. In April 1559, for example, during the peace talks with Spain at Amiens, Monluc had a terrible fever, which left him ice-chilled or burning hot and deprived him of sleep for three nights. He attributed his recovery to the robust strength of his body, hardened by years of war.[47] Monluc visited mineral baths when he could, particularly those at Barbotan in the Gers, though they rarely provided him much relief. He also liked to self-medicate when feverish by drinking Greek wine.

Many of the medical treatments Monluc underwent proved almost as searing as his original traumas. After his face wound, he gives a matter-of-fact description of feeling the fingers of a surgeon named Maître Simon dig out the bits of bone, flesh, and teeth from what were once his cheeks, mentioning along the way that this same surgeon later fashioned him a leather nosepiece that he wore whenever he went out in public. Monluc's wife enters the narrative only when she appears to take Monluc back home to recover from wounds or illnesses; he does credit her with saving his life more often than doctors or surgeons who treated him. Monluc does not have much positive to say about medical professionals. He relates, in particular, how French surgeons nearly amputated his arm in 1528 but for the sage advice of a captured young Italian surgeon. Better to risk his life with the prospect of possibly resuming his career, his body intact, than end it with only one arm. As a result, Monluc spent the next three months bed-ridden, suffering excruciating back pain while his arm healed through the use of poultices and bandage wraps. It was three full years before his arm was completely functional again. In 1554, when he lay wasting away from dysentery at Siena, the doctors in effect gave up on him, going so far as to relay word to the French ambassador to Venice, Odet de Selve, that Monluc's replacement should be arranged for since he was as good as dead.[48] Doctors cauterized his leg in 1568, which so wracked him with pain that he wished someone would cut his throat to end his misery. It became infected, which only prolonged his torment. Finally, in 1569 a boil on his right nipple had to be lanced, leaving his chest so tender that he could not wear a shirt. He underscored the importance of providing medical care for his men, going so far in his memoir as to prepare a memorandum for the king to establish veterans' hospitals to care for disabled soldiers.[49]

Anti-courtier rhetoric infuses much of Monluc's text and his own self-image as a warrior, contrasting his life of privation and endurance with the soft, voluptuous life at court. The coded language of masculinity and femininity lends a gendered edge to this contrast. Monluc even dared his enemies at court to shed their clothes before the king so the world could see whose body bore the signs of service to the monarchy.[50] In recent years, gender studies have analyzed masculinity in all its various manifestations and facets, not denying but certainly moving beyond the longstanding feminist fixation on its relationship to patriarchy now to study the dynamics of male presentation with all its complexities and inconsistencies. Some historians have recently posited a crisis in masculinity in France during the sixteenth century as a result of widespread violence and the breakdown of royal authority.[51] Does Blaise de Monluc fit into this scenario? Focus on the male body relates to understandings of corporeality, notions of masculine honor, and its interconnections with discourses of effeminacy, misogyny, sodomy, and masculinity.[52] Personal relationships among men during this period usually adhered to the coded behaviors of hierarchical dominance and subordination or the platonic (and occasionally homoerotic) dictates of friendship (amitié).[53]

Wounds and their treatment held metaphorical meaning for Monluc. Young noblemen like him rushed to join the fray in the Italian wars, these "belles occasions," and to prove their valor and thus their value by performing signal actions of daring and sacrifice for others to see. Happy were those who carried on their bodies the visible marks of such exploits, the more spectacular (in the literal sense of "seeable") the better. In defending his honor, Monluc spoke of his willingness to lose an arm in the king's service. His wounds proved his steadfast loyalty to the crown and could be healed only by the gentle balm of royal gratitude. Yet, he bitterly allowed that the main recompense he received was not the king's support but a musket blast to the face. Monluc was not above self-pity, even despair, admitting on a few bleak occasions his desire, when facing the calumnies of his enemies, to contemplate suicide.[54] His wounds became an inscription of manhood, for giving and receiving body blows formed the core experience of warfare between warriors, as opposed to the less manly combat that used projectiles of various sorts. Echoing a commonplace of the time, Monluc decried firearms as a coward's weapon because the shooter dared not look upon his adversary's face. Through bodily suffering, the battle moved inside the warrior to the very recesses of his being. Pain shaped his identity, while scars carved his body with marks of honor.

Monluc spoke a tough-guy sixteenth-century Franco-Gascon idiom replete with references to the body, both real and metaphorical, literally from head to toe. Eyes and ears figure quite prominently, for Monluc led a life of intense observation. Head, heart, and stomach also served as metaphors for prudence, courage, and perseverance. While Monluc retailed commonplaces about the need for bodily self-control, he described the physical exhilaration he experienced after combat, when he felt so light that he hardly touched the earth.[55] Yet Monluc was quite willing to admit the stark terror he felt in battle as the blood drained from the face, the heart constricted, and hands and legs shook uncontrollably. On one occasion, he almost soiled his pants from fear. The only times he did not feel the painful throbbing of his various war wounds was when he was afraid. He even noted, with irony, that the fear of death often cures diseases.[56] Yet dishonor was to be feared over death, as the fate of the body was not paramount.

Monluc's memoirs provide gripping, almost cinematic descriptions of his body in full motion amid a kaleidoscope of sights and sounds, such as the feeling of galloping on a horse suddenly shot out from under you, or the sensation of the heaving clash of row upon row of pike men, or the sickening but telltale sound of lead shot smashing through a comrade's body. Along with the cinematic came thespian moments, for Monluc occasionally impersonated different characters, including ironically himself. He took pride in his ability to converse in several tongues, including Italian, Spanish, English, and Dutch. At the siege of Perpignan in 1542, for example, he pretended to be a Basque cook as he reconnoitered the city for the French, while at the 1544 siege of Boulogne he fooled English sentries when he answered their calls of "who goes there" with the assuring answer "a friend." Perhaps Monluc's most remarkable performance occurred during the siege of Siena, where he served as the embattled city's governor. He took up his post in August 1554 already seriously ill with dysentery. High fever so sapped his strength he could hardly get out of bed. Never one to underestimate his own importance, Monluc relates how the courage of the Sienese sank as his condition worsened. In an effort to rally the city's spirits in January 1555, Monluc decided to make an unannounced appearance incognito at the Palazzo Communale, where he conversed with others in Italian. He eschewed his usual somber black garb and instead dressed up as an amorous courtly fop, with a feathered hat, frilly silken shirt, and bejeweled fingers. He also vigorously washed his face with wine and rubbed a crust of bread on his cheeks and lips to give them a ruddy color. His disguise proved so convincing that when he saw himself in the mirror, it appeared as if God had given him a totally new face; later at the meeting when someone asked him if he knew where Monluc was, barely containing his laughter Monluc said he knew not. It was a splendid example of Renaissance self-fashioning if ever there was one.[57]

Monluc evinces little of an interior life in his narrative, though he apparently dropped his mask of Stoic calm when he conversed with good friends like Montaigne, who relates the heartbreaking sadness and regret that Monluc felt over the death of his second son, Pierre-Bertrand, whose death Monluc likened tellingly to the loss of one of his arms.[58] Not only his body, but also his somnolent imagination spoke to Monluc in the form of vivid, prophetic dreams. Perhaps a clumsy literary device or an actual experience (Monluc would swear and offer witnesses for the latter), dreams figured prominently in the *Commentaires*. For as long as he could remember, Monluc was plagued with a dream in which he fought against imaginary foes, a condition that doctors could not explain. The dreams he relates in detail, however, marked key moments in public affairs. Three nights before Henri II's fatal jousting accident in early July 1559, for example, Monluc recounts a dream in which he saw the king sitting on his throne, his face dripping blood much like Christ at Calvary, hearing the plaintive murmuring of attending doctors as the king neared death. In his dream Monluc began crying, only to awaken with tears streaming down his cheeks.[59] The resemblance to Monluc's own facial injury at the time he dictated his account is of course obvious, as were occasional references likening Monluc's suffering to that of Christ. In September 1567, as civil war again loomed, Monluc relates another dream in which he saw France torn apart by fratricidal strife; a battle raged atop a wall, from which he fell headlong into a river, becoming drenched. He awoke so bathed in sweat that his valet brought him a new nightgown and changed his sopping bed linens.[60] The last dream he details is the one presaging his face wound at Rabastens. Like his other dream tales, his description of the actions leading up to his injury, while horrifying, contains much melodrama.[61] Monluc's retelling, seen from a Freudian perspective, illustrates a "repetition compulsion" whereby the mind re-enacts traumatic events in order to deal with them. By contrast, Julia Kristeva identified the human tendency to be passionately (*avec jouissance*) drawn to the abject, with pain and violence attending.[62] For her, purification of the abject lay in catharsis produced by literature, especially poetry (the sublime) or, more usually, religion (salvation). The abject reveals the disintegration of accepted boundaries and structures of meaning predicated on the distinction between subject and object.

Observing pain and inflicting it on others shaped Monluc's attitude toward his own physical pain. Atrocities were commonplace and usually personal. In 1544, for example, Monluc relates an incident when his men set fire to a stable in which enemy soldiers had taken refuge. Crying "mercy, mercy," a number of them leapt from the burning building only to be massacred by his waiting men out of vengeance for the recent deaths of two comrades.[63] Massacres became standard policy for Monluc when he led Catholic soldiers against Huguenots in the 1560s, receiving with relish the sobriquet "the Butcher of Guyenne." Monluc famously distinguished between the kind of violence found in foreign wars and the more perverse sort in civil wars.[64] Consider what befell a certain Jean Parisot, the son of a *charcutier* in Montauban and fervent Protestant, who terrorized Bigorre in 1574, during which time he fell into a quagmire on his horse. Monluc's men seized and killed both him and horse, cutting off his ears (the horse's we're not sure about) before tossing their carcasses into the town's ditch.

Religion also figures in Monluc's narrative. Monluc sent one of his lieutenants to Loretto in his stead to fulfill a vow to aid his recovery in 1528, for example. He asked God's pardon in praising the French alliance with the Turks, arguing that in war one must make arrows out of any wood at hand.[65] He regularly invoked God and Fortune almost interchangeably to explain his haphazard life, employing deus ex machina devices to underscore man's limited power to control events. How else to account for who dies and who lives in combat? As one example among many, Monluc describes a ferocious battle in September 1542 when out of the 35 men he led, only he and 5 others returned unscathed. Like unto Job, God communicated to Monluc through the pain of wounds and illnesses.[66] Yet by contrast, Monluc also spoke of the need for commanders to prepare assiduously for every eventuality if they hoped to be successful. In this respect, his was a very Machiavellian view of human affairs, though without the political theorizing. Monluc regularly thanked God for preserving his good judgment when faced with the unknown, for it helped ensure he preserved both his life and his sense of honor.[67] Monluc seemed to embody Machiavelli's adage that a man of *virtù* controlled at most only half of what happened in his life.

Monluc meditates quite frequently on death, which had been perhaps his closest companion during so much of his life. At the end of his memoirs there appear, like apparitions, the names of all the dead comrades-in-arms with whom he had had the privilege to serve over the years. He took comfort in the thought he would soon join them after his death. War was death, he remarked on more than one occasion, and a leader's main duty was to lead his men into the jaws of death. As he grew older, he noted his fear of death increased, though he found solace thinking it would bring an end to the constant and excruciating leg pain that only became worse as he aged. In his dedication, Raemond described Monluc as composing his memoir suspended midway between his bed and the coffin. Monluc often expressed his desire to be buried as a man of honor, for his bones to go into the ground while his reputation endured among men. Just like Christ, who shed his blood for mankind, so Monluc shed his for the king; and he hoped, too, on the day of resurrection to have every drop, along with his mutilated face and body, restored in full just as he had come into the world from his mother's womb.[68]

Blaise de Monluc's graphic, often gripping account of his bodily travails raises important questions about the relationships between pain, suffering, and physicality. His conscious construction of a sense of self marked by his scarified wounds rendered his body a complex inscribed text. It defined and integrated the different facets of his personality, what it meant to be a warrior, a man, a Catholic, and a comrade. His world of pain and suffering offered these sensations as forms of redemption, as scourging experiences that heightened his worthiness, at least as he saw it in his one remaining eye. Monluc displays his broken body, a monument fashioned by a life of reckless action and pitiless duty, to us, his readers, who cannot help but also feel— if only with a passing *frisson* in our own bodies—the sum total of a life fully lived.

Notes

1. See Margaret Lock and Judith Farquhar, eds., *Beyond the Body Proper: Reading the Anthropology of Material Life* (Durham: Duke University Press, 2007).
2. Michel Foucault, *Discipline and Punish* (New York: Vintage Books, 1975), 125.
3. Judith Butler, *Undoing Gender* (New York: Routledge, 2004).
4. In addition to her landmark study, *Holy Feast and Holy Fast: The Religious Significance of Food to Medieval Women* (Berkeley: University of California Press, 1987), see Carolyn Walker Bynum's *Fragmentation and Redemption: Essays on Gender and the Human Body in Medieval Religion* (New York: Zone Books, 1991).
5. The notion of "complexity" forms a central feature of the essays in *Responding to Men in Crisis: Masculinities, Distress and the Postmodern Political Landscape*, Brian Taylor, ed. (New York: Routledge, 2006).
6. See Mary-Jo DelVecchio Good, *et al.*, eds., *Pain as Human Experience: An Anthropological Perspective* (Berkeley: University of California Press, 1994), and Arthur Kleinman, *et al.*, eds., *Social Suffering* (Berkeley: University of California Press, 1997).
7. For intriguing ways of approaching disabilities, see *Corporealities: Discourses of Disability*, David T. Mitchell and Sharon L. Snyder, eds. (Ann Arbor: University of Michigan Press, 2001).
8. Margaret A. Winzer, "Disability and Society before the Eighteenth Century: Dread and Despair," in Lennard J. Davis, ed., *The Disabilities Studies Reader* (New York: Routledge, 1997), 75–109.
9. See Henri-Jacques Stiker, *Corps infirmes et sociétés: essais d'anthropologie historique*, 3rd ed. (Paris: Dunod, 2005).
10. See Luc Botanski, *La Souffrance à distance* (Paris: Métailié, 1993).
11. Elaine Scarry, *The Body in Pain: The Making and Unmaking of the World* (New York: Oxford University Press, 1985).
12. Joseph A. Amato, *Victims and Values: A History of a Theory of Suffering* (New York: Praeger, 1990). By contrast, John Bowker points out the heterogeneity of meanings and modes of suffering outside of the Christian tradition in *Problems of Suffering in Religions of the World* (Cambridge: Cambridge University Press, 1970).

13. Thomas J. Csordas, ed., *Embodiment and Experience: The Existential Ground of Culture and Self* (Cambridge: Cambridge University Press, 2005).

14. Jean Tagault, *De chirugica institutione libri quinque* (Paris, 1543). The book, in which Tagault treats tumors, ulcers, wounds, fractures, and dislocations, was republished some twenty-two times over the next century.

15. I. Eisenberg, "A History of Rhinoplasty," *South African Medical Journal*, 62:9 (1982): 286–92.

16. André Thevenet, "Guy de Chauliac (1300–1370): The 'Father of Surgery,'" *Annals of Vascular Surgery*, 7:2 (1993): 208–12.

17. P. Santoni-Rugiu and R. Mazzola, "Leonardo Fioravanti (1517–1588): A Barber-Surgeon Who Influenced the Development of Reconstructive Surgery, *Plastic Reconstructive Surgery*, 99:2 (1997): 570–75.

18. Gaspare Tagliacozzi, *De curtorum chirurgia per insitionem, libri due* (Venice, 1597). See M. T. Gnudi and J. P. Webster, *The Life and Times of Gaspare Tagliacozzi, Surgeon of Bologna, 1545–1599* (New York: Herbert Reichner, 1950).

19. Ambroise Paré, *La Méthode de traicter les playes faictes par hacquebutes et aultres bastons à feu et de celles qui sont faictes par flèches, dardz, et semblables, aussi de combustions spécialement faictes par la pouldre à canon* (Paris: Vivant Gaulterot, 1545).

20. See Dominick LaCapra, *Writing History, Writing Trauma* (Baltimore: The Johns Hopkins University Press, 2000).

21. Nikola Grahek, *Feeling Pain and Being in Pain*, 2nd ed. (Cambridge, MA: MIT Press, 2007).

22. Dalia Judovitz, *The Culture of the Body: Genealogies of Modernity* (Ann Arbor: University of Michigan Press, 2001).

23. Montaigne's body recurs as a motif throughout the *Essais*. See his interesting if excruciating discussion of kidney stones in "De l'experience," book 3, ch. 13.

24. Glending Olson, *Literature as Recreation in the Later Middle Ages* (Ithaca: Cornell University Press, 1982).

25. Joël Blanchard, "Commynes et la nouvelle histoire," *Poétique*, 79 (September 1989): 287–98, and Jean Dufournet, "Commynes et l'invention d'un nouveau genre historique: les mémoires," in Danielle Buschinger, ed., *Chroniques nationales et chroniques universelles: actes du colloque d'Amiens* (Göppingen, 1990), 59–77.

26. Marc Fumaroli, "Mémoires et histoire: le dilemma de l'historiographie humaniste au XVIᵉ siècle," in Noémi Hepp and Jacques Hennequin, eds., *Les Valeurs chez les mémorialistes français du XVIIᵉ siècle avant la Fronde* (Paris: Klincksieck, 1979), 21–45.

27. Pierre Nora, "Les Mémoires d'État de Commynes à de Gaulle," in Pierre Nora, ed., *Les Lieux de mémoire*, II, *La Nation*, 3 volumes (Paris: Gallimard, 1997) and Nadine Kuperty, *Se dire à la Renaissance: les mémoires au XVIe siècle* (Paris: Vrin, 1997).

28. Yuval Noah Harari, *Renaissance Military Memoirs: War, History, and Identity, 1450–1600* (Woodbridge, Suffolk: Boydell Press, 2004). Harari identifies eleven criteria that generally characterize Renaissance memoirs and offers six earmarks of military ones (7, 17–18). While at times interesting, Harari's book suffers from an overweening desire to connect the Renaissance literary representation of war with modern exemplars.

29. David Rivault de Fleurance, *Les Estats, esquels il est discouru du prince, du noble, et du tiers estat, conformément à notre temps* (Lyons: B. Rigaud, 1596), 313.

30. See the still-central studies by Ellery Schalk, *From Valor to Pedigree: Ideas of Nobility in Sixteenth- and Seventeenth-Century France* (Princeton: Princeton University Press, 1986), and Kristen Neuschel, *Word of Honor: Interpreting Noble Culture in Sixteenth-Century France* (Ithaca: Cornell University Press, 1989).

31. Florimond de Raemond invoked this tradition, as well as added references to celebrated Greek and Roman military commanders, particularly Julius Caesar, in his dedication of the first edition in 1592 addressed to the nobility of Gascony.

32. The craze for things classical can also be seen in the ornamentation and remodeled spaces in Gascon castles and manor houses, as well as in the libraries, some modest, others quite substantial (such as Montaigne's), built by Gascon nobles of Monluc's generation. Examples of such places include the château de Crabé near Lectoure, that of Tauzia near Valence-sur-Baïse, Plieux near Miradoux, Manlèche at Pergain-Taillac, and the château d'Aymeric de Voisins in Gramont. Pilasters and classicized capitals proved the most common additions in this style.

33. See Livy 6.14.6, and Cicero, *De oratore*, 2.124 and 195–96, for examples of such displays, though Quintilian thought the practice too ostentatious (*Institutio oratoria*, 6.3.100). For a discussion of the significance of such public exhibitions of wounds, see Yan Thomas, "Se venger au forum: solidarité familiale et procès criminel à Rome," in Raymond Verdier and Jean-Pierre Poly, eds., *La Vengeance* (Paris: Édition Cujas, 1984), 3:71–72.

34. The wildly popular *Amadis de Gaule*, translated into French in 1540, and Ariosto's celebrated *Orlando Furioso*, translated into French in 1543, both left influences, too.

35. The first account of Bayard's life and death was published in 1527.

36. Jean Jacquart, *Bayard* (Paris: Fayard, 1987).

37. *Les Tragiques*, "Aux lecteurs," *Œuvres d'Agrippa d'Aubigné*, Henri Weber, Jacques Bailbé and Marguerite Soulié, eds., (Paris: Gallimard, 1969), 5.

38. Robert III de la Marck, Sire de Florange, *Mémoires du Maréchal de Florange dit le Jeune Adventureux*, Robert Goubaux and P.-André Lemoisne (Paris: H. Laurens, 1923–24), 1:127–28.

39. Götz von Berlichingen, *The Autobiography of Götz von Berlichingen*, H. S. M. Stuart, ed., (London: Duckworth, 1956), 25–26. Like Monluc's memoir, this one was never originally intended for publication, appearing in German for the first time in 1731.

40. Véronique Larcade, *Orphelins d'une Amérique: les capitaines gascons à l'époque des guerres de religion* (Paris: Éditions Christian, 1999), examines the careers and topoi associated with this "type."

41. Pierre de Bourdeille, seigneur de Brantôme, "Les Vies des grands capitaines et estrangers," in *Œuvres completes*, Ludovic Lalanne, ed. (Paris: Société de l'histoire de France, 1864–1882), 1:314–19.

42. Pierre de Ronsard, "Les Hymnes, Hymne de la mort," in *Œuvres complètes*, Paul Laumonier, ed. (Paris: Hachette-Droz-Didier, 1953), 8:178–79.
    Ne me laisse longtemps languir en maladie,
    Tourmenté dans mon lict: mais puisqu'il faut mourir,
    Donne moi que soudain jet e puisse encourir,
    Ou pour l'honneur de Dieu, ou pour servir mon Prince,
    Navré d'une grand'playe au bort de ma province.

43. Pierre de Bourdeille, seigneur de Brantôme, *Œuvres complètes*, Léon Lecestre, ed., 2 volumes, (Paris: Librairie Renouard, 1889), 2:53.

44. *Commentaires*, 46–47.

45. *Commentaires*, 75.

46 Yet Monluc's was a portable litter, as his men carried him about so that he could fulfill his duties while lying prone: "Ceux qui liront ma vie pourront veoir de combien de sortes de maux j'ay esté assailly; et neantmoins je n'ay jamais pour cela esté oisif ny retif aux commandmens de mes maistres ou en ma charge. Cela n'est bien séant à un guerrier de croupier dans le lit pour un peu de mal." *Commentaires*, 683.

47. "Il m'a bien servi d'estre fort et robuste; car j'ay mis autant mon corps à l'espreuve que soldat ait fait de mon temps." *Commentaires*, 461.

48. *Commentaires*, 274.

49. *Commentaires*, 447–48. It was also not uncommon for captains to pay for the medical treatment of men wounded in their service. Thus, a certain captain Cabos paid 29 *écus* to a

surgeon and apothecary who cared for a soldier named Molinis, who was wounded in the leg by a musket shot and had to convalesce for a week with a local peasant (Larcade, 83).

50. *Commentaires*, 12–13.
51. Kathleen P. Long, ed., *High Anxiety: Masculinity in Crisis in Early Modern France* (Kirksville, MO: Truman State University Press, 2002).
52. These problems have been studied more fully in the English context. See Tim Hitchcock and Michèle Cohen, *English Masculinities, 1660–1800* (London: Addison Wesley Longman, 1999).
53. On the latter was the celebrated friendship between Michel de Montaigne and Étienne de la Boétie, which joined paeans to love and equality with a repertory of misogynistic commonplaces. See Marc D. Schachter, "'That Friendship Which Possesses the Soul': Montaigne Loves La Boétie," in Jeffrey Merrick and Michael Sibalis, eds., *Homosexuality in French History and Culture* (New York: Haworth, 2001), 5–21. Male friendship, however, occasionally invited suspicions or even charges of sodomy, as occurred with the circle of young noblemen, derisively called *mignons*, whom Henri III chose as his boon companions.
54. "à quoy j'avois autant pence comme à me donner la mort moy-mesmes; mais je suis né sur ceste planette d'estre tousjours subject aux calomnies" (*Commentaires*, 727).
55. Bodily control also meant regulating desires for food and sex, though again Monluc's own behavior contrasted with his admonitory advice to captains of war. While underscoring the need to put up with privations during a campaign, Monluc also bragged about how his was the most lavishing provisioned tent in the entire French army, as he offered sumptuous banquets with fine wines to entertain—and no doubt seek to influence—members of the high aristocracy. "Il me sembloit, en ces banquets, que mon corps ne pesoit pas une once et que je ne touchois pas en terres" (*Commentaires*, 245).
56. "La peur ou l'affection me fait aller plus droict et plus viste, de sorte que je ne sentois guières mon mal" (*Commentaires*, 240). No doubt an adrenaline rush. And "J'ay ouy dire que l'apprehension de la mort a guery des maladies" (*Commentaires*, 313).
57. *Commentaires*, 294–97. On this sense of individual agency, see Stephen Greenblatt, *Renaissance Self-Fashioning: From More to Shakespeare* (Chicago: University of Chicago Press, 1983).
58. Montaigne, *Essais* (Paris: Gallimard, 1958), book 2, essay 8, "De l'affection des pères aux enfants," 375–76. Peyrot, as his father called him, died from a septic wound to the thigh, received during a naval encounter with the Portuguese off the coast of Madera.
59. *Commentaires*, 467–69.
60. *Commentaires*, 593–94.
61. Take, for example, Monluc's plea to his men to continue to fight on no matter what happens to him, even though he knew what would: "et si je suis tué ou blecé, ne vous en souciez point, et me laissez là, et poussez seulement outré, et faictes que la victoire en demeure au Roy" (*Commentaires*, 772). His relation of the whole incident goes on for fifteen pages.
62. Julia Kristeva, *Powers of Horror: An Essay on Abjection*, trans. Leon S. Roudiez (New York: Columbia University Press, 1982), especially chapter one.
63. *Commentaires*, 94.
64. "Ce n'est pas comme aux guerres estrangères, où l'on combat comme pour l'amour et l'honneur; mais aux civiles il faut ester maistre ou valet, veu qu'on demeure sous le mesme toit. Et ainsi il faut venir à la rigueur et à la cruauté" (*Commentaires*, 463).
65. *Commentaires*, 81–82.
66. "si n'ay-je esté exempt de grandes blessures et de grandes maladies: car j'en ay autant eu que homme du monde sçauroit avoir sans mourir, m'ayant Dieu tousjours voulu donner une bride, pour me faire cognoistre que le bien et le mal depend de luy, quand il luy plaist" (*Commentaires*, 49–50).

67. "C'est la requeste principale que vous devez faire à Dieu de vous garder l'entendement" (*Commentaires*, 177–78).
68. "Et quand je seray mort, à grand peyne dira on que j'en apporte le jour de la resurrection en paradis tout le sang, oz, nerf et voyens que j'ay porté au monde du ventre de ma mère" (*Commentaires*, 14).

9

From the Chemin des Dames to Verdun: The Memory of the First World War in War Memorials in the Red Zone

Alexandre Niess

War memorials commemorating combatants killed in action during the Great War were erected throughout the world, on a scale corresponding to the immense slaughter of *la der des der* (World War I). The phenomenon of commemoration occurred in Great Britain, the United States, Australia, but also in Germany and especially in France, where the number of monuments is far greater than elsewhere. Consequently, the following analysis will focus specifically on French space.

Springing from popular will, supervised by the state that frequently passed laws concerning memorials immediately following the First World War (some 42 laws between 1918 and 1925), war memorials commemorating soldiers killed in action during the first global conflict appear as the first large-scale commemorative and artistic enterprise of the twentieth century, as Annette Becker rightly notes in all her major work on the question since 1988.[1] The duty to remember, resulting from the debt incurred by the living toward the dead, quickly became a pressing one in the eyes of contemporaries. *La Reconnaissance nationale*, for example, an association founded on the initiative of Jean-Baptiste Belloc, was established a few weeks after the beginning of hostilities. Its function was to supply communes with marble plaques on which the list of the dead could be inscribed. In 1916, Jean Ajalbert launched an inquiry involving several personalities in the world of the arts, letters, science, and politics (Paul Fort, Alexandre Millerand, Paul Doumer, Henri Bergson, Simon Pozzi, Adolphe Willette, Steinlen, Abel Faivre, etc.) to obtain their opinion about commemorating the dead.[2]

In the communes, the duty to remember is present everywhere and with equal intensity, even if it is not always expressed in similar fashion. Various studies have already shown that commemoration resulting from the Great War, witnessed in the form of communal war memorials, was a powerful symbolic and political symbol.[3] Located on the frontline during four long years, several French communes underwent their own particular wartime experience, whether they were occupied or were within range of the French, Allied, or German canons. This experience of war marks both space and spirits. As Hugh D. Clout has established in his material assessment of the Great War's impact on the communes that were destroyed, the war marked this space forever[4] We can use J. Guicherd and C. Matriot's mapping of the region immediately after the end of the war to illustrate the idea of the "Red Zone," a region devastated by the violent battles of the Great War and more than 90 percent destroyed (figure 9.1).[5]

This particular situation raises questions we shall aim to answer as precisely as possible based on the sources available. Nevertheless, since the Red Zone stretches across a vast expanse from Flanders to the Vosges, we have deliberately chosen to focus exclusively on the central part of the area of the French front, made up of battlefields with the evocative names of the Chemin des dames, Champagne, Argonne, and Verdun.

The area we will investigate stretches from the Chemin des Dames to Verdun. It corresponds to the actual shape the frontline took during the conflict. In fact, the various assaults often resulted in taking only minimal amounts of new ground. This zone stretches for almost 180 kilometers, and to assess the impact of the war itself on the commemorations of French communes we have included 15 kilometers on each side of the frontline in order to investigate the shape that the duty to remember actually takes in the spaces of war. We will focus our analysis on a space of 5,400 square kilometers in which 445 communes are located, some of which still today are devastated and in ruins, including communes in Marne such as Perthes-les-Hurlus or Ripont, or communes in Meuse such as Louvemont-Côte-du-Poivre or Douaumont. Among these 445 communes we have inventoried nearly 379 public commemorations to date. These 379 war memorials thus constitute our corpus, allowing us to investigate whether significant differences exist distinguishing the situation of the Red Zone from the rest of metropolitan France. We shall ask whether commemoration takes on similar forms in this presumably similar space (the Red Zone) or whether we should consider that symbols, placement, and costs (which constitute revelatory elements of the practice of memory) vary according to the individual commune's wartime experience, the latter being determined by the commune's location and whether it was occupied throughout the four years or was situated behind the Allied frontlines.

In the following analysis, we shall show first that the duty to remember in this particular space is accompanied by the will to display publicly the commune's participation at the cost of national blood. We will see that these symbolic representations possess singular characteristics with recurring themes. Then we will examine the extraordinary solidarity that developed around the French communes, making the communal war memorials into sometimes international works. Finally we will analyze the impact of these facts (the display of pain and suffering, original symbolic representations, and sponsored works) on the cost itself of commemoration.

Display

The duty to remember was apparent early on in France, as seen in the projects undertaken by Jean Ajalbert or by the *Reconnaissance Nationale* led by Jean Richepin and Jean Des Vignes-Rouges. Following the First World War, war memorials sprang up in the French countryside and were placed in villages as if they were milliary columns. Almost every commune commemorated the sons of the village who died on the battlefield, yet the ways they chose to do so are varied and revealing.

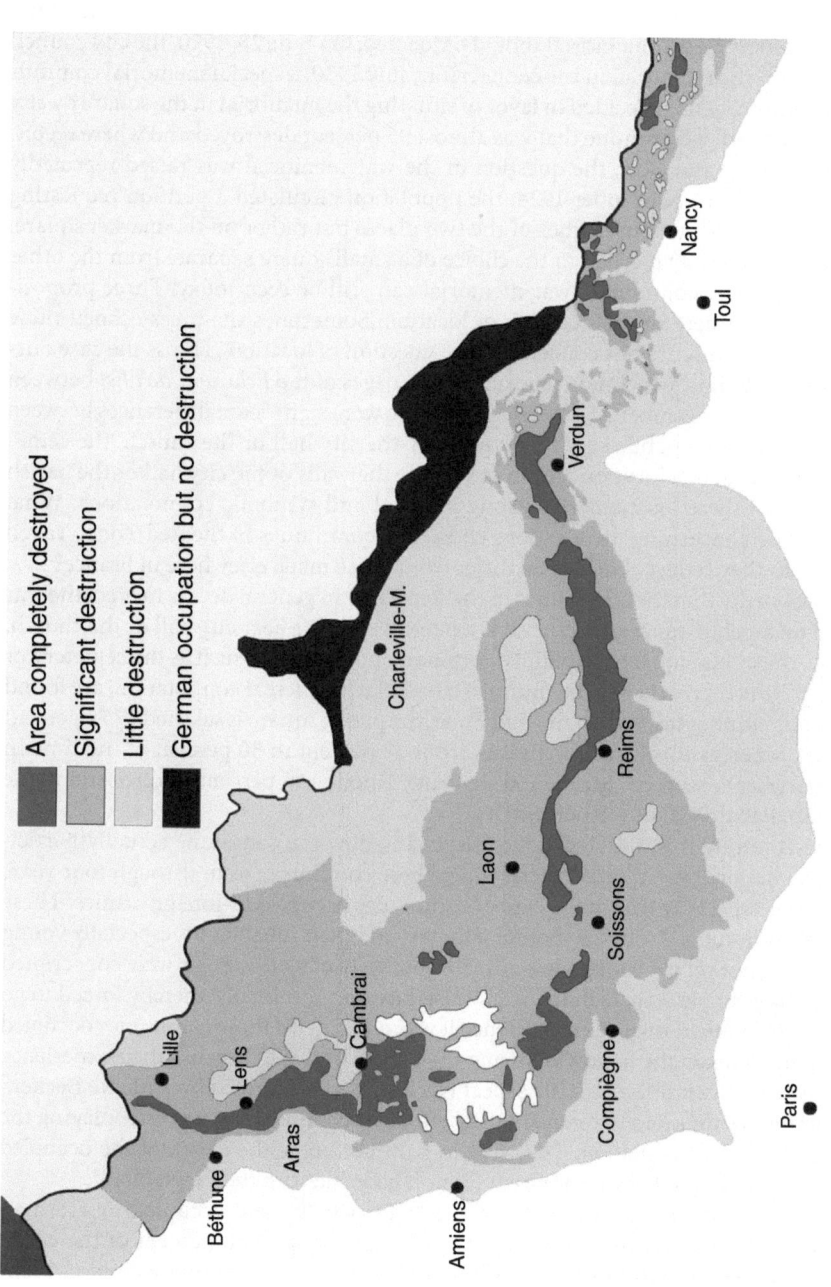

Figure 9.1 Relative Destruction in Northern and Eastern France (in J. Guicherd and D. Matriot).

Legend:
- Area completely destroyed
- Significant destruction
- Little destruction
- German occupation but no destruction

Béthune · Arras · Lille · Lens · Cambrai · Amiens · Compiègne · Soissons · Laon · Reims · Charleville-M. · Verdun · Toul · Nancy · Paris

Such is the case involving the choice of locations for the war memorials. Sometimes the decision created problems and led to intense debates. Thus in Bazancourt in Marne, the municipality encountered resistance from a large part of the population concerning its choice of where the war memorial should be located. On June 28, 1920, the city council decided to erect the memorial in the cemetery. In July 1920, a special memorial committee was constituted, and it decided in favor of situating the memorial in the square by city hall. In Bazancourt, a commune that was almost 95 percent destroyed and where reconstruction was long and slow, the question of the war memorial was raised repeatedly and intermittently. In November 1924, the population circulated a petition requesting that the project be located in neither of the two places but rather on the market square. A consensus was finally reached in the choice of a small square separate from the other locations, where the commune's war memorial can still be seen today. Three propositional forces clash here in the question of location. Sometimes the press echoed these occasionally lively discussions concerning the question of location. This is the case during the crisis in Reims, which was reported in the pages of the *Éclaireur de l'Est* between June 15, 1921, and November 22, 1922. Clearly there were significant differences between the various options: a public square, a square by the city hall or the church, the cemetery or churchyard, or a location within or outside the walls of the city hall or the parish church. Each of these locations has strong political and symbolic connotations. What then is the case concerning the locations chosen by communes in the Red Zone? Taken as a whole, do they reflect choices that differ from those made elsewhere in France?

It is noteworthy that the communes in the Red Zone in general decide to erect the war memorial on a public square, whether it is a neutral space or near city hall or the church. Elsewhere, the public square is certainly dominant, but locations such as the cemetery or churchyard, which give the war memorial a particularly funereal connotation, are found in significant numbers as well. In the Red Zone, the public square is selected 77.79 percent of the time, whereas this location appears from 32 percent to 50 percent of the time in other departments studied: Marne (50 percent), Rhône (46 percent), Loire-Atlantique (32 percent), Pas-de-Calais (48 percent).[6]

For the communes in the Red Zone, displaying how the commune actually participated in the national sacrifice is crucial, since these communes went through four years of war. This is especially true in the case of communes occupied by foreign armies. These communes were cut off from the mother country, and their inhabitants, especially young men who were not yet old enough to be drafted but were of working age, were conscripted by the Germans to participate in the war effort, however minimally, thereby forced to go against the will of their own country. Hence it was difficult for these communes occupied by the enemy to rejoin the nation following the First World War because their experience of the Great War was different. This aspect has been well examined by Annette Becker.[7] These communes thus had to intensify their efforts at commemoration by displaying the war's impact on the population. They had to hammer home the idea that the occupied communes too, like all the communes of France, had paid with their own blood.

This is why the communes north of the front, which were occupied by German troops, located their war memorials on a public square in 85.88 percent of the cases, as opposed to 68.90 percent of the cases south of the front. Cemeteries and churchyards were selected more often by communes south of the front, their active and actual participation throughout the war being above question. The selection of this location corresponds to the desire to stress mourning more than the heroism of the commune and of its youth who died for France. A funerary aspect is present in only 4.12 percent of the cases in the northern part of the zone, whereas it is stressed in 14.35 percent of the cases in the southern part. It is worth noting moreover that this situation is more frequent in the southernmost communes of the zone examined here.

Significant differences can be observed within the Red Zone. The commune of Jouy-en-Argonne (55) represents a fairly atypical case that differs in two respects from our conclusions since this commune's war memorial (although located south of the front) was erected in front of the cemetery and away from the village. Such differences are even more numerous when we try to generalize, comparing the Red Zone with the rest of metropolitan communes. The cemetery or churchyard was selected by communes in the following proportions according to department: Marne, 32 percent; Rhône, 40 percent; Loire-Atlantique, 60 percent; and Pas-de-Calais, 38 percent.[8] It is interesting to show that the smallest figures are found in departments that were destroyed and partially occupied. This suggests that the conclusions we draw for the battle space stretching from the Chemin des Dames to Verdun can be applied unrestrictedly to all the battlefield areas. Finally, it is not surprising that commemoration after the First World War that was taking place in Loire-Atlantique, which belonged to the Catholic west, was located above all in cemeteries or churchyards. Burial is a deeply Judeo-Christian act, and the Church could not give over its essential defining feature, the *cura animarum* (care of souls), to the Republic alone. Hence, locating the war memorial in a cemetery or churchyard is a cultural reflex harkening back to Judeo-Christian origins. We should not be surprised to encounter a greater number of monuments in such locations in the Catholic west.

Specific Symbolic Representations

The choice of location expressing the wartime experience led communes to select specific symbolic elements to implement their duty to remember. We first shall examine the symbolic elements appearing on the war memorial itself. As elsewhere, it is not surprising to find that the elements that return repeatedly are the military cross and palm cluster. In this sense the Red Zone communes resemble other French communes, using the military cross in 68.10 percent of the cases and the palm cluster in 70.24 percent, a ratio encountered elsewhere. By contrast, the battlefield communes use other much more significant elements to a greater or lesser extent than do other communes.

Soldiers: The first such element is the poilu, the figure of the World War One soldier, which is present in 10 percent of the war monuments. In the Red Zone communes, however, this proportion rises to 23.32 percent. Almost one-quarter of the war monuments in the Red Zone display soldiers, whether they are presented in the form of a medallion as in the war memorial at Sommepy (51), a bust as at Sivry-la-Perche (55) or free-standing sculpture at Corbeny (02).

However, noteworthy differences exist as well, depending on where the commune is located relative to the front and its position within the area under consideration. This makes the difference between north and south particularly noteworthy. The communes in the northern zone stress the figure of the soldier, as if to replace those whom the commune could not send to the front or else to express gratitude to those who fought with glory and courage to prevent the commune from falling into the hands of the nation's enemy. The soldier is present in 16.75 percent of the war memorials south of the frontline and in 30.59 percent of those erected to the north. Besides this difference between the northern and southern parts of the Red Zone, importance differences exist between departments. In the areas studied, this symbol is present in 11.81 percent of the cases in the department of Aisne, 30.30 percent in Ardennes, 19.86 percent in Marne, and 43.59 percent in Meuse.

This typology could be even more precise since several different kinds of soldiers are represented. In such places as Ludes, a commune located behind the French front, the memorial was erected in front of vineyards and facing the fort of Pompelle, a center of French army resistance around Reims, which makes the soldier's body into a shield against the invader. In other places such as Saint-Étienne-à-Arnes, figures of the soldier express the meaning of the sacrifice by clutching the flag, the sign of their commitment to the values of the Republic according to the expression that was established during the conflict—for Law and Liberty. The soldiers represented in the communal war memorials thus symbolize patriotic commitment.

Only rarely do war memorial sculptures of soldiers represent the reality of the war, something unnecessary for populations that knew it all too well. Thus in Samogneux, although the soldier is shown preparing for a gas attack, his face reveals nothing of the anxiety preceding the confrontation or of the difficult conditions experienced in the trenches (see figure 9.2).

Moreover, what pertains to the realities of the war also pertains to how the reality of death is represented. Death and the dead soldier are almost never represented (with the exception of the war memorial at Pontfaverger, where the dead soldier is represented but hidden by a bronze shroud), and the dying soldier, as in the memorials at Pévy, Monfaucon-en-Argonne, and Urcel, is purified to the point of becoming unrealistic.[9] The dying soldier resembles a clean-cut, well-groomed young man ready for a New Year's Eve party rather than for death. The hand on his heart suggests that he is dying from a bullet that has just struck him—a clean and proper death that contrasts with the reality of the wounds produced by bombings, machine guns, and shrapnel.

Civilians: Besides the figure of the soldier, civilians are also honored in the Red Zone. Civilian victims hold a special place among them since their names appear on numerous memorials (approximately 65 percent). Nevertheless, civilian victims are of secondary importance since only rarely are their names presented on the front of the memorial; this list of martyrs is presented instead on the sides of the memorial. In the great majority of monument cases, what is most important is the sacrifice made by soldiers Even in the case of destroyed villages where civilian victims were numerous, the civilians figured are often those who remained; the combatants are the soldiers, and the destruction appears only in third place, as seen in the war memorial in the commune of Ornes (55). The village's destruction is commemorated on the sides and in the background of the main motif, which is on the back of the war memorial. But what attracts attention above all is a tearful family, protected after the conflict by the Republic and during the conflict by the unassailable actions of the frontline soldiers, represented here on the attack in the middle distance.

The presence of civilians (tearful families, those who remain) and feminine allegories (whether they represent the nation in arms or decorate its dead, the Republic crowning its heroes or villages cherishing their dead) are found in only 10 percent of the cases. In the Red Zone, however, the appearance of the civilians and feminine allegory should be emphasized, since they are present there in 14.75 percent of the war

Figure 9.2   A Soldier Before the Fight, Samogneux.

memorials. Here, too, significant north-south differences exist since these characteristics are more present in the south (in 16.27 percent of the cases), interestingly enough
with allegories and the presence of civilians occurring in strictly identical proportions.
This semiology is employed less in the north than in the south, but more than the
national average (12.35 percent), according a larger place to civilians, who are often
feminine. These civilians have several roles. They show the suffering and sadness of
those who remain, especially the widows, fiancées, and mothers. This can be seen in
the war memorials at Senuc (08), Isles-sur-Suippe or Époye (51), which use the same
subject, or at Sillery (51) and La Cheppe (51).

Contemplation sometimes offers a way to show that life goes on after all, as is suggested by the war memorials at Suippes (51), Fromeréville (55), and Esnes-en-Argonne
(55). The aim here is to show that despite conflict, the destruction of the means of
production, the ravages of war, and the absence of most of the men, life continues and
is even entitled to after several deadly years. Hope in a better future is often expressed

through the presence of suns (the Senuc model, the back of the Fromeréville monument, etc.) or by field work.

Ruin, desolation, and mourning are also useful in showing the dramatic side of war to the children of future generations. The war memorial plays a highly important educational role following the conflict. This theme is taken up both in the semiology of the monuments and in the speeches given during their inauguration. In Pouillon, Robert Lefèvre, chair of the war memorial construction committee, explained that "placed in front of the school where students learned lessons of devotion, courage, and sacrifice, [it] will recall the undying memory of the generation that safeguarded the freedom of our homes."[10] The mayor of the commune of Heutrégiville, Camille Brisset, reminded those present at his inaugural speech that the war memorial would "remind future generations of the great acts that set so many of our countrymen apart as the most heroic of men."[11] Even in the war memorial itself, mourning leads to civic education, as can be seen at Presles-et-Boves (02), Reims (51), and Fismes (51) (figure 9.3).

Feminine allegories: The role of feminine allegories is to show the debt of the nation or the Republic to the soldiers who died on the battlefield and to the civilians who fought as well as they could against enemy oppression and occupation. This role is seen in the memorials at Vauquois (55), La Neuville-en-Argonne (55), Pontavert (02), and Roucy (02).

We can conclude that in the Red Zone a high degree of correlation exists between commemorations and the duty to remember. Consequently, monuments are found that

Figure 9.3   A Grandfather and his Grandson: Educational Lesson at Presles-et-Boves.

use the same semiological elements. For instance the figure of the dying man entitled "Le Dormeur du val" is found at Urcel, Pévy, and Montfaucon-d'Argonne, in other words in the three departments from which are drawn the most cases for our study. Nevertheless, a certain internal variation needs to be taken into consideration, and it is probably incorrect to treat the Red Zone as a single uniform area. The experience of the Red Zone is certainly different from that of the nation, and thus it is important to investigate the specific nature of this space. But in the Red Zone itself, a distinction must be made between north and south, since the wartime experience in each was different, as were subsequent forms of commemoration. This individual experience also produced another particular characteristic of the war memorials, their "extracommunality."

Works Outside the Commune, Canton, or Department

These war monuments are communal, that is, they commemorate the children of the commune lost during the Great War. They are related to a communal or municipal will, since without political authority nothing is really possible. But the monies collected by mayors, treasurers of specially constituted committees, and others weren't received solely from the people of the commune or even the canton. A pen pal system, like the one that provided so much support to fighters during the war, was also set up between French communes following the war, as well as between French and foreign communes, to finance the reconstruction needed in the devastated regions. Thus, the war memorials of certain French communes located in the Red Zone were not just local works but truly international ones, especially when a private solidarity was developed.

At Poilcourt (08), for instance, the war memorial and the entire reconstruction of the commune were partially financed through the support of the Australian city of Sydney, which explains why the commune's name today is Poilcourt-Sydney. The city of Orléans also participated in establishing the commemoration of the commune of Vauquois (55). Only rarely did a community outside the village community assume responsibility for the war monument, although this did occur at La Neuville-en-Argonne (55), whose children were commemorated through the financial support of Guadeloupe.

In Pévy, the war memorial was financed through the reimbursement of expenses incurred when English troops were stationed around the area of the commune during the war. Some outside private individuals gave considerable sums of money to facilitate the reconstruction of northeastern France. Robert Carnegie, for instance, took part in the reconstruction of the city of Reims, and the library bearing his name today was financed with his support. Foreign investment, which often came from the United States, was involved in all the works. In the small commune of Samogneux (55), the building of its war memorial and the reconstruction of the bridge across the Meuse were financed in part through the generosity of a certain Miss Gray from Boston.

Larger Investments

In the war memorials in the Red Zone communes, the most significant commemorations can be understood in terms of their location in the village, their symbolic features, and the artistic quality of the projects realized, which were financed in part from abroad. It seems clear, then, that commemoration in the Red Zone should be more expensive. This hypothesis can be verified with available sources, which confirm that the average cost of war memorials in Marne was 8,947 francs. Considering only the zone in the same department represented in our study, we see that the average project cost 15,397 francs, in other words almost double. Commemoration was actually more expensive in the Red Zone.

In this regard as well, the Red Zone is far from homogeneous since the project promoters were influenced, either consciously or unconsciously, by the presence of the frontline. In the southern zone, the communes and their benefactors spent on average 10,668 francs (Reims not included since the project there alone cost 362,824 francs, an amount that would have skewed our figures), whereas the communes in the north spent on average 13,014 francs.

While these figures are suggestive, we see two main pitfalls in their use. The first is. that we are considering only the figures from the Marne department, which must be compared with other regions. The second is that we are relying only on raw figures, and it is obviously impossible to compare the 362,284 francs spent by Reims with the 5,100 francs spent by the commune of Grateuil. We will examine these two aspects successively.

To establish points of comparison, we will use the studies of Monique Luirard (Loire)[12] and Bénédicte Grailles (Pas-de-Calais).[13] The following table shows the cost of war memorials according to price, comparing Loire, Pas-de-Calais, Marne, and the Red Zone.

Price in Francs	Red Zone	Loire	Marne	Pas-de-Calais
< 2,499	10.0	4.7	22.2	9.9
2,500–4,999	10.0	17.2	26.7	29.4
5,000–9,999	36.7	37.5	29.8	24.7
10,000–24,999	31.1	30.5	16.0	32.6
25,000–49,999	8.9	6.2	3.9	
50,000–99,999	2.2	2.3	0.6	2.6
> 100,000	1.1	1.6	0.8	0.8

As surprising as it may seem, this table shows that the cost breakdown of war memorials in the Red Zone is much closer to that in Loire than that in the two departments of Marne and Pas-de-Calais, which were affected by the events of the war. Nonetheless the main difference is important since it concerns small budgets. In fact, 10 percent of the communes examined here have a war memorial costing less than 2,500 francs, so in this sense these communes resemble those of Pas-de-Calais. This proves at least that the commemorations were designed to be more expensive than usual and that the large majority of the war memorials took on obligations for expenses between 5,000 and 50,000 francs, as was the case in no less than 76.7 percent of the communes in the Red Zone and 74.2 percent in Loire, as opposed to only 57.3 percent in Pas-de-Calais and 49.7 percent in Marne.

As was briefly noted above though, we must remember that raw data have only the importance we wish to give them. To obtain relatively accurate data, we decided to relate the cost of war memorials to the recorded number of inhabitants in the commune in 1921. This process allows us to compare the efforts the living undertook to honor the commune's dead. From this perspective, the cost of certain memorials differs significantly. Most important is the Reims war memorial, which did cost 362,824 francs, but in a commune of 76,645 recorded inhabitants in 1921. In other words the expense undertaken cost 4.73

francs per inhabitant. By comparison, the effort made by the commune of Grateuil with its war memorial costing 5,100 francs appears enormous, since in 1921 this commune had only 49 inhabitants, making for a ratio of 104.08 francs per inhabitant. Comparing the Red Zone communes reveals considerable differences, which are even greater if we extend the area studied to include the entire Marne department. The difference between the northern and southern zone is consistent since the average cost per inhabitant is 37.40 francs in the north, as opposed to only 22.16 francs in the south. But the efforts undertaken by the communes in the south of the Red Zone must be situated in a larger context in order to see their particularity, since on average the cost per inhabitant in all the Marne communes (including the Red Zone communes) is no more than 11.31 francs per inhabitant. The range of the two categories is revealing since that of the northern part of the zone is larger (because of the exceptional case of Rouvroy) than that of the southern part, which in turn is smaller than the range of the entire Marne department (11.25–433.33, 3.46–85.06, 1.57–433.33). It should be noted that the commune that is least involved in its war memorial project is Saint-Just-Sauvage, a commune of 1,256 inhabitants in 1921 that spends only 1,977 francs for its war memorial. This commune is one of the southernmost communes in the department, and one of the most distant from the combat zone. By contrast, the commune that spent the most in terms of its capacity is Rouvroy, at the center of the Red Zone and whose inhabitants honored their dead with an expenditure of 2,600 francs.

Consequently, we can claim with certainty that the communes that underwent a significant wartime experience developed greater financial capacities in order to commemorate the dead of the Great War. The effort made was larger the more significant the village's destruction and the more important it was to heal the wounds of the war before turning to the dead.

Conclusions

War memorials in the Red Zone possess particular characteristics showing that the spaces involved in significant wartime events are marked to such an extent that the very commemoration of the human sacrifice the inhabitants accepted takes on a different dimension and nature. More visible, less directed toward a funereal aspect, stressing the figure of the soldier and the civilian, works of a national and international nature, more expensive monuments and requiring from the local populations significant financial efforts—these are the characteristics of the war memorials of the Great War located between the Chemin des Dames and Verdun. It could also be shown that these memorials were erected much later than elsewhere since it was necessary to rebuild housing, restart agricultural and industrial production, rebuild waterways, transportation, and electric networks before turning to honor those who died for the Fatherland.

Notes

Chapter translated by Daniel Brewer.

1. Annette Becker, *Les Monuments aux morts: mémoire de la Grande Guerre* (Paris: Errance, 1988); Annette Becker and Philippe Rivé, *Monuments de mémoire: monuments aux morts de la Grande Guerre* (Paris: Mission permanente aux commémorations et à l'information historique, Secrétariat d'État aux Anciens Combattants et Victimes de guerre, 1991).

2. Jean Ajalbert, *Comment glorifier les morts pour la Patrie* (Paris: Crès, 1916).

3. Besides the well-known work of Annette Becker, see Bernard Cousin and Geneviève Richier, "Les Monuments aux morts de la guerre 1914–1918 dans les Bouches-du-Rhône," in *Iconographie et histoire des mentalités* (Paris: CNRS, 1979), 124–30; Jean Giroud, Maryse Michel, and Raymond Michel, *Les Monuments aux morts de la guerre 1914–1918 dans le Vaucluse* (Isle-sur-la-Sorgue: SCRIBA, 1991); Alexandre Niess, *Les Monuments aux morts communaux de la Première Guerre mondiale dans la Marne, mémoire de maîtrise* in contemporary history, directed by Maurice Vaïsse (Reims: Université de Reims, 2001); Alexandre Niess, "Monuments aux morts et politique: l'exemple marnais," *Guerres mondiales et conflits contemporains* (December 2003): 17–32; Alexandre Niess, "La Mort du soldat de la Grande Guerre: représentations dans les monuments aux morts de la Marne," *Des vivants et des morts: de la construction de la "bonne" mort*, Simone Pennec, ed. (Brest: Université de Bretagne Occidentale, 2004), 323–36; Alexandre Niess, *Cimetières militaires et monuments aux morts de la Grande Guerre—Marne*, Inventaire général-ADAGP, "Itinéraires du Patrimoine," 2005; Yves Pilven Le Sevellec, "Les Monuments aux morts en Loire-Atlantique," *Visions contemporaines*, 3 (1989), 7–70; and 4 (1990), 7–132.

4. Hugh. D. Clout, *After the Ruins: Restoring the Countryside of North France after the Great War* (Exeter: University Exeter Press, 1996).

5. J. Guicherd and C. Matriot, "Les Régions dévastées," *Journal d'agriculture pratique*, 34 (1921), 155.

6. Alexandre Niess, *Les Monuments*; Kim Danière, *Les Monuments aux morts du Rhône, mémoire de maîtrise* in history, Université de Lyon (1996); Pilven Le Sevellec, "Les Monuments aux morts"; Bénédicte Grailles, *Mémoires de pierre: les monuments aux morts de la Première Guerre mondiale dans le Pas-de-Calais* (Arras: Archives départementales du Pas-de-Calais, 1992).

7. Annette Becker, *Les Oubliés de la Grande Guerre: humanitaire et culture de guerre, populations occupées, déportés civils, prisonniers de guerre* (Paris: Noësis, 1998).

8. Alexandre Niess, *Les Monuments*; Kim Danière, *Les Monuments*; Yves Pilven Le Sevellec, "Les Monuments"; Bénédicte Grailles, *Mémoires de pierre*.

9. Alexandre Niess, "La Mort du soldat de la Grande Guerre," in Pennec, ed., *Des vivants et des morts*, 323–36.

10. *Le Nord-Est*, September 28, 1925.

11. *La Dépêche du Nord-Est*, November 20, 1922.

12. Monique Luirard, *Les Monuments commémoratifs de la Loire* (Le Puy: Université de Saint-Étienne, Centre interdisciplinaire d'études et de recherches sur les structures régionales, 1977).

13. Bénédicte Grailles, *Mémoires de pierre*.

National Sacrifices, Local Losses: Politics and Commemoration in Interwar Alsace

Christopher J. Fischer

As French troops marched into Alsace in November 1918, they were met by adoring crowds, windows bedecked in tricolor bunting, and a regional press that jubilantly declared on behalf of all Alsatians, "We are again French!"[1] For Alsatians and the newly arrived French troops, the end of war, suffering, and privation gave cause for celebration. For many, the return of Alsace to the embrace of France likewise provided a source of joy. Yet this "Spirit of 1918" belied the difficulties that would soon beset the region.

Conflicted ideas about the Great War sprang, in large part, from divergent conceptions of the goals, and price, of the war. In the French narrative, Alsace—after nearly five decades of unlawful annexation by Germany—had been regained at the tremendous sacrifice of nearly 1.4 million men. Alsatians, indisputably French, therefore rightfully welcomed back French troops. French policies, French designs for commemoration of the conflict, and reactions to Alsatian complaints about French rule were filtered through this narrative of sacrifice.

While Alsatians certainly celebrated the war's end and mourned their losses, their understanding of the war differed from French national memory. Most Alsatians who served—and died—did so in German uniforms, not French ones. Moreover, Alsatians at home had suffered both the normal privations of war and German martial law. For Alsatians, the experience of loss as both victor and vanquished imparted a more ambiguous meaning to the conflict. How could Alsatians recognize their losses in war within the context of the French narrative of sacrifice and victory?

Divided memories also clashed with confused expectations and disillusionment over French policies. The newly regained province fitted rather uneasily in France. A host of economic and political issues gave rise to a strident autonomist movement and strengthened regionalist tendencies among the area's existing political parties. In response, French officials and more nationally-minded locals cried foul; Alsatians, at least the autonomists, denigrated the heroic sacrifices of French poilus in liberating Alsace from the German yoke.

This article lays out some of the ways the memories of the war became enmeshed in interwar developments in Alsace. In particular, it will examine how the memory of the war shaped French policies, how it led to ruptures over proper modes of commemoration, and how it sharpened political differences between region and nation, especially in the latter half of the 1920s. For French authorities and the most intensely nationalist Alsatians, the sacrifice of the war demanded the swift assimilation of the region, and more importantly the rapid eradication of German influence in Alsace. For many Alsatians, the wartime suffering reinforced the notion that Alsace was unique; French infringements on Alsatian memories and perceived rights, therefore, seemed an extension of wartime losses.

The Contours of the Alsatian Experience

Before analyzing the impact of the memory of war on Alsace, some of the circumstances framing Alsatian and French attitudes must be laid out. During the years of German rule (1871–1918), a strong and vibrant regionalism evolved in Alsace. Alsatian regionalism was by no means monolithic. Some Alsatians favored a vision of the region with a decidedly German cast; others strongly defended the region's culture, traditions, and German dialect within the context of the German Empire, but remained ambiguous on the issue of national belonging. Finally, some Alsatians evoked the notion of a regional "dual culture" to defend French culture and memory, an often thinly disguised version of a French vision for the region.

While the pro-French variant of Alsatian regionalism resonated poorly with German administrators and nationalists, all the variants of Alsatian regionalism were fostered, if not encouraged, by a German conception of a federalized nation in which regional identities could exist concomitantly with loyalty to the German nation-state.[2] Annexation and heavy-handed German policies, in conjunction with a federal German nation, helped to forge Alsatian regionalism; the constitutional debates of 1910–11, the so-called Zabern Affair, and the First World War honed Alsatian regionalism to a fine edge.[3] Such intense feelings of regionalism, of course, survived into the period following the armistice.

These sharpened feelings of regional uniqueness were further tempered in the crucible of war. The place of Alsatians in both armies reflected their unusual status within the two nations. French officials wanted Alsatians to serve, but feared their fate at German hands if captured, and, in some cases, were uncertain of Alsatian loyalty. Though opposed by some intensely fervent Alsatians, the French solution saw most Alsatian troops largely shipped to the colonies to free up additional soldiers for the front.[4]

German military leaders, in marked contrast, questioned Alsatian loyalty with greater intensity. For many German officials, Alsatians remained little better than disguised Frenchmen who would betray German troops at the first opportunity or, barring that, desert. Such fears had been amplified not only by the ongoing tribulations between civilian and military leaders before 1914, but more proximately by many leading Francophile Alsatians who slipped across the border in the summer of 1914 to serve in the French army and government. Reports of preferential French treatment of Alsatian POWs further exacerbated such fears.[5] To avoid desertions and betrayals, German military officials took action. Beginning in 1915, troops from the region were often sent to the Eastern Front, put under special regulations or limited in their travels during leave.[6]

Such measures reflected a broader feeling of mistrust toward the Alsatians. Such doubts, combined with the region's strategic location, led to the imposition of a much stricter form of martial law in Alsace-Lorraine than in the rest of Germany. Those deemed to have "anti-German sentiments," a history of criticizing German rule, were placed in "protective custody" in the German interior. In the meantime, German military authorities attempted to accelerate the pace of Germanization. Although local politicians complained in both the regional and national legislatures, their pleas fell upon deaf ears. Taken in conjunction with the broader wartime suffering, it is not surprising that Alsatians increasingly lost faith in Germany and at war's end embraced a return to the French national fold. Such experiences would color Alsatian memories of the war and its meaning.

This ready acceptance seemed reflected in those early weeks at war's end with the cheering masses lining the streets to welcome the French back to Alsace and Alsace back into the *mère-patrie*. President Raymond Poincaré, visiting Strasbourg in December 1919, affirmed this situation in a speech dwelling on the warm acceptance of French troops in the region as a sign of Alsatian national loyalties.[7] Yet within months, local papers would write of an Alsatian *malaise*, and in less than a decade Alsace would witness the rebirth of a strong autonomist movement. Although French authorities looked for the menacing hand of Germany behind these assumed Alsatian puppets, the problems in the border region stemmed more from a collision of French and local hopes and perceptions, as well as from French policies that flew in the face of local traditions.

Many of the difficulties that arose between French authorities and the Alsatian populace sprang from myriad, often mutual misunderstandings and misconceptions. For almost five decades, Alsace along with Lorraine had been the focus of French myth. The declarations made by the representatives at the 1871 National Assembly at Bordeaux, the literary works of Léon Daudet and Maurice Barrès, and the activities of Alsatians such as the caricaturist Hansi—joined to wartime propaganda—all fed the perception that Alsatians had remained French, valiantly resisting German efforts to coerce them into loyalty to the German nation. Such conceptions of Alsace did not quite square with reality. While some Alsatians had quite vociferously opposed German rule, many had acclimated themselves to German administration and had worked to reform the place of the region within the *Kaiserreich*. In addition, the overwhelming majority of Alsatians spoke German (and little French), were more religious than the French, and had grown accustomed to administratively adept (if politically incompetent) civil servants.[8]

If the French had a mistaken image of Alsace, then the Alsatians also misunderstood what it meant to return to the French fold. The France that reclaimed sovereignty over the region in 1918 was not that of 1871. The French nation-state was far more unified spiritually, culturally, and administratively,[9] thus greatly different from the federalist German Empire under which Alsatian regionalism had come of age. In addition, regional representatives possessed only a vague awareness of the implications of the French separation of church and state for Alsatian affairs; the full and logical consequences of the French separation law of 1905 were not foreseen. Finally, Alsatians believed the wartime promises of Joseph Joffre and Poincaré, among others, to respect Alsatian customs and traditions implicitly protected the status quo in church-state relations, preserved German laws deemed more beneficial (e.g., social welfare provisions and municipal administration) than the French equivalents, and maintained the German language as a constituent element of Alsatian culture.[10]

Thus, Alsatians and their new French compatriots brought not only different experiences of the war into their new union, but also vastly different conceptions of one another and of the relationship between region and state. These differences would cause no end of difficulty in Alsatian interwar politics.

Excising German Ghosts: The Reintegration of Alsace into France

In 1919, French authorities sought to integrate the region rapidly while simultaneously erasing German influences in the area. While perfectly logical, especially in light of French wartime sacrifices, such policies would prove controversial. Already during the war, the *Conférence d'Alsace-Lorraine*, a collection of French politicians and exiled Alsatians in Paris charged with postwar planning for the two provinces, began laying out a plan of economic, administrative, and cultural assimilation. Despite warnings from Alsatians in the conference, French administrators assumed that the local population would easily be brought into the secularized, centralized French state, and that most of the practical matters involving reintegration would be worked out in due course.[11]

From the outset, practical issues undermined local confidence in French authorities. Local civil servants and state employees, for example, found their pay, seniority, and pensions reduced, issues that would bedevil French authorities throughout the interwar period.[12] Many Alsatians moreover blamed the general economic slowdown after the war on the French government. As a part of the drive to shed the German influence in the area and reap the benefits of victory, for example, German companies were sold off to French concerns; locals perceived these fire sales as detrimental to the local economy. What seemed like logical steps to French authorities in cutting ties between the region and Germany felt like unfair punishment to locals. Indeed, the economic situation left long-lasting bitterness and contributed to the birth of the Alsatian Communist Party.[13]

French authorities, however, did not simply want to bring Alsace (as well as Lorraine) back into the French state, but also wished to purge the region of its Germanic elements.[14] Although many of the 300,000 Germans in the region left voluntarily at war's end, French suspicions of the remaining Germans, and their putative local supporters, led to the creation of—*commissions de triage*, commissions meant to sort through the local population to weed out suspect members of the local community. Labor leaders, former German civil servants, and potentially troubling regionalists, all came under suspicion. The process suffered from numerous flaws as those summoned were allowed no legal representation, charges could be brought on the basis of denunciations rather than hard proof of wrongdoing, and the commissioners themselves often came from upper class Alsatians returning to the region for the first time in decades.[15] Between November 1918 and October 1919 the commissions processed over 11,000 cases, handing out punishments ranging from surveillance for those deemed "suspect" to expulsion to Germany.

The 1919 *épuration* or purification ultimately affected few Alsatians. However, combined with political and economic missteps, this process contributed to a growing unhappiness with French rule, a feeling termed by contemporaries to be a "malaise alsacien." This dissatisfaction would not grow into full blown anger until the later half of the 1920s, when Alsatians once again turned to the language of regionalism to express their complaints over French rule. More immediately, divergent French and Alsatian experiences found expression in the commemorations of the First World War.

Commemorating the War: Monuments, Armistice Day, and Muted Conflicts

As in France and across Europe, Alsatians sought to come to grips with the massive losses incurred in the war. As in much of Europe, the process of finding proper modes of mourning proved difficult. The "normal" debates over placement, costs, and the aesthetics of monuments, explored in great detail by such scholars as Daniel Sherman and Jay Winter, of course played a role in Alsace.[16] The need to create public narratives of sacrifice and symbols of commemoration, moreover, clashed with private memories. Most Alsatians experienced the war in the German army and not the French. Victory suffused with sacrifice, therefore, was at best experienced vicariously for most Alsatians. Nonetheless, French authorities and local politicians set about the difficult, often contentious task of commemorating the war. The traditional autonomy of Alsatian communes exacerbated such conflicts, as local officials felt that they had the right to represent all their constituents in the various monuments, not only those or not only in a manner of which higher ranking officials approved.[17]

Such tensions played out in several ways. In Mulhouse, for example, veterans of the French and German armies clashed over the proper form of commemoration and the proper inscription for a future monument; the involvement of Souvenir Français, which did not want to honor the fallen enemy, only complicated matters. The prospect of a dual monument and a split commemoration was, as Shane Story has pointed out, anathema to French officials who wanted a single French memory of the war. The controversy was obviated when the statue was redesigned to include two figures, one of which glorified the war effort, the other of which honored the sacrifices of the soldiers.[18]

In other towns, the local community tried to show its loyalty to France while also remaining mindful of the divergent experiences of the war. For example, the town of Quatzenheim dedicated its monument in 1919. The inscription on the monument, fully endorsed by the departmental head of monuments as well as by local officials from the Interior Ministry, read "In remembrance of the liberation of Alsace, to the young men of Quatzenheim, victims of the war, 1914–1918."[19] Thus, it did not differentiate between those who fell for the Germans and those who fell for the French.

Unfortunately for French administrators, such easy ways to approve monuments were the exception rather than the rule. Other municipalities submitted projects that did not, according to French officials, possess the appropriate artistic merit. Independent-minded communes often submitted their plans as faits accomplis to departmental authorities.[20]

Strasbourg proved an especially difficult case in the quest to commemorate the fallen of the First World War. The reluctance of the city council to accept monuments dedicated to the French nation delayed the erection of a war monument.[21] For example, in the municipal elections of May 1929, a strange alliance of autonomists, Catholic regionalists, and autonomist-minded communists won a narrow majority in the city council. The new administration led by the autonomist-communist politician Karl Hueber, a vociferous critic of the French administration in Alsace, hindered progress on the moment and more generally tried to avoid celebrating French national holidays in municipal spaces.[22] The displacement of the Hueber government by Democrat Charles Frey in the 1935 elections created an opportunity to rectify the lack of a monument in Strasbourg.

A private commission, headed by Henri Lévy, raised several hundred thousand francs for the monument.[23] The monument itself received a place of honor in the city, the center of the Place de la République, the former *Kaiserplatz* and home to a statue of Wilhelm I.[24] Sculptor Léon Drivier designed a *Pietà*-like statue. However, instead of holding one fallen son, the mother cradled two figures, their hands outstretched to one another and touching; they represented Alsatians who had fought on both sides of the war. For the inscription, the committee chose the relatively neutral "A nos morts: 1914–1918"; a German translation, however, was rejected as it would only cause a "demonstration."[25] Inaugurated in October 1936, the sculpture was erected at the center of the Place de la République.[26] By choosing a monument that could commemorate the memories of all sides in Alsace, the steering committee for the war memorial managed to bridge some of the differences dividing the local community.

While local monuments could exacerbate local divisions and create tensions between communal and departmental officials, the only national monument erected in the region, the memorial complex at Hartmannwillerskopf, appears to have enjoyed local support. The ridge and surrounding slopes of Hartmannwillerskopf, located near the town of Cernay, had seen hard combat, especially in 1915, and over 20,000 soldiers from the French and German armies had died along this section of the front.[27] The monument stood alongside those at Douaumont, Notre Dame de Lorette, and Dormans as central sites of commemoration.

Even as efforts got underway in 1920 to collect donations for the monument, French officials had to worry about the physical preservation of the site, especially as the battlefield became a tourist destination. For example, in 1920 the regional military commander demanded tighter control of the site after a large Swiss group had crossed into French territory to "profane" the battlefield with a rendition of the *Internationale*.[28] Similarly, General Commissioner Alapetite had to admonish members of the Touring-Club de France that photograph stands at the site could not be expanded; moreover, postcards were to be in French, though the captions for pictures could have translations under the French text.[29] To judge by the absence of complaints in subsequent years, French authorities seemed to have regulated visits to Hartmannwillerskopf to their satisfaction.

Meanwhile, the committee in charge of raising funds for the monument went about the arduous task of seeking donations. Led by General Tabouis, a veteran of the campaigns in the area and a commander in the garrison at Mulhouse, the steering committee sought both public and private support.[30] Some subsidies came for the local administration. Some donations were in kind, as the local government offered material from leftover war stocks, especially bronze, for the construction of the site.[31] The committee also gained some popular support; for example, the *Société de secours aux blessés militaires*, led by a Strasbourg socialite, helped secure the support of the prefect and raise money among Strasbourg's Francophile elite.[32]

The committee campaigned for broader public help by appealing to Alsatians' sense of duty and history. After reminding Alsatians of their desire to return to France, the committee turned to examine the sacrifice of French soldiers and Alsatian duty to their memory:

> Some of them, many even, arrived and you saw them march before you in those unforgettable days of November 1918. It is in their name that we appeal to you who watched them.
>
> At issue is honoring and glorifying the memory of those who did not arrive because they fell while thinking of you… banish from your heart, from your spirit indifference, ingratitude, and forgetfulness. Give, give for the soldiers of France, who died for your deliverance.[33]

The extent to which such calls actually won over Alsatians is not clear. Much of the support in the region seems to have come from prominent industrialists and business leaders.[34] Regardless of the financial bases for the project, by 1925 the committee was ready to present a model of the future monument, designed by Robert Danis, at a Parisian exposition.

The inauguration of the monument sparked neither loud protest from the Alsatian autonomists nor abundant enthusiasm from local nationalists. Rather, a measured support characterized President Albert Lebrun's October 1932 visit to Alsace, during which he led the dedication of the memorial. The ceremonies included a short prayer service and were attended by political leaders, members of patriotic organizations, and the Bishop of Strasbourg as well as by local luminaries.[35]

In keeping with the spirit of commemoration, both the speeches and press coverage adopted a subdued tone. In particular, politicians and newspapers stressed two elements. First, the horrors of war—and the subsequent desire for peace—were expressed in the ossuary, crypt, cemetery, and monument atop the "bloody mountain." Second, the motif of French sacrifice for the liberation of Alsace assumed a central place in the rhetoric of the day. Indeed, President Lebrun went so far as to tie the events of 1870–71 to the sacrifices of 1914–18; not only had France avenged, albeit at a high cost, the loss of 1870–71, it had also reclaimed Alsace and its populace, which had fought valiantly against German domination for decades.[36] The elision of Alsatian sacrifices in the German army did not spark protests, at least not as evidenced by the major Alsatian dailies.

If monuments marked commemorative space, holidays created commemorative time. While some national holidays opened up political debates over the nature of church-state relations (Fête de Jeanne d'Arc) or offered opportunities to debate the nature of national belonging (Quatorze Juillet), Armistice Day (Fête de la Paix) offered a more muted day of commemoration. In somber tones, the Alsatian press annually called on its readers to commemorate with them the terrible losses of the First World War. At times, Alsatian attitudes mimicked those of the larger French polity. For example, in 1921, when the French parliament tried to shift the holiday to Sunday, November 13, Alsatian editors decried the move just as had veterans across the country.[37]

Beyond a communal moment of sorrow, the day had divergent meanings. The nationalist daily *Journal d'Alsace et de Lorraine*, not surprisingly, used the occasion to celebrate the return of Alsace to France and recall French suffering at the hands of the Germans. Moreover, the paper at times reminded its readers that the Germans had yet to fulfill their treaty obligations and even went so far as to charge Germans with revanchist intents with regard to Alsace.[38] The *Journal de L'Est* espoused a similar view, but underscored more clearly the links between Alsace and France by recalling the victorious days of November 1918. The issue of national sacrifice also came to the fore; for example, in 1925, the paper stressed how French soldiers had "subordinated regional and personal interests" to the war, an obvious jab at the growing regionalist movement.[39]

The socialist press, in contrast, took an internationalist track and focused upon the need for greater peace in Europe. In part, such papers as the *Le Républicain* and the *Freie Presse* wanted to see disarmament, increased moral investment in the League of Nations, and revisions to the Versailles Treaty, all of which in turn would increase the likelihood of opportunities for European peace. As such, the articles of the later 1920s praised the idea of the Locarno Treaties. On occasion, the Socialists also used the issues of peace, war, and commemoration to hammer on right-wing/nationalist politicians such as Raymond Poincaré and Georges Clémenceau whose "politics of hate" disturbed European peace.[40]

Much like the pro-French and socialist press, the moderate regionalist press adopted a reverent air for its coverage of Armistice Day. Catholic, regionalist-oriented papers pushed for greater efforts toward peace; unlike the Socialists, such dailies as the *Elsässer Kurier* and the *Elsässer* focused on the need to avoid a renewal of specifically Alsatian suffering by calling attention to the unique position of the region as a borderland between two great powers. Only in the late 1920s did a critical stance creep into the commemorative articles. For example, the November 11, 1929 edition of the *Elsässer* recalled that Alsatians fought on both sides of the war. The French government needed to realize that all Alsatians, not just those in the uniform of the poilu, "had died for the Fatherland. Only when the current administration took steps to include all Alsatians could the reintegration of Alsace into France be completed."[41]

If the moderate Catholic regionalist press demonstrated a measured respect for the holiday, then the autonomist press was conspicuous in its failure to offer commentary concerning Armistice Day. Such an omission is not easily explained, though the limited number of years of publication of autonomist papers in the 1920s (the leading papers, *Die Zukunft*, *Die Volkswille*, and *Die Volksstimme* enjoyed approximately two-year press runs apiece) offer only a small range of possible material to examine. Why the silence? One can venture several potential explanations. The autonomists, some of whom harbored separatist longings, may not have wanted to recognize a day marking German defeat. More likely, the autonomists, though often outrageous in rhetoric and harsh in style, shared with their compatriots the sorrow of the day and had the decency, or political acumen, not to make political hay out of a harvest of destruction.

Alsatians and Frenchmen both sought to come to terms with the immense losses endured during the Great War. Commemoration and mourning though proved at times difficult to reconcile. French narratives of sacrifice failed to square easily with Alsatian losses during the war, especially those in the German army. French officials recognized the need for locals to honor fallen soldiers but sought to excise memory of

Alsatian sacrifice in the German military from local memory. Thus, the public spaces of commemoration at times became arenas of conflict between regional and national, private and public memory. Yet commemorative time as embodied in the November 11 ceremonies offered more universalizing moments. Political debates were set aside, French sacrifice and Alsatian loss for a time overlapped, and, for a day at least, staunch nationalists and fierce regionalists mourned together.

Languages of Perfidy and Sacrifice

The memory of the war also cast its long shadow over political developments in the region. As Alsatians grew increasingly frustrated with French policies, and as French officials became angered over local unrest, the experience of war provided an increasingly important lens, although a blurred one, through which to understand events. Alsatians could point to their suffering during the war (as soldiers, under German martial law, and in some cases as interned civilians in French concentration camps), but this held little purchase with French authorities. In contrast, French authorities and local French nationalists repeatedly evoked the memory of the French sacrifice during the war; Alsatian resistance was at best ingratitude, at worst a form of treason to the 1.4 million dead. The memory of war, therefore, precluded Alsatians from making claims based upon their suffering, and it further reinforced French aims to integrate the region.

By 1925, large segments of the population remained unhappy with developments since 1919. For most Alsatians, the combination of traditional political rhetoric (as invoked by the local parties, Catholic, Socialist, Radical, etc.) with strong doses of regionalism sufficed to frame Alsatian demands. Local politicians, especially after the 1924 decision of the Herriot government to introduce the full range of the *lois laïques* or secular laws into Alsace, sought a more vigorous regional administration to defend regional culture and traditions. Such demands evoked opposition, if not outright horror, from French administrators from the interior.

Yet calls for greater decentralization throughout France, and for a special set of statutes for Alsace in particular, came to represent a rather moderate line in Alsatian politics. More alarming to French authorities was the appearance, beginning in 1925, of new groups in the region that framed their desires in terms of a rhetoric born in the war, that of national minorities.

Three main and closely interrelated groups became the standard bearers of Alsatian autonomy and began to evoke an almost Wilsonian rhetoric to further their regionalist and later separatist agenda. First, a new paper, *Die Zukunft*, appeared in May 1925. In polemical terms, the paper demanded respect for Alsatian culture and autonomy for the region; it questioned whether liberty, fraternity, and equality had been achieved in Alsace; and it queried when the wartime promises of Poincaré and General Joffre to respect Alsatian traditions would be fulfilled. Over the following months, as the paper grew in popularity, it also began to run numerous articles asking whether Alsatians were a national minority, often drawing comparisons between Alsace and the Italian, formerly Austrian, region of South Tyrol, and presenting the League of Nations as one potential guarantor of Alsace's "national character [*Volkstum*] and language."[42]

In 1926, the editors of the *Die Zukunft*, along with select members of the regional Catholic Party and others disillusioned with French rule, formed a new group, the so-called *Heimatbund* or Homeland Union. Intended as an overarching organization to direct efforts to fight for Alsatian rights rather than as a political party, the Bund called upon all "true Alsatians" to "no longer bear the circumstances" and fight "as a national minority for complete autonomy in the framework of France." In particular, the Heimatbund wanted a regional legislative and administrative body with budgetary authority, maintenance of the current situation with regard to church-state relations, preservation of the "Christian character of the Alsatian people," and education in German.[43] The Heimatbund program not only evoked Alsace's right to greater autonomy as a national minority, but did so within the "spirit of Locarno." As Heimatbund leaders would later elaborate (most notably in the pages of *Die Zukunft*), the Treaty of Locarno allowed France to deal with the problems in Alsace more favorably. Since Germany had renounced border changes in the West, the French no longer had to fear Alsatian autonomy as a Trojan horse for German revanchism.[44] The Heimatbund manifesto concluded by stating that its proponents were "enthusiastic supporters of the idea of freedom and international cooperation and opponents of imperialism and militarism." Alsace, in this conception, could return to its rightful role as a "meeting place between two cultures" and would help foster "reconciliation between Germany and France."[45]

If *Die Zukunft* had raised the issue of national minorities, and if the Heimatbund (in conjunction with *Die Zukunft*) elaborated on the idea, then the foundation in 1927 of the Landespartei marked the transition of autonomism into the realm of separatism. This new party, largely comprising the more radical supporters of *Die Zukunft* and the Heimatbund, demanded autonomy within France; yet it also utilized the idea of an Alsatian national minority in a manner that surpassed such claims. After asserting that France had retaken Alsace and Lorraine "under the pretext of liberating them" in 1918, Landespartei co-founder Karl Roos openly asserted the Alsatian right to "self-determination" (Selbstbestimmung) as a national minority and claimed that Alsatians could, in the event autonomy was not granted, appeal to the League of Nations for relief.[46]

Some in the autonomist camp even went so far as to test the waters for support for the Alsatian claim to national minority status. Camille Dahlet, the leader of *Fortschrittliche Partei* (Progressive Party) and deputy in the Chamber of Deputies, met in Geneva with the leaders of the Minorities Congress in February 1929. Dahlet hoped to gain support for the broader autonomist movement in Alsace from a group that promoted national minority issues to the League of Nations. The French government kept close tabs on Dahlet's activities. A French consul would happily report that the Congress leaders, composed mainly of representatives of German minorities and allegedly tied to the German Foreign Ministry, showed little interest in supporting Alsatian aspirations to minority status.[47]

While the claim that Alsatians should be considered a national minority represented a new element of Alsatian regionalism, one that would incite official government countermeasures and nationalist opprobrium, the underlying aims and sentiments of the autonomist movement descended directly from pre-1914 developments. Although the specific demands of autonomists in the Heimatbund and Landespartei pushed prior demands, it was the intellectual heritage of the concept of dual culture and earlier moves for autonomy that provided the backbone for the interwar autonomist movement. The addition of Wilsonian idealism to these existing traditions was meant to add weight to autonomists' demands to protect and preserve Alsace's cultural, linguistic, and religious traditions.

French officials and French nationalists in Alsace responded to the radicalized turn in Alsatian regionalism on a number of levels. The French government took measures directly aimed at quashing the most radical regionalists. For example, state employees who signed the manifesto of the Heimatbund, one of the early incarnations of the autonomist movement, faced disciplinary measures or dismissal. In November 1927, the French government banned certain autonomist newspapers and shortly thereafter began arresting key leaders of the movement. These arrests were followed in May 1928 by a long trial of some 15 defendants accused of conspiring to separate Alsace from France. The trial, though pointing to ties between some of the leading autonomists and Germany, was largely ineffective as the only defendants found guilty would be pardoned in July 1928. More importantly, the publicity surrounding the trial strengthened, if only for a limited time, the autonomist movement.

For the nationalist press, the very nature of the demands of the Heimatbund, especially the claim for a special regime for Alsace and Lorraine, went against the French conception of the nation-state. Autonomist demands would, in effect, undermine "France indivisible" and end the "union sacrée."[48] Moreover, the papers had long argued in favor of linguistic and cultural assimilation into the French nation; the argument that Alsatians were a German minority, or even that the German language was integral to being Alsatian, was anathema to such groups. The claims to national minority status, therefore, were dismissed as the first step toward separatism.[49] As Anselm Laugel, a key member of the regionalist movement before 1914, argued, no other region such as Picardy or Brittany would ever invoke a claim to the League of Nations. The Heimatbund, Laugel continued, was catering to the "vengeful and imperialistic aspirations" of Germany under the guise of a "false appeal" to minority status and the spirit of Locarno.[50] However, it was not simply the "anti-national" element of national minority rhetoric that fueled the nationalist response, but also the less threatening claims to a cultural uniqueness for Alsace and the attendant hope to preserve some of the regional heritage. As the Socialist *Freie Presse* stated, the Heimatbund claim that it wished to protect "Alsatian customs and traditions" was nothing more than "a Chinese wall which would separate the provinces of Alsace and Lorraine" from France.[51]

Much of the rage directed at the autonomists sprang from the belief that autonomism equaled not simply separatism but support for the recently vanquished Germans. Leaders of the various autonomist factions such as Eugène Ricklin, a former leader in the local Catholic party, and Georges Wolf, a liberal before the war and radical-socialist thereafter, were believed to be sympathetic to Germany. Suspicions of German financial support for the autonomist movement slowly surfaced, and although relatively trivial they did cast the claims of autonomists in a yet more dubious light. Finally, a general suspicion of Germans, grounded in the experience of the Great War, underlay nationalist concerns. The nationalist press in Alsace demonstrated little faith in Locarno and saw a German puppet master behind every autonomist marionette; yet such concerns also blinded the nationalist press, and the government as well, to the legitimate complaints of Alsatians under French rule.[52]

The suspicion and hatred of Germans was married to the recollection of sacrifices offered by both France and Alsace to secure the return of the "lost provinces" to the *mère-patrie*. This can be seen in an open letter from a local branch of the *Union nationale des combattants* in response to the Heimatbund in which the veterans stated that they, "in true memory of the sacrifice of our sons, fathers, and brothers who fell in the war, and the suffering of the Alsatians under German occupation, condemn the campaign in Alsace led by a few fanatics ... [and] oblige [the government] to fight every attempt to endanger national unity."[53]

Perhaps the most ardent evocation of French sacrifice, however, came not from the Alsatian nationalist press but rather from Raymond Poincaré. In late January and early February 1929, a wide-ranging and often contentious debate took place in the Chamber of Deputies concerning the problems in Alsace. In a three-day, ten-hour speech, Poincaré lambasted the autonomists as a threat to the French state and as agents of Germany, denounced the introduction of the idea of "national minorities" in Alsace, argued that France already had made concessions to Alsatian particularities, and repeatedly pointed to the heroic sacrifices of French soldiers to free Alsace of German domination. Poincaré ended, stating his cause powerfully, that 1.4 million Frenchmen "had died in the hope that the victory of our country would secure the liberation of enslaved Alsace."[54]

In the ensuing years, French rhetoric and policies became more measured. In part, French officials decided to slow the pace of integration. Alsatians witnessed an evolution of the local political scene. The autonomist movement by 1932 had largely fractured into a variety of splinter groups ranging from agricultural fascists to autonomist communists to a nascent National Socialist Party. More importantly, the Great Depression in the early 1930s, the looming shadow of Nazism, and in the later 1930s the gathering storms of war all led to a desire for peace, and for many Alsatians a firmer embrace of the French nation.

Conclusion

In keeping with the spirit of the volume, we can ask ourselves: how did the war and the memory of the war transform Alsace and Alsatian spaces? The region, at a fundamental level, was both subject and object. Alsatians suffered through the war: their sons served on both sides of the conflict; martial rule and material privation marked the home front. But at the same time, Alsace as a territory was a prize of war, and the "liberation" of its people a French war aim. The French liberation of Alsatians, Alsatians who had loyally resisted the Germans, became imbricated in the meaning of France's sacrifices to the hated German foe. Alternative stories of loss and sacrifice based in the region's unique experiences of the war were therefore subsumed to the meanings the French imparted to the conflict. Alsace was no longer simply French territory, but space imbued with the memory of sacrifice, French soil regained in the dear currency of French lives. Such understandings of Alsace brought a sharper edge to the contention wrought by the creation of memorials in the region's public squares.

Ultimately, the war's legacy went beyond shaping commemorative spaces in Alsace. On one level, the war defined the tolerated boundaries of political discourse. Autonomists, especially those evoking a Wilsonian demand of self-determination, stepped outside the borders of acceptable political discourse and provoked a swift rhetorical, and sometimes legal, response. On a yet more profound level, the war's legacy created a dissonance between regional and national spaces. For Alsatians, regional space and national space could be congruous; however, such melding had to be on terms that comported with their experiences. For the French, imagining the region as national space was much simpler; Alsatians who refused to conform failed to appreciate French sacrifices. The war may have allowed the realignment of French national frontiers; ironically, it also disrupted the integration of regional into national space.

Notes

1. Charles Frey, "Der 22. November," *Strassburger Neue Zeitung* (November 23, 1918) clipped in ADHR 2 J 231 "Collection Heitz." See generally Archives Départementales du Haut-Rhin 2 J 231 "Collection Heitz," a collection of newspaper clippings concerning the entrance of French troops into Alsace.

2. Celia Applegate, *A Nation of Provincials: The German Idea of Heimat* (Berkeley: University of California Press, 1990); Alon Confino, *The Nation as Local Metaphor: Wuerttemberg, Imperial Germany, and National Memory, 1871-1918* (Chapel Hill: University of North Carolina Press, 1997); Caroline Ford, *Creating the Nation in Provincial France: Religion and Political Identity in Brittany* (Princeton: Princeton University Press, 1993).

3. Jean-Marie Mayeur, *Autonomie et politique en Alsace: la constitution de 1911* (Paris: Armand Colin, 1970); David Schoenbaum, *Zabern 1913: Consensus Politics in Imperial Germany* (London: George Allen and Unwin, 1982); Christopher Fischer, "War Weariness or National Reunion?" *Proceedings of the Western Society for French History* (Spring 2003): 223-31.

4. ADHR AJ 30/85 (Purg. 11745), "Etude sur la Question des engagements des alsaciens-lorrains dans l'armée française," and ADHR AJ 30/85 (Purg. 11745), "Rapport à Général en chef" signed "Masselin."

5. Christopher Fischer, *Alsace to the Alsatians?: Visions and Divisions of Alsatian Regionalism, 1890-1930* (Ph. D. Dissertation, University of North Carolina, 2003), 188-91.

6. Alan Kramer, "*Wackes* at War: Alsace-Lorraine and the Failure of German National Mobilization, 1914-1918," in *State, Society, and Mobilization in the Europe during the First World War,* John Horne, ed. (Cambridge: Cambridge University Press, 1997), 105-21.

7. "Discours de M. Poincaré," *Elsässer,* December 12, 1918.

8. Baechler, *Le Parti catholique*; Jean-Jacques Becker, "L'Opinion publique francaise et l'Alsace en 1914," *Revue d'Alsace,* 109 (1983), 125-38; Laird Boswell, "From Liberation to Purge Trials in the 'Mythic Provinces': Recasting French Identities in Alsace and Lorraine, 1918-1920," *French Historical Studies,* 23:1 (2000), 129-62; François Dreyfus, *La Vie politique en Alsace, 1919-1936* (Paris: Armand Colin, 1969), 24-27; Stefan Fisch, "Dimensionen einer historischen Systemstransformation: zur Verwaltung des Elsass nach seiner Ruckkehr zu Frankreich," in *Staat Verwaltung: Fünfzig Jahre Hochschule für Verwaltungswissenschaften Speyer,* Klaus Luedtke, ed. (Berlin: 1997), 381-98; Stefan Fisch, "Assimilation und Eigenständigkeit: Zur Wiedervereinigung des Elsass mit dem Frankreich der dritten Republik nach 1918," *Historisches Jahrbuch,* 117:1 (1997): 111-28; Julia Schroda, "Der Mythos der 'provinces perdues' in Frankreich," *Konstrukte nationaler Identität: Deutschland, Frankreich, und Grossbritannien (19. und 20. Jahrhundert)* (Würzburg: Ergon, 2002), 115-33; William Shane Story, "Constructing French Alsace: A State, Region, and Nation" (Ph. D. Dissertation, Rice University, 2001), 18-33.

9. Eugen Weber, *Peasants into Frenchmen: The Modernization of Rural France, 1870-1914* (Stanford: Stanford University Press, 1976).

10. Ibid.

11. Rothenberger, 37.

12. Harvey, *Constructing Class,* 131-52; Rothenberger, 46-52.

13. Harvey, *Constructing Class*, 131–52.
14. *Conférence d'Alsace-Lorraine*, 70 ff, 94 ff.
15. See Boswell, "From Liberation," 129–62; Irmgard Grünwald, *Die Elsass-Lothringer im Reich* (Frankfurt/Main: Peter Lang, 1984), 29–34; David Allen Harvey, "Lost Children or Enemy Aliens?: Classifying the Population of Alsace after the First World War," *Journal of Contemporary History*, 34: 4 (1999): 537–54.
16. Examples of literature on this phenomenon include Mark Connelly, *The Great War, Memory and Ritual: Commemoration in the City and East London, 1916–1939* (Rochester, NY: Boydell. 2002); William Kidd, "From the Moselle to the Pyrenees: Commemoration, Cultural Memory, and the 'Debatable' Lands," *Journal of European Studies*, 35:1 (March 2005): 114–31; Catherine Moriarty, "Private Grief and Public Remembrance: British First World War Memorials," *War and Memory in the Twentieth Century*, Martin Evans and Ken Lunn, eds. (Oxford: Berg, 1997), 123–45; Daniel Sherman, *Construction*; *War and Remembrance in the Twentieth Century*, Jay Winter and Emmanuel Sivan, eds. (New York: Cambridge University Press, 1999); Jay Winter, *Sites of Memory, Sites of Mourning: The Great War in European Cultural History* (Cambridge: Cambridge University Press, 1995).
17. Story, 228: also see generally ADBR 121 AL 584.
18. Story, 229–30.
19. ADBR 69 AL 70, Memo from Directeur de l'Architecture et des Beaux-arts dated 25 July 1919.
20. Story, 229–35.
21. Story, 235–37.
22. See Goodfellow, *Between*, 75–77; AN F7/13392, Report signed "Bauer" dated 27 September 1929.
23. Archives Municipales de Strasbourg Div. IV 178/971, Report entitled "Aufstellung eines Gefallenendenkmales auf der Place de la République" dated 1 March 1935.
24. Efforts had also been made in the 1920s to erect a statue first to Jeanne d'Arc and later to Victor Hugo and Alphonse Lamartine at the Place de la République. Both failed to garner support in the Strasbourg city council and had to be placed elsewhere in the city.
25. AMS Div VI 178/971, Letter from Henry Lévy to Charles Frey (19 June 1935).
26. Besides the years of the First World War, the years for the Second World War and the Vietnamese and Algerian wars were added to the monument.
27. Hartmannwilleskopf (le Vieil Armand), Official Website, http://www.abri-memoire.org/hwk.php (site accessed January 11, 2003).
28. ADBR 121 AL 1091, Letter from General Boisseau to the General Commissioner dated August 20, 1920.
29. ADBR 121 AL 1091, Letter from Alapetite to the Touring-Club de France dated December 8, 1920.
30. ADBR 121 AL 1091, "Monument du Hartmannwillerskopf."
31. ADBR 121 AL 1091, "Monument du Hartmannwillerskopf."
32. ADBR 286 D 304, Projet du comité du monument national de l'Hartmannswillerkopf (1927).
33. ADBR 121 AL 1091, "Monument et ossuaire du Hartmannswillerkopf."
34. ADBR 286 D 304, Projet du comité du monument national de l'Hartmannswillerkopf
35. "Feierlichkeiten im Haut-Rhin," *SNZ* (October 10, 1932).
36. "Le Président de la République sera demain l'hôte du Haut-Rhin," *La France de l'Est* (October 8, 1932); "L'Inauguration de crypte du Vieil-Armand," *La France de l'Est* (October 9, 1932); "Feierlichkeiten im Haut-Rhin," *SNZ* (October 10, 1932); "Lebrun ehrt die Kriegsgefallenen," *Elsässer Kurier* (October 10, 1932); *Journal d'Alsace et de Lorraine* (10 October 1932); *Strassburger Neueste Nachrichten* (October 10/11, 1932).

37. Antoine Prost, *In the Wake of War: 'les anciens combattants' and French Society, 1914–1939*, Helen McPhail, trans. (Berg: Providence, 1992), 58–61. For Alsace, see the press clippings in ADBR 286 D 304.

38. "11 November 1925," *Journal d'Alsace et de Lorraine* (November 11, 1925); "Il y a dix ans," *Journal d'Alsace et de Lorraine* (November 11, 1928); "L'Alsace et le 11 Novembre," *Journal d'Alsace et de Lorraine* (November 12, 1929).

39. *Journal de L'Est* (November 11, 1925); "11 Novembre," *Journal de L'Est* (11 November 1926); "Une date," *Journal de L'Est* (November 11, 1927); "Strasbourg célèbre avec éclat la Fête de la Victoire," (November 12, 1927); "L'Arrondissement de Hagenau célèbre l'anniversaire de l'armistice," *Journal de L'Est* (November 13, 1927); "Le Commémoration de l'armistice," *La France de l'Est* (November 11, 1929).

40. *Freie Presse* (November 11, 1925); *Freie Presse* (November 11, 1926); *Freie Presse* (November 11, 1927); *Der Republikaner* (November 11, 1925); *Der Republikaner* (11 November 1926); *Der Republikaner* (11 November 1927).

41. *Elsässer* (November 12, 1925); *Elsässer* (November 12, 1926); *Elsässer* (November 10, 1927); *Elsässer* (November 9, 1929); *Elsässer Kurier* (November 10, 1926); *Elsässer Kurier* (November 12, 1927); *Elsässer Kurier* (November 12, 1928).

42. *Die Zukunft* (May 9, 1925); *Die Zukunft* (November 7, 1925); *Die Zukunft* (May 29, 1926); *Die Zukunft* (November 13, 1927).

43. "Manifesto of the Heimatbund," June 8, 1926 in *Documents de l'Histoire*, 455.

44. Karl-Heinz Rothenberger, *Die Elsass-lothringische Heimat- und Autonomiebewegung* (Frankfurt: Peter Lang, 1975), 103–04.

45. "Manifesto of the Heimatbund," June 8, 1926, in *Documents de l'Histoire*, 455.

46. AN F7/13395, Report dated November 25, 1927 signed "Bauer." See also Rothenberger, 133–36.

47. ADBR 98 AL 691.

48. "La France a le devoir de se défendre," *Journal d'Alsace et de Lorraine* (August 3, 1926); "La Condamnation de l'autonomisme," *Journal d'Alsace et de Lorraine* (August 6, 1926); "Le Projet de loi contre les menées anti-françaises et l'Alsace-Lorraine," *Journal d'Alsace et de Lorraine* (August 8, 1926); "La Propagande séparatiste en progression," *Journal d'Alsace et de Lorraine* (August 22, 1926); "Que fait-on contre les agents de l'Allemagne en Alsace et en Lorraine," *Journal d'Alsace et de Lorraine* (July 21, 1926).

49. J. A. Jaeger, "Un manifeste autonomiste," *Journal de l'Est* (June 8, 1926); "De quoi se plaignent-ils," *Journal d'Alsace et de Lorraine* (June 10, 1926); "Échos de ce qui se passe en Alsace et en Lorraine," *La Revue d'Alsace et de Lorraine* (July 1926), 133–34.

50. Anselm Laugel, "Le Véritable danger des menées autonomistes," *La Revue d'Alsace et de Lorraine* (August 1926).

51. "Der Manifest der Heimattreuen," *Freie Presse* (June 8, 1926); "Klerikale und Heimatbund," *Freie Presse* (24 June 1926); "Der neue Kurs im klerikalen Lager," *Freie Presse* (7 July 1926).

52. *Le Nouvelliste* of Strasbourg, the former paper of the Francophile regionalist Emile Wetterlé, warned of this fact in 1926. While decrying the separo-autonomists as anti-national and simultaneously labeling the appeal to the League of Nations as "humiliating" to France, the paper also went to lengths to dissect the many legitimate complaints of Alsatians against the French government. See *Le Nouvelliste* (June 23, 1926). In contrast, papers such as *S'Elsass* and *Das Bollwark*, nationalist papers meant to counter the influence of the Heimatbund and *Die Zukunft*, saw a German behind every autonomist door. A report from early 1927, built on dozens of earlier police reports, points to the constricted French perspective. See AN F7/13395, "La Propagande anti-française en Alsace-Lorraine [marked secret]" dated January 7, 1927.

53. ADBR 98 AL 671, Letter from UNC-Wissembourg to the Prefect of Bas-Rhin dated June 15, 1926.
54. Rothenberger, 172–175; *Elsässer Kurier* (February 6, 1929); *Elsässer* (January 30, 1929); "Nachklänge zur Rede des H. Poincaré," *Elsässer* (February 11, 1929); *Journal de L'Est* (2 Feb. 1929); *Journal de L'Est* (Feburary 9, 1929); *Freie Presse* (February 2, 1929); *SNZ* (February 2,1929); *Volkswille* (February 2, 1929).

II

A Disembodied Memory: The Contemporary Legacy of the Great War Viewed through the Lens of the Political

Julien Fragnon

On the eve of the twenty-first century, the memory of the war of 1914–18 experienced a popular revival.[1] From the literary and media success of *Paroles de poilus*[2] to films by Jean-Pierre Jeunet[3] and Christian Caron[4] or the comic strips by Tardi, the memory of the first global conflict has hardly ever been more present.[5] The eightieth anniversary of the armistice led to an intense debate between Prime Minister Lionel Jospin and President of the Republic Jacques Chirac concerning the rehabilitation of the mutineers of 1917. Moving beyond this controversy, which historians have already closely examined,[6] our objective is to consider the keys to interpreting the past offered by this "historical memory."[7] To this end, we will approach this memory as a political object in its own right by showing how memory meets the conditions of mobilization and identification that define political discourse. Based on a quantitative[8] and qualitative analysis, we will then display the narrative structure of commemorative discourse produced by governmental representatives.[9] The resulting semantic structure enables us to understand the political uses of the past as ritual processes that carve out a "national habitus."[10] The wide-spread use of the interpretive models of the nation, Europe, or citizenship is accompanied by a denunciation of war-like violence that leads to an affective appropriation and a naturalization of the past.

Memory as a Political Object

Conceiving of historical memory as a political object seems like an exercise in tautology since we are examining cases of commemorative discourse that are produced by representatives of the political arena. In this approach, reference to the past is a political act simply because it is made by an individual within this space, which is even more the case if the individual occupies a position of power (president, prime minister, or minister). The politicization of memory goes even farther in that the pertinent requalification of memory reinforces its intrinsic functions.[11] For Maurice Halbwachs, memory is located in a specific space and social group whose recollection was shaped by the present.[12] Memory corresponds to a past reconstructed in a present that constrains the past in ways that lastingly modify its meaning. "It can be taken for granted that memory is a form of relation to the past whose final cause is not knowledge, reality or intelligibility of the past but the truth of the present, the construction or strengthening of a shared identity."[13] Just as personal memory maintains a continuous individual identity, historical memory provides the group with a temporal durability; compensating for the fragility of identity, it functions as a mechanism of integration.[14] "Memory in fact is the only mechanism today for recognizing the unity and legitimacy of 'France', as will and as representation, which in the past could be known only through its identification with the state, the expression of a great power throughout its long period of greatness."[15] Historical memory constitutes a social legacy whose public expression conveys a set of references within a given collective unit. Its recognition, however minimal it might be, is decisive for determining belonging to the collectivity. This historical memory[16] is thus not a natural given but rather a dynamic social construction, the unfinished product of a historical process of unifying and selecting past experiences.[17]

Expressed by the representatives of the collectivity, memory intensifies its exemplary and integrative ability. For Jean Leca, the political serves to resolve conflicts by using transcendental principles that are representative of society's ideals.[18] This heightened generality characterizes the politicization of discourse, especially in situations when individuals are in conflict.[19] This quasi-philosophical perspective concerning politics intersects here with an anthropological viewpoint in which the political is the organization of communities that are in situations of conflict. "The political refers in general to the constitution of political communities, and thus it involves the relations that the governed form among themselves and those that they establish with the governing powers. These conceptions consider that, in complex, pluralist, and inegalitarian societies, the political is similar to a mode of expressing and taking on conflicts."[20] Because of its identity-forming capacity, the political production of memory plays a central role in constituting communities and regulating their relations.[21] Remembered historical episodes supply a body of references that delimit an interpretive grid for the event related, which becomes a symbol of condensation.[22] Through this process, the political transmission of a national memory crystallizes a set of framing values, which play a constitutive role in producing national identity. However, this work of organizing and selecting what will be remembered and forgotten is performed by diverse social agents whose interests and abilities to make things public can vary and are at time contradictory (politicians, journalists, association leaders, historians, fiction writers, etc.). Thus, historical memory, as expressed by the political, displays the symbols of the representation of power, whereas its expression in the area of the media or scholarly research is shaped by other objectives (popularization, truthfulness, etc.). As it traverses these different spaces, historical memory becomes vested with different representations,[23] but bearing the political authority of their producers, historical memory enjoys considerable influence in the individual appropriation of collective memory, such as the description of soldiers' suffering.[24] Although political uses of the past thus constitute an object in which we can perceive the transformation of contemporary political space, these uses are not unilateral and can modify our relation to the past.[25]

The Political Use of the Past

The politicization of historical memory does not mean that these uses are necessarily untrue, but only that they include a strategic finality. Questioning the reasons for the interest in the past, Moses Finley asks:

> Interest towards what end, to fulfill what function? The past was studied with the didactic and moral goal of illustrating the fundamentally bad character of humanity and of guiding future political action. It fulfilled the socio-psychological function that consists of giving a society coherence and a goal, of reinforcing morale and encouraging patriotism. It can be used romantically, as it has been. And so forth. Each of these interests requires a different approach and way of knowing. None of the forms of interest just listed requires a systematic presentation of the past.[26]

By way of example, recollections of the Great War transmit a relatively believable past. Whereas the description of soldiers' lives squares with that of contemporary historians, the passages concerning what battles meant and how soldiers endured an unheard of violence relate more to the inscription in a coherent historical narrative or to a current political situation than to historiographical truthfulness.[27]

Historical memory involves certain discursive strategies. Among the distinguishing features of political language,[28] social origin no longer has the power to differentiate that it had prior to the 1980s.[29] Although the remarks made by current politicians concerning their origins and social trajectory are more or less similar, introducing biography sometimes offers an original way of explaining things.[30] Wishing to reintegrate "those who were shot as an example" in national collective memory, Lionel Jospin paid homage to his father, a radical pacifist during the interwar period and who was born in the territories occupied before 1914.[31] One might think then that party membership is a more pertinent variable: the period in question covers three different legislatures[32] during the single presidency of Jacques Chirac, offering a fairly representative grasp of the French political landscape (6 speeches by figures from the left, 12 from those on the right). However, content analysis turns up no striking differences: the principal themes are identical—(France,[33] time,[34] and conflict[35])—as are techniques of modalization employed (essentially techniques of intensification designed to dramatize the speech[36]). To see any differences, one must consider the fourth discursive theme, which indicates a greater ideological differential (6.6 references to the army on the right and 11.75 references to politics on the left).[37] But the readings of history are far from identical even within parties. Annette Becker has noted that for quite some time the various parties of the left have been divided over the legacy of the First World War between pacifism, antifascism, and assent to mourning.[38] In fact, Lionel Jospin, Jean-Pierre Masseret, and Jacques Floch represent a socialist left majority that favors a moderate patriotism willing to call certain things into question, but that remains distant from communist or ecologist positions.

Finally, institutional functions weigh heavily. Politicians' speeches are highly defined; despite the variety of speakers, reference to this historical period remains very specialized.[39] The majority of speeches (13 out of 20) are from leaders in the executive branch (2 speeches by the president and the chief of the army, and 3 by the prime ministers[40]) and from governmental positions related to the army (8 speeches given by the three secretaries of state for veterans affairs). Institutional homogeneity is also seen in the similarity of narrative structure of Lionel Jospin and Jacques Chirac (narrative style, modalization, references employed), who represent two opposing parties but aspire to the same presidential office.[41] Lastly, the appearance of certain occurrences is shaped by political circumstances. Despite the regular commemorating of November 11, leaders intervene only sporadically, endorsing particular political interests. Speeches are divided between two events, the armistice (10 speeches) and the battle of Verdun (6 speeches).[42] Although the latter maintains a strong symbolic legitimacy,[43] it is not celebrated on a fixed date: the speeches occurred on February 21,[44] June 16, June 23, and June 25.[45] The variable of circumstances is strikingly apparent in the case of integrating veterans from former African and Asian colonies and from overseas departments. These soldiers were mentioned in celebrations in response to increasingly strong social pressure, which stemmed from a colonial memory that for long had been overlooked and repressed.[46]

Having identified the variables linked to the social conditions of production of the political discourse under consideration, let us now turn to figures of discourse in these speeches.

Defending the Land: The Great War or the Acme of the Nation

Evoking a defensive war that would protect a country invaded by the enemy, commemorative speeches contain a high number of territorial references (nouns or place names), especially in the 2 presidential speeches where they are the discursive units most frequently used 15 instances in 1996 and 20 in 2006. Jacques Chirac mentions the city of Verdun metonymically: the site of the battle becomes an identifying marker, and the "soldier of Verdun" stands for the French soldier.[47] Discursive proximity is coupled with a relation to resistance: the relation "soldier-Verdun" is linked to "symbol," "stakes," "possession," "resistance," or "battlefield." June 16, 1996, Jacques Chirac said, "more than any other battle, Verdun is the symbol of resistance and patriotism. For the soldier of Verdun defends his country the way he would defend his family." In his second speech, the metonymic process becomes more encompassing: "during the endless year of 1916, all of France was at Verdun, and Verdun had become all of France." As a crucible for national unity, the territory became sacred. At the Chemin des Dames, Lionel Jospin began his speech by repeating three times the phrase "sacred place." The name of the city, Craonne, and the noun "place" were mentioned ten times each. In the French republican way of thinking, the expression of patriotism is identified with defending the territory.

Raoul Girardet has shown that the development of the nation was deeply rooted in a territorial idea.[48] Commemorations thus praise the defense of the land, the land of France that had been devastated by bombings. "In a few hours, an entire landscape was mutilated, becoming an appalling chaos" (Chirac, June 25, 2006). Ground confers upon men remarkable qualities of resistance. "Driven as if by instinct, even before they received the order that would become legendary, the soldiers defending Verdun immediately decided to 'hold'" (Mekachera, *France-Soir*, February 25, 2006). The earth is a metaphor for national unity. For the president of the senate, Christian Poncelet, only the recovery of Alsace and Lorraine allow the French nation to be reconstituted. "The end of the appalling war of 1914–1918...in a land deeply affected and devastated by the ravages of war, marked the victory of France and her allies and the completion of French unity. Alsace and Lorraine...returned once and for all to the country's bosom thanks to the sacrifice of thousands of nameless heroes. This very sacrifice, which touched the entire nation, even in the remotest village, was the ultimate test that unified national consciousness following the wrenching experiences of the previous century" (November 10, 1998). Rooted in the land, commemorative discourses sanctify national unity.

In recalling the memory of the Great War, references that promote political unity are intensified, transcending partisan division. Although the word "nation" and the various terms linked to it ("country," "people") make up the first group of references employed by the right (81 instances in 10 speeches, making on average 8.1 references per speech), while it is only the fifth group in leftist speeches, it is more frequent in the latter (33 instances in four speeches, making 8.25 references per speech). The first global conflict constitutes the acme of a national mobilization that has been sought since the French Revolution. This identificatory claim transcends particular differences and social or regional splits in the name of a shared endpoint, the defense of invaded territory. The Secretary of State for Veterans Affairs, Hamlaoui Mekachera, is delighted to see "together, these Frenchmen who are still so different, the peasant and the city dweller, the school teacher and the priest, the Breton and the inhabitant of Marseille, people from Vendée and Lorraine" (February 21, 2006),[49] while Lionel Jospin evokes an "army of professional soldiers, but also of intellectuals, peasants, workers and colonists, raised in the unity of the Nation" (November 5, 1998). Jacques Chirac uses a double rhythm to stress the unifying moment of Verdun: "The city dweller and the peasant, the aristocrat and the worker, the school teacher and the priest, the republican and the monarchist, the believers in heaven and those who didn't—all social conditions, all opinions, all religions were at Verdun" (June 25, 2006). Unlike the beginning of the twentieth century,[50] the republican nationalism that is found in speeches does not construct a menacing figure of the Other,[51] and it has only one internal unifying aim. Since contemporary French and European political culture cannot accept anything other than a moderate patriotism (precisely because of the tragic example of the two World Wars), the reminder of the past must be extended beyond national borders.

Constructing Europe from the Great War

In commemorative speeches, the basis of unification extends to the construction of Europe, whose fundamental pacifism assures legitimacy. Although in a few instances victory is ascribed to the promotion of universal values,[52] recalling the epic tones of Georges Clemenceau,[53] these references remain marginal compared to references to Europe. In a more consensual perspective, victory is presented as triggering a tortuous yet unstoppable movement along the path to peace.

The Great War is claimed to be the base for European construction, whose rootedness in history speechmakers express by means of memory. "[The war] convinces us of the importance of peace that later prevailed on our continent. The progress of the idea of Europe is also one of the delayed, paradoxical, and fortunate consequences of this merciless 'Great War'" (Mekachera, November 11, 2002). The peace process occurring between former belligerents, the product of European unification, is the direct extension of the aspiration for peace that emerged at the armistice. "The priceless advantage of the construction of a united Europe is that it preserved our continent and our peoples from the return of war" (Jospin, November 5, 1998). Memory is reappropriated as an undivided reading of the past in order to pacify the history of relations between France and Germany. "[The] context of a Franco-German rapprochement opens the way to a new interpretation of the past: the memory of Verdun becomes the witnessing on the part of all combatants, both French and German. No longer is the issue about condemning and calling for revenge, but rather about producing a single reconciliatory narrative. Soldiers of both sides are included in a single form of homage."[54] Politicians use the universal idea of suffering, moreover, so that the memory of the armistice will be shared.[55] The importance of peace is born in a common suffering between former adversaries. "The war that was supposed to 'bleed dry' one of the sides left them both bloodless. This ordeal made people see things differently" (Jospin, June 23, 2001). Although pacifism had a real effect upon the origins of the European Union, making it the key for interpreting the representations of soldiers amounts to reproducing an element of this mythical discourse. Historical memory overvalues the importance of interwar pacifism,[56] and it separates the rise of Nazism and the outbreak of World War II[57] from the consequences of the armistice treaties. In masking facts with logical connections, narrative introduces causality where there is only contingency, through a process that characterizes a mythological narrative structure. "What allows the reader to consume myth innocently is that he does not see it as a semiological system but as an inductive one. Where there is only equivalences, he sees a kind of causal process."[58] Language, which is the basis of myth, is confirmed by fragmented and repetitive commemorative practices (praise of the flag, a minute of silence, the placing of wreaths, the presence of local and national political representatives, the presence of military detachments, commemorative monuments). Characteristic of ritual as Lévi-Strauss describes it,[59] this double movement frames a memorial discourse corresponding to traditional ritual aspects (sacredness, territory, the primacy of collective values and symbols). By valorizing territory and collectivity, the political uses of the past ensure power's representativity and authority.[60] Reappropriated as political representation, historical memory also includes a denunciatory strategy.

A Disencarnated Denunciation of Violence

The valorization of peace is accompanied by a virulent accusation of war's violence. The obsessive presence of death,[61] the ravaged countryside, and the ferociousness of industrial war are recurrent themes. The numerous references to mud conjure up the image of a world inevitably foundering in the cesspool of desolation, "a swamp where the clay was as much as a meter thick, an endless bog where here and there still floated the remnants of the forests the bombs had cut to pieces, along with human remains" (Chirac, June 16, 1996). This realist approach explains the massacre of men and landscape by the tormenting bombing. Artillery fire is presented as the primary cause of excessive mortality,[62] as in Lionel Jospin's speech where he repeats the terms of a military speech whose author was distressed at the sight of "French troops... melted by the fire of enemy artillery" (November 5, 1998). This mechanization of death represents a qualitative shift, a passage to "industrial death" (Chirac, June 25, 2006). Just as the memory of individual battles was marginalized, men are erased by the mechanization of violence, as in this passage in which artillerymen are absent in the description of their weapons at work: "The fire of enemy artillery...senses relief troops, launches flares, and seeks out men to massacre" (Chirac, June 16, 1996). Although the unleashing of violence results from human interactions, it is evoked by metaphors of nature, synonymous with atmospheric disturbances (storms, thunder). Jacques Chirac mentions "a maelstrom of fire and steel" and a "flood of fire." The element of intentionality in the violence of war is overlaid with metaphors that mechanize and naturalize the fury of combat.[63] Violence strikes indiscriminately and at random, producing "[a] death...all the more sudden because it resulted from blind strikes: with World War I mechanical death appears, striking nameless bodies at random and with a violence theretofore unknown" (Jospin, November 5, 1998). War is understood as a disembodied object forced upon soldiers, whereas it is the result of human actions and political and military decisions. Through a double movement of dehumanization (the mechanical register) and naturalization (atmospheric metaphors), commemorative speeches evacuate the human and history. Once again, this naturalization is related to a mythological process. "In passing from history to nature, myth acts economically: it abolishes the complexity of human acts, it gives them the simplicity of essences; it does away with all dialectics, with any going back beyond what is immediately visible, it organizes a world which is without contradictions because it is without depth, a world wide open and wallowing in the evident, it establishes a blissful clarity: things appear to mean something on their own."[64] Although they naturalize the use of violence, commemorative speeches highlight the individual because of his potential status as victim.

A New Collective Image: The Suffering Individual

Contemporary speeches praise not only politicians or glorious generals of the period, but also the anonymous soldier who embodies fighting citizenship. Although fighters are valorized, it is not for their military qualities but their excessive will to hold out against the enemy and war itself. Secretary of State for Veterans Affairs Jean-Pierre Masseret of the Socialist Party praised the soldiers' collective disinterestedness and sacrifice compared to contemporary individualism. "During four long years, in mud, fear, and horror, they did their duty, whose call they received from the Republic. In the face of rising individualism, the growth of categorical interests, they remain the reference point for fighting citizenship" (Le Monde, November 11, 1998). There is but a step from fighting citizenship to the suffering individual, since from the extreme hardship of battle, heroism is born. Affliction is displayed according to a strategy of proximity, which attempts to get as close as possible to the soldiers' conditions of daily living and survival. The description of psychological suffering (waiting in the trenches, the fear of bombing and assaults) and physical suffering (illness and wounds) is clear, as the president of the Republic emphatically expresses: "In a Dante-like world, these men will live and fight, chilled to the bone, then under a leaden sun, gnawed at by vermin, tortured by hunger and thirst, without sleep, with bombings that cease only to give way to disorganized, unrelenting battles ending in fearsome hand-to-hand fighting" (June 16, 1996). This didactic procedure, which is found repeatedly in Chirac's speeches on history,[65] will be used by others. On November 5, 1998, Jospin passionately describes "soldiers exhausted by attacks that were doomed to failure, slipping in blood-soaked mud, plunged in a deep hopelessness." Brigitte Girardin urges that suffering be recalled: "let us remember the atrocity of these combats in the trenches: wetness, diseases, cold, hunger, and especially the mud all became part of the soldiers' waiting for the shelling, the mine explosion or the next bloody offensive" (November 11, 2004). Speakers introduce the effects of the real, such as precise place names and times, in order to anchor their speeches in reality, as does Hamlaoui Mekachera who identifies with the fighters, laying claim to a singular memory that alone can transmit suffering.[66] Politicians favor processes of identification as they humanize their speeches, which they construct as if they too were close to the sufferers, capable of empathy and affection. These discursive techniques belong to a broader pattern marked by a shift in the way political legitimacy is constructed. The logic of proximity created by eliminating distance with respect to the citizen has already been observed.[67] Its actualization in commemorative discourse takes the form of recognizing the fighters' suffering. However, the repeated use of affects in commemorative discourse tends to "victimize" the fighters. The victim-like figure can even constitute a criterion for identificatory mobilization like European pacifist memory or Lionel Jospin's 1998 remarks concerning the mutineers. In the face of acts of disobedience, Jospin could not celebrate insubordination, and so he employed a different strategy for maintaining collective identity: connectedness is not founded on hierarchical subordination but upon the sharing of a common suffering. It is in the name of suffering endured that the soldiers, even if they mutinied, most be reintegrated into national collective memory.[68] This tone was used by the first secretary of the Socialist Party in what finally was contradictory support for the mutineers: "[they] were not all bad Frenchmen, but simply men whose strength was gone, who had lost their way in a hell of fire and blood."[69] The claim of humanity makes it possible to discredit the criticisms of dishonor that are connoted by the act of disobedience. These discursive strategies have influenced how the memory of the First World War has been read. They amount to introducing into the political sector deep-seated changes in the historiographical field, certain representatives of which play a role in writing speeches (Jean-Jacques Becker, for instance, for Jospin's 1998 speech).

The war is read through the unique prism of the fighters, especially by centering memory on pain. Lexically, affliction refers to the figure of the victim. The soldiers are read as people suffering, as victims of war's violence, sacrificed on the altar of the defense of fatherland and territory. Historical memory is broken down into units here, where soldiers are presented only as individuals capable of being killed but incapable of killing. Annette Becker notes the perverse effects this victim memory has on the historiography of the conflict,[70] while Valérie Rosoux exposes the inherent absolution in a victimizing reading of the past. "A position like this makes everyone short on credibility. Laying claims makes it possible to assert one's rights. As if one's suffering absolved one of all kinds of sins."[71]

By providing keys to interpreting the past, the historical memory that politicians produce shapes a "national habitus" made up of traditional representations and political circumstances, which advances the defense of the homeland, national unity, denunciation of war's violence, and the recognition of individual suffering. Although these political uses of the past are not an accurate view of it, their integration into a historical continuum belongs to a constant process of constructing a collective identity (with a national or partisan end). However, the denunciation of war and its related victimizing of combatants lead to a discourse that eliminates all responsibility. Like the mythical narrative, victim discourse implies a naturalizing of the social world.[72] Debates over the mutineers are characterized by the unifying use of the figure of the victim and its inherent compassion. The critical speeches of praise given by Lionel Jospin cannot abandon a feeling of humanity toward soldiers, as in the case of François Léotard (center-right) who "also understands that it is possible to display humanity towards these soldiers."[73] The prime minister defends himself by claiming that "what he said quite simply was humane, just, and necessary."[74] By means of this victim claim, the speaker locates himself on the level of values of justice and ethics, which are politically legitimate and unassailable. The leader situates himself within a logic of proximity whose positive value no longer needs to be questioned in the political arena.[75] This logic leads to singularizing the ways social conflicts are apprehended. This contemporary feature of political expression remains troublesome once the victim becomes the central figure of penal policy[76] or of the grateful nation.[77]

Notes

Chapter translated by Daniel Brewer.

1. The French historian Stéphane Audoin-Rouzeau has written of "a new life" given to the event of 1914–18 in "La Grande Guerre, le deuil interminable," *Le Débat* 104 (1999): 123.
2. In March 1998, on the 80th anniversary of the armistice, a journalist with Radio France launched an appeal to recover letters soldiers sent from the front. The radio broadcast was a tremendous success (more than 8,000 letters were received), and the book subsequently sold over a million copies.
3. *Un long dimanche de fiançailles* (2003).
4. *Joyeux Noël* (2005).
5. This is a period marked by the death of the last soldier, Lazare Ponticelli, on March 17, 2008.
6. Especially Nicolas Offenstadt, *Les Fusillés de la Grande Guerre et la mémoire (1914–1999)* (Paris: Odile Jacob, 1999); and Nicolas Offenstadt, "Les Mutins de 1917 dans l'espace public ou les temporalités d'une controverse (1998– ?)," in Maryline Crivello, Patrick Garcia, and Nicolas Offenstadt, eds., *Concurrence des passés: usages politiques du passé dans la France contemporaines* (Aix-en-Provence: Publications de l'Université de Provence, 2006).
7. The expression "historical memory" is used here in the sense given to it by Marie-Claire Lavabre: "Historical memory is taken to mean the uses of the past and of history as it is taken over by social groups, parties, churches, nations or states. These are appropriations by those in power or by those subjected to power, appropriations that are multiple and selective in any case, marked by anachronism and by the resemblance between past and present, whereas history properly speaking is directed in principle, if not towards unity, at least towards the critique of historical memories and the establishing of differences between the past and the present." Marie-Claire Lavabre, "De la notion de mémoire à la production des mémoires collectives," in Daniel Cefaï, ed., *Culture politique* (Paris: PUF, 2001), 242.

8. The quantitative analysis for this project was done using the analytic content software Tropes V7 (2004). It can be used to extract concepts from a given corpus and produce a network of relations between these concepts in graph form in order to explain the meanings of a text. It identifies the most frequent units of meaning, with their characteristics, interrelations, and evolutions. It thus makes it possible to verify empirically the analyst's interpretations of a text.

9. Our corpus includes 20 speeches or press interviews given by governmental leaders (President of the Republic, Prime Minister, Minister of Defense, Secretary of State for Veterans, etc.) under three different legislatures (1996–97, 1997–2002, 2002–06).

10. "The social units that we call nations differ widely in the personality structure of their members, in the schemata by which the emotional life of the individual is molded under the pressure of institutionalized tradition and of the present situation." Norbert Elias, *The Civilizing Process*, Edmund Jephcott, trans. (Cambridge, MA: Blackwell, 1994), 27.

11. For Jacques Lagroye, politicization is a process of "requalifying the most varied kinds of social activities, a requalification leading to a practical agreement between social agents who are inclined, for numerous reasons, to transgress or call into question the differentiation of the spaces of activities." "Les Processus de politisation," in Jacques Lagroye, ed., *La Politisation* (Paris: Belin, 2003), 360–61. This definition could also apply to the strategies used by associations of war veterans to politicize the stakes of memory.

12. "It is not sufficient, in effect, to show that individuals always use social frameworks when they remember. It is necessary to place oneself in the perspective of the group or groups. The two problems, moreover, are not only related: they are in effect one. One may say that the individual remembers by placing himself in the perspective of the group, but one may also affirm that the memory of the group realizes and manifests itself in individual memories." Maurice Halbwachs, *On Collective Memory*, Lewis A. Coser, ed. and trans. (Chicago: University of Chicago Press, 1992), 40.

13. Marie-Claire Lavabre, "Usages du passé, usages de la mémoire," *Revue Française de Science Politique* 44:3 (1994): 487.

14. Paul Ricœur, *Memory, History, Forgetting*, Kathleen Blamey and David Pellauer, trans. (Chicago: University of Chicago Press, 2004).

15. Pierre Nora, "La Nation-mémoire," in Pierre Nora, ed., *Les Lieux de mémoire*, vol. 2 (Paris: Gallimard, 1986), 2210.

16. Improperly called collective memory, national memory or history in public debate.

17. "It is work, the socialized work of reducing the number of possible representations, homogenizing memories, bringing together the political expressions of memory in the way that groups, in other words spokespersons, authorized witnesses, individuals of note or memory agents, express them as memories of lived experience." Marie-Claire Lavabre, "Peut-on agir sur la mémoire," in "La Mémoire entre histoire et politique," *Les Cahiers français* 303 (2001): 9.

18. Jean Leca, "Le Repérage du politique," *Projets* 71 (1973): 11–24.

19. Luc Boltanski, *L'Amour et la justice comme compétence* (Paris: Métailié, 1990).

20. Sophie Duchesne and Florence Haegel, "La Politisation des discussions, au croisement des logiques de spécialisation et de conflictualisation," *Revue Française de Science Politique* 54:6 (2004): 880.

21. "[T]he work of memory, that is, the organization of things remembered and forgotten, the order given after the fact to the past, operates in a political group the way it does on the level of individuals: it shapes the preservation of identity." Henry Rousso, *Vichy: l'événement, la mémoire, l'histoire* (Paris: Gallimard, 2001), 352.

22. Murray Edelman, *Constructing the Political Spectacle* (Chicago: University of Chicago Press, 1988).

23. We have observed that the registers of commemorative speeches and the main references of the novelists of the Great War were similar (involving the trenches, mud, fire, bodies in

pieces, and devasted countrysides). See Micheline Kessler-Claudet, *La Guerre de quatorze dans le roman occidental* (Paris: Armand Colin, 1995). This analogy indicates that literary greatness is preserved in constructing political legitimacy.

24. If studies in the sociology of reception have taught us to be cautious in revealing the fragmentation and plurivocity in the appropriation of representations, we think that the individual acquisition of historical memory results from an interconnection between "exterior" objective constraints and their private appropriation on the part of the individual. Our identity can be seen as a variable and changing mixture of "objectivized history" (institutional and social structures, historical precedents, etc.) and "subjectivized history" (personal reinterpretation and reappropriation of exterior elements).

25. The weight given to political readings of the past can in turn modify historiographical or media practices, as in the view of wartime violence.

26. Moses Finley, *Mythe, mémoire, histoire: les usages du passé,* Jeannie Carlier and Yvonne Llavador, trans. (Paris: Flammarion, 1981), 26.

27. The mixture is formed when the politician prides himself in recalling the truth of the past as did Prime Minister François Fillon, who during the ceremony of July 22, 2007, commemorating the Vél d'Hiv roundup, spoke of the "cult of truth" (http://www.nouvelobs.com, accessed July 23, 2007).

28. Christian Le Bart, *Le Discours politique* (Paris: PUF, 1998).

29. Daniel Gaxie, *La Démocratie représentative* (Paris: Montchrestien, 2003).

30. Despite this contemporary homogeneity, one must not overlook the symbolic struggles that engendered this situation of domination. Pierre Bourdieu has clearly shown how political legitimacy is accompanied by linguistic legitimacy: the mastery of "speaking well" determines the general conditions for participation in the political arena. Pierre Bourdieu, *Ce que parler veut dire: l'économie des échanges linguistiques* (Paris: Fayard, 1982).

31. "Robert Jospin, le pacifiste," *Le Monde,* November 7, 1998.

32. 1995–1997, the rightist government of Alain Juppé; 1997–2002, the leftist government of Lionel Jospin; 2002–2006, the rightist government of Jean-Pierre Raffarin, followed by Dominique De Villepin in June 2005.

33. Under this theme we find toponyms linked to France (cities, departments, etc.) or to a use of France as prosopopeia. The partisan division is balanced: 16.3 instances per rightist speech, 20 on the left.

34. This theme refers to the temporal inscription of the narrative: 12.7 instances on the right, 21.5 on the left.

35. 10.2 instances on the right and 18 on the left.

36. On average, in the speeches of politicians on the right, 39.2% of the modalizers are stress adverbs, compared with 30.25 percent on the left.

37. This theme includes all the nouns related to institutions.

38. Annette Becker, "La Gauche et l'héritage de la Grande Guerre," in Jean-Jacques Becker and Gilles Candar, eds., *Histoire des gauches en France,* vol. 2, *20ème siècle: à l'épreuve de l'histoire* (Paris: La Découverte, 2004), 330.

39. Seven different governmental offices (President of the Republic, Prime Minister, Defense Minister, Secretary of State for Veterans Affairs, and Minister for Overseas Affairs) are represented by a total of 20 speeches.

40. Two for Lionel Jospin alone during the period of cohabitation during which he seriously challenged the "preserves" of the President of the Republic (collective memory, foreign policy). Conversely, we have not located any speech by Alain Juppé (1995–1997), and the only mention by Jean-Pierre Raffarin during this period is part of a longer speech on governmental policy regarding veterans.

41. This fact has already been mentioned by Damon Mayaffre. "[From] a lexical point of view, the two men blend into each other, and their speeches come to resemble each other.... We

see here the effect on the left/right split produced by a cohabitation that tends to dissolve differences. Without ever employing the same language, between 1997 and 2002 Jospin and Chirac gradually lost their own identities and increasingly shared a common vocabulary," Damon Mayaffre, *Paroles de président: Jacques Chirac (1995–2003) et le discours présidentiel sous la Ve République* (Paris: Honoré Champion, 2004), 233–34.

42. The only other episode of a conflict that was the occasion of a ministerial speech was the celebration of the 90th anniversary of the victory of the Marne, September 28, 2004, during which the Minister for Veterans Affairs spoke.

43. Since the 1960s, Verdun has been the place where presidents speak about the Great War, and it is here that President Chirac gave his two speeches. One might advance the theory that the Prime Minister's speech at Verdun on June 23, 2001, coming at the end of a long period of cohabitation and one year before the presidential elections of 2002, was part of a strategy for constructing a presidential profile. Moreover, Verdun has been one of the symbols of Franco-German reconciliation since François Mitterrand and Helmut Kohl met in 1984.

44. Beginning of the German bombing of French positions at Verdun.

45. The end of June marked the beginning of the German troops' summer offensive. The end of the battle of Verdun, dating from General Mangin's offensive that took the Hardaumont hill, December 15, 1916, gave rise to no speeches by ministerial leaders.

46. Precisely when we found emphatic praise on the part of President Valéry Giscard d'Estaing in his speech of November 11, 1978: "We can give similar praise to soldiers from overseas and to Allied soldiers who gave their life for France and whose delegations I greet here today.

47. "Soldiers" and "Verdun" are the two words most frequently related.

48. Raoul Girardet, "A closed domain, tightly enclosed upon itself and that first had to be constituted and that now must be watched over, the idea of the French fatherland thus tends to be conflated with the archetypal figure of the patrimonial domain," *Mythes et mythologies politiques* (Paris: Seuil, 1986), 161.

49. In one speech, the minister begins a sentence 11 times in a row with the word "together."

50. Michael Jeisman, *La Patrie de l'ennemi: la notion d'ennemi national et la représentation de la nation en Allemagne et en France de 1792 à 1918* (Paris: Editions du CNRS, 1992).

51. In a passage in his Verdun speech on June 23, 2006, Jacques Chirac referred to the republican perspective guiding the combatants' actions as being distinct from any kind of bellicose nationalism: "these men who fought relentlessly were not driven by nationalism or by hatred for the enemy. Their soul was not militaristic. Their soul was patriotic. It was republican."

52. According to Defense Minister Michèle Alliot-Marie (UMP), "November 11, 1918, symbolizes the victory of peace, liberty, democracy" (Rethondes, November 11, 2004), whereas her colleague for Overseas Affairs, Brigitte Garardin (UMP), borrows a quote from maréchal Joffre consecrating the victory of the camp at Good: "You won the greatest battle in history and saved the most sacred cause—the world's freedom" (Paris, November 11, 2004).

53. On the day of the armistice, in the National Assembly he said, "Thanks to our great dead, France, which used to be God's soldier and is now the soldier of humanity, will always be the soldier of the ideal," quoted by Jean-Jacques Becker, *Clemenceau: l'intraitable* (Paris: Liana Levi, 1998), 138.

54. Valérie Rosoux, "Les Usages du passé dans la politique étrangère de la France," in Claire Andrieu, Marie-Claire Lavabre, and Danielle Tartakowsky, eds., *Politiques du passé: usages politiques du passé dans la France contemporaine* (Aix-en-Provence: Publications de l'Université de Provence, 2006), 179.

55. In his speech at Verdun in 1996, President of the Republic Jacques Chirac referred to a "shared martyrdom." *Le Monde*, June 18, 1996.

56. Lionel Jospin mentions the idea of a coherent and irrepressible will for peace that was born of the Great War: "The idea of a reconciliation between France and Germany was momentarily swept aside by World War II. But it reemerged immediately afterwards, even more forcefully, so as to prevent such an explosion of violence from ever occurring again" (Verdun, June 23, 2001).

57. The memory of the second conflict remains inseparable from the first. This is seen with Jacques Chirac who praises Pétain in 1917 yet denounces his policy of collaboration with the Germans. "A man who succeeded in making decisions that led to victory. He remained the conqueror of Verdun. This man is Philippe Pétain. Alas, in June 1940, the same man, entering the winter of his life, bestowed his glory on the fateful choice of the armistice and the dishonor of collaboration. This French tragedy is a part of our history. Today we can face it directly" (Verdun, June 23, 2006). Along with others, we can even question the autonomous nature of the memory of the Great War. Nicolas Offenstadt observes the displacement of arguments towards more burning subjects (collaboration or the Algerian war) during the 1998 debates over the mutineers. "If collective memory seems to have calmed down concerning the mutineers and the men who were executed, a displacement must occur, in order to intensify the charge, towards higher stakes." *Les Fusillés de la Grande Guerre et la mémoire*, 188.

58. Roland Barthes, *Mythologies*, Annette Lavers, trans. (New York: Hill and Wang, 1957), 131.

59. "In all cases, ritual makes constant use of two procedures: parcelling out and repetition." Claude Lévi-Strauss, *The Naked Man: Introduction to a Science of Mythology, 4*, John and Doreen Weightman, trans. (New York: Harper and Row, 1981), 672.

60. "These rituals provide material for a dual political operation: first, the expression of a strong cohesion among the governed who show their attachment to values, symbols, and a common history; second, the reaffirmation of collective consent to established power and those who embody it." Marc Abélès and Henri-Pierre Jeudy, ed., *Anthropologie du politique* (Paris: Armand Colin, 1997), 254.

61. Jacques Chirac describes "a gigantic mass grave…with bodies as far as the eye can see" (Verdun, June 25, 2006).

62. A fact that historians have confirmed.

63. Stéphane Audoin-Rouzeau and Annette Becker already observed this device concerning the perception of brutality in the accounts of veterans: "What's stressed is always anonymous and blind brutality, in other words a violence without identified responsibility, and thus a violence that is exonerated." *14–18: retrouver la guerre* (Paris: Gallimard, 2000), 53.

64. Barthes, *Mythologies*, 143.

65. For Jean-François Tanguy, the President gives greater importance to realism through methodological descriptions in order to denounce the absurdity of war. "Le Discours 'chiraquien' sur l'histoire," in Claire Andrieu, Marie-Claire Lavabre, and Danielle Tartakowsky, ed., *Politiques du passé*, 133–45.

66. "The first shell lit up the sky and cast a dismal shadow over the land at 7:15, February 21, 1916. The ensuing silence did not last long. None could imagine that they were living the last moment of respite for more than 300 days. Three hundred days with the flash of explosions and the noise of cannons—no more nights, no more rest. With fire and gas, mud and dust, with the cold and the heat, suffering and fear, it was hell" (Verdun, February 25, 2006).

67. "The ideology of nearness should be understood as a way of believing that over-determines what in the moment is politically thinkable, with the time variable suppressing the partisan variable." Christian Le Bart and Rémi Lefebvre, ed., *La Proximité en politique: usages, rhétoriques, pratiques* (Rennes: Presses Universitaires de Rennes, 2005), 19.

68. "Some of these soldiers, exhausted by attacks that were doomed to failure, falling down in blood-soaked mud, plunged in deep hopelessness, refused to be sacrificed. May these

soldiers, who were 'shot as an example' by an order whose inflexibility was equaled only by the harshness of the fighting, be reintegrated fully in our national collective memory."

69. *Le Monde*, November 9, 1998. Thus some were bad, and the others who weren't had strayed nonetheless from the narrow path of obedience.

70. "As painful as that may be, it is easier to accept that one's grandfather or father was killed in combat than to admit that he could have killed. In memorial consciousness, it is better to be a victim than an agent of suffering and death; always received, always anonymous, death is never given; one is always its victim." Annette Becker, "Politique culturelle, commémorations, et leurs usages politiques: l'exemple de la Grande Guerre dans les années 1990," in Claire Andrieu *et al.*, ed., *Politiques du passé*, 36.

71. Valérie Rosoux, *Les Usages de la mémoire dans les relations internationales: le recours au passé dans la politique étrangère de la France à l'égard de l'Allemagne et de l'Algérie de 1962 à nos jours* (Brussels: Bruylant, 2001), 263–64.

72. "Generalizing the status of the victim—since henceforth the combatants are the victims of having fought—leads to depoliticizing the stakes of the past. The sense of history runs the risk of being erased in the face of growing sensitivity towards the victims' misfortune." Claire Andrieu, "La Commémoration des dernières guerres françaises: l'élaboration de politiques symboliques, 1945–2003," in Claire Andrieu *et al.*, ed., *Politiques du passé*, 45.

73. *Libération*, November 9, 1998.

74. *Libération*, November 10, 1998.

75. One can wonder whether the political field, by valorizing nearness and singularity, does not come to resemble modes of legitimation pertaining to literary creation. Nathalie Heinich refers to a "regime of singularity" that functions as an axiological regime structuring representations in literature. *Être écrivain: création et identité* (Paris: La Découverte, 2000), 342.

76. Following Nicolas Sarkozy's proposal, the government of François Fillon plans to create a "victims judge" responsible for following the victim's case throughout the judicial process. See Denis Salas, "L'Inquiétant avènement de la victime," in "Violences," *Sciences Humaines*, special issue 47 (December 2004–January 2005): 90–91.

77. "The second message the President wishes to send is an appeal to the Nation, which must extend a hand to victims. Several hundred French people have been invited to the garden party who, as the President put it, 'placed a knee on the ground' this year. His wish is to show that, for these people, the Nation is there. The victims will be invited to both the parade and the garden party, accompanied by heroes who are quite often anonymous heroes and who merit the Nation's special homage on the day of the national holiday." Press conference given by Elysée Palace spokesman David Martinon, July 12, 2007 (http://www.elysee.fr, site accessed July 23, 2007).

12

"Pedestals Dedicated to Absence": The Symbolic Impact of the Wartime Destruction of French Bronze Statuary

Kirrily Freeman

Paris was populated all of a sudden with pedestals dedicated to absence.

(Paul Claudel)[1]

On a tree-lined Parisian boulevard near place Denfert-Rochereau stands a pedestal inscribed "F. Arago, 1786–1853, *Souscription Nationale*." Few passersby, if they notice it at all, would pause to wonder whether a statue ever graced its top. Some long-time resident of the neighborhood might remember that the base once supported a bronze homage to the scientist and politician François Arago raised by national public subscription in 1893.[2] Some may even recall the statue's removal in December 1941.[3] But for most, the story behind this empty pedestal remains unknown. The same is true for scores of bronze monuments all over France.

On October 11, 1941, France's collaborationist government legislated the destruction of bronze statuary in the public domain. Crippled by copper shortages and bound by the terms of the Franco-German Armistice, the Vichy regime sought to "mobilize" all potential sources of nonferrous metals, including statues.[4] Between October 1941 and August 1944, French municipalities lost the majority of their monuments. Estimates of the number of works destroyed range between 1,527 and 1,750 decorative and commemorative statues in the public domain (war memorials and monuments on church property were excluded).[5] This widespread removal of statuary touched almost every community and significantly undermined a form of civic artwork that dominated French landscapes.

Despite public forgetfulness, the fact that bronze statuary was melted during the Second World War is well known and appears anecdotally in a number of histories and memoirs of the period, as well as in art historical studies of French monuments.[6] References to melted statues are frequently marshaled to invoke the impact of Nazi occupation, the extent of Vichy's social and moral revolution, and the turbulent climate of wartime France. But the campaign itself has only recently been the topic of any sustained scholarly inquiry.[7] In particular, the symbolic, political, and cultural implications of the loss of this artwork have remained largely unexplored.[8]

One of the most interesting and hitherto neglected aspects of the story of French monuments is the mixed public reaction to their wartime destruction. The question of public statuary in Paris was already highly politicized, and, for many, Vichy's mobilization of Parisian monuments was a laudable correction of Third Republic excess. In the French provinces, however, the campaign was loudly condemned. There was widespread protest ranging from poignant, though futile, gestures such as leaving wreaths and bouquets on empty pedestals, to the more perilous and proactive theft and concealment of statues slated for demolition. There was also extensive administrative opposition to the campaign from local and regional authorities who took the considerable risk of written protest, petitions, and resignations.[9]

This reaction to the removal and destruction of bronze statuary illustrates the importance of these symbols to French communities. The French population rejected Vichy's position that monuments primarily represented a source of strategic raw materials. As one official in Marseilles lamented to the prefect of Bouches-du-Rhône, "[people] don't understand how icons worthy of such respect can be sacrificed to purely material needs."[10] In their protest, the French population testified to the myriad ways this public art touched local life. Their reaction reveals the symbolic impact of Vichy's bronze mobilization campaign, and ensured that the campaign would leave a legacy for French communities.

Monuments play a significant role in the forging of identity and community. Central to this process was the dedication of statues to figures of local significance who had contributed to the advancement of the nation.[11] This statuary fostered local patriotism and linked municipalities symbolically to French national patrimony. A statue in Lyon dedicated to the scientist André-Marie Ampère demonstrated that the city fostered intelligence and innovation and contributed to France's international renown through the inventions of its talented son. The town of Dole erected a statue to Louis Pasteur "in order to remind the world 'of the fact that Dole is the birthplace of the illustrious scientist.'"[12] The statue to François Millet in Cherbourg demonstrated that the town was "a source of 'national glory of which all of France should be proud.'"[13] The role played by public statuary in fostering local patriotism, linking the glories of French communities to national patrimony and ensuring the inclusion of French municipalities in the imagined community of the nation, assured that the removal of public statuary—an assault on the memories and identities statues embodied—had a clearly symbolic dimension.

The symbolic impact of the bronze mobilization campaign was reinforced by its legacy to French communities—a palpable absence created by the profusion of empty pedestals that marked French municipal landscapes after October 1941. Rosalind Krauss describes the symbolic function of plinths as mediators between the site of a statue and the work itself. According to this "logic of the monument" any sculpture is commemorative because "it sits in a particular place and speaks in a symbolical tongue about the meaning or use of that place."[14] An empty pedestal divested of its "symbolical tongue" becomes an emblem of absence. The two most striking characteristics of absence are that it highlights uncertainty and that it creates a need for resolution:

> Once an absence has been made conspicuous...we are forced to accommodate some degree of uncertainty in our interactions with that larger entity of which the absence is a part....The perception that something that could or should be present is not there becomes vaguely threatening, requiring some sort of resolution or closure.[15]

The ways that French communities dealt with the loss of their statuary and the gaping empty spaces that were opened in their public places during the war and in the Liberation and postwar periods were all attempts to cope with the uneasiness generated by absence, and to achieve resolution by redefining the function of these empty spaces. The significance of these uneasy spaces was, however, closely linked to shifting social and political contexts and historical moments. Before August 1944, the empty pedestals had a memorial function: as residue of war, they were evidence of a range of losses and sacrifices. After the Liberation, however, the empty pedestals became uncomfortable reminders of a difficult, divisive period. The responses of French communities to this absence and the uneasiness it generated illustrates the negotiation between remembering and forgetting that is the central dynamic of memory.

Decapitated Pedestals

As with the removal of the bronzes themselves, a combination of economic constraint and pragmatism determined the fate of their pedestals. Initially, a number of municipalities envisioned removing the bases at the same time that they dismantled their statues. In Toulouse the removal of bases was scheduled for December 22, 1942.[16] A note from the prefecture of Bouches-du-Rhône in April 1942 explained to municipalities that "if the bases are damaged, their demolition is justified and will be at the expense of the state."[17] Dismantling the bases proved to be prohibitively expensive, however. An estimate for the demolition of pedestals in Paris was 280,000 francs.[18] The expense and the increasing difficulty of securing labor and transportation prevented the removal of bases after early 1943.[19]

Bases were also left intact because, in some cases, Vichy intended to replace melted bronze statuary with new works in stone. Vichy's project of erecting stone replacements was pursued slowly between early 1943 and August 1944, with meager results. Those that appeared in French municipalities before the end of the war were mostly small busts. Cost, lack of materials, and protest from municipalities contributed to the modest size and number of the replacement statuary and the slow pace of their completion. As a result, many of the stone replacements commissioned by Vichy were not completed until well into the 1950s. The replacement statue of Corneille in Paris, for example, was the first on Vichy's list in February 1943. It was not inaugurated until 1952.[20]

The majority of pedestals in French municipalities remained, therefore, both bare and intact. These "decapitated" pedestals inspired some witticisms in the press: "We hope that the administration will excuse our irreverence, but we are forced to point out that it is a terrible gardener. While it pulled some of the weeds—and not all, by the way—it left the roots. And what roots by God!"[21] One contributor to *Aujourd'hui* wrote in 1942, "we knew that a pedestal was important to a statue, but who would have guessed how essential a statue is to a pedestal?"[22] The prevailing reaction to the abundance of naked plinths, however, was a combination of grief and anger. A 1942 article in *Université Libre* insisted that "each empty plinth become a new monument to remind the French people of the crimes we must avenge."[23]

Michael Taussig describes the controversial removal of a public sculpture from the shores of Lake Burley Griffin in the Australian Capital Territory. Weeks after the sculpture had been removed people still flocked to see where it had been, to witness "the perfectly empty space filled with what was."[24] The decapitated pedestals throughout France created just such an empty space. While Vichy's minister of industrial production expressed the hope that the sight of bare pedestals would "help the population better understand that they are not alone in making sacrifices,"[25] the impact of so many empty bases dotting the landscape was the opposite. The prefect in Marseilles suggested that the empty bases were really sufficient monuments in themselves, claiming that "often the inscriptions remain, and these are the essence of memory."[26] Inscriptions on bare pedestals did not suffice, however, for the French public whose attention was instead focused on the glaring absence above them.

For the French population during the war, the "empty space filled with what was" took on a memorial function. Instead of signaling the greater sacrifices made by the state, these empty bases instead reminded French communities of Vichy's Faustian bargain and of all that they had subsequently lost. The impact of war on material culture is largely discursive and symbolic, and wartime damage to objects serves as a record of collective trauma.[27] The trauma inscribed on empty pedestals, for Charles Maurras, was national shame:

> We hadn't really fully understood what the defeat meant....The departure of these heavy masses of non-ferrous metal made it clear; clear enough, here and there, to cause tears. And so it was really true, we had been defeated. And this obliges us to part with these [symbols] whose meaning is revealed suddenly with eloquence—the honour of our country, our honour![28]

Andreas Huyssen has explored the ways in which monuments function as substitute sites of mourning.[29] Similarly, Elizabeth Hallam and Jenny Hockey describe how material culture can mediate between the present and the absent: "Material cultures do not simply operate as means through which memories may be retrieved and sustained.... Rather, objects might stand as painfully isolated vestiges of those persons with whom they were once surrounded or associated."[30] Hallam and Hockey offer a compelling example of how an unusual object can become a memorial, standing for the grief and loss of an entire community. They describe a melted watch face retrieved from the wreckage of a troop train derailment at Gretna in May 1915. The impact of the disaster and ultimately of the Great War is, they argue, materialized on the face of the watch. "The watch is linked to a single death, yet it carries the weight of multiple deaths, not just at the site of the train disaster but also those resulting from war."[31] In a similar way, the empty pedestals in French town squares resonated with personal and collective loss. They served simultaneously as reminders of the missing and the dead, of individual sacrifices, but also of the collective trauma suffered by French communities, and of values and achievements subsequently undermined or lost.

With their new resonance as symbols of loss, the plinths were incorporated in the rituals that surrounded First World War memorials—symbols of absence and loss par excellence. Great War memorials are the architectural expression of absence. Their arches and walls "delineate a space within which absence is allowed to impose itself as the dominant feature."[32] In expressing this absence, Great War memorials also enable its resolution. Daniel Sherman describes the construction of war memorials in the interwar period as a way for the French nation, as well as for the bereaved families of fallen soldiers, to cope with the absence of a generation. Sherman stresses that the memorializing impetus was first manifested at the local level: "towns and villages throughout the country sought to compensate for physical absence with the symbolic presence a monument could project."[33] In the period between the removal of statuary and the Liberation, the empty bases of lost bronzes were incorporated into the same local commemorative practices as those associated with Great War memorials. Once again, in this time of crisis, expressions of grief and loss came to the fore. The laying of wreaths and flowers is an overtly funerary practice,[34] and these offerings appeared on innumerable monuments and empty bases in French towns. A police report on Bastille Day 1941 described the discovery of a placard and wreath on a statue in Maurs.[35] Flowers were left on the naked base of La Savoyarde in Chambéry, and wreaths were placed on the empty pedestals of Mistral in Arles and Étienne Dolet in Paris.[36]

Taussig's theory of defacement gives us a useful framework for the incorporation of empty pedestals into local expressions of grief through commemoration. Taussig sees the state of desecration as one of immense symbolic power: "When...a public statue is defaced, a strange surplus of negative energy is likely to be aroused from within the defaced thing itself. It is now in a state of desecration, the closest many of us are going to get to the sacred in this modern world."[37] For Taussig, the act of defacement is paradoxically an act of revelation—"it brings insides outside, unearthing knowledge, and revealing mystery."[38] The mystery it reveals is what Taussig calls the public secret—that which is generally known but cannot be articulated. In the case of the naked plinths, the public secret they revealed was twofold—the fact that the statue metal was destined to German munitions factories and, behind that fact, Vichy's culpability. These bases testified not only to the physical loss of monuments, but to the symbolic rape of France and to the many more concrete losses and privations to which the French population responded with anger and sorrow.

Manifestations of grief and loss continued to find expression through public commemorative monuments, and empty bases came to take on this function because they, like war memorials, marked a poignant absence for French communities. In their state of desecration, these empty bases were daily reminders of the French defeat, of occupation, of daily trials and hardships, and ultimately of the French government's complicity with Nazi Germany. As the political and social context shifted after the war, however, the meaning of this absence changed fundamentally, and the symbolic power of the bases' desecration was lost.

"From the Municipality of 41–44, to the Municipality of the Liberation"

A celebration took place in Amiens on March 4, 1945. The statue *Pierre l'Hermite* was replaced on its base, "marking the official transfer of the monument from the municipality of 41–44 to the municipality of the Liberation."[39] This ceremony involved much more than the re-inauguration of a statue. It marked a symbolic transition for Amiens by filling one of the many absences created during the war. For many municipalities, however, their bronze statues were not only removed but destroyed, and a re-inauguration such as that in Amiens was impossible.

The fact that few statues were chosen for state commissioned replacements, and that the program of replacements was subsequently abandoned, left the issue of dealing with the legacy of Vichy's bronze campaign largely up to French municipalities. What communities did with their bare pedestals, the choices they made about replacements, and the considerations that shaped these decisions also illustrate the issues French communities faced as they embarked on the process of reconstruction after the war. Here we see how communities rebuilt themselves physically and symbolically, and how they chose to recreate and redefine their urban landscapes and their memories of the past.

The choices that French municipalities made in dealing with this legacy varied widely. Statues were replaced sometimes by replicas, sometimes by new works; bases were removed, recycled or left empty. Municipalities sometimes accepted the Vichy replacement statuary, and sometimes did not. But all of their choices stemmed from the need to deal with the absence created by empty pedestals, and the uneasiness that this absence created. A report prepared by the prefect of the Seine outlines the state of Parisian statuary in 1950, illustrating the absences left in the wake of the bronze mobilization, and the variety of ways these voids were filled:

> One can't help being struck by the number of empty pedestals...in the streets, sad reminders of the years of occupation. In Paris and in the *banlieue*, 140 and 61 bronzes respectively were removed or melted down....[Vichy] had planned to replace some of the bronzes that were removed by works in stone. This was continued by the governments born of the Liberation, and several statues have already been produced including: *Claude Bernard*,...*Berlioz* by Saupique inaugurated on December 1, 1946, and *Villon* by Collamarini, placed in square Monge on March 24, 1947. Other statues are finished and awaiting installation: *Voltaire*, by Drivier; *Béranger*, by Lagriffoul; *Corneille*, by Rispal; *Rousseau* by Bizette-Lindet; *Lamartine* by Niclausse. Still others are in the process of completion: *Shakespeare*, by Landowski, *Étienne Dolet*, by Couturier. In addition to these statues commissioned by the state, a number of committees have offered monuments to the City of Paris. To them we owe the monument to Thomas Paine raised in 1948; *La France* by Bourdelle, and the statue of Maréchal Foch, erected in 1949. Other projects are still being studied. For example, a fountain will replace the inventors of quinine, Pelletier and Caventou, on the boulevard St. Michel.[40]

Filling the empty spaces left by lost statues and dealing with the uneasiness created by these "sad reminders of the years of occupation" was an integral part of the process of recovery in Paris. A similar impetus to fill or redefine these spaces is evident throughout France.

Many municipalities elected to replace their lost statuary with identical replicas by recasting the original bronzes. This suggests that people were attached to the character of their statues as well as the memories they invoked. The creation of replicas was even demanded when a lost statue had been replaced by a new work in stone. The monument to Flaubert in Rouen was on Vichy's list of works to be replaced, and a stone effigy had taken the place of the lost bronze. This new statue was so unpopular it was moved to the municipal museum and the original was recast in 1955. The city was also able to re-create *Georges Dubosc* when the statue's cast was discovered in an antique shop.[41] The statue of Mistral in Arles was recast and re-inaugurated in 1948. Likewise, monuments to Steinlein and Condorcet in Paris were recast in 1962 and 1989 respectively.[42]

When a bronze replica was not possible, some municipalities chose to recombine surviving elements of lost statues to create a new work. The monument to Jean Jaurès in Toulouse is an amalgamation of elements of the old statue, which had been raised in 1929.[43] Since 1982, Paris' 20th district has housed a fragment of the enormous monument to Gambetta that was formerly the main feature of the Carousel du Louvre.[44] If nothing of the previous monument remained, many municipalities chose to commission a stone copy of the original bronze. The monument to Eugène Carrière in Paris, for example, was originally created in bronze by his son Jean Réné. In 1959, Jean Réné produced a stone copy of the original.[45] The statue of Rameau in Dijon is, likewise, a stone copy of the original bronze, as is the monument to Maréchal Niel in Muret, which was re-inaugurated in 1949.

While municipalities struggled to replace some of their lost statues with copies in bronze or stone, they sometimes elected to opt for entirely new monuments. The monument to Sergeant Bobillot, which once stood at the intersection of Boulevard Voltaire and Boulevard Lenoir in Paris was replaced in 1959 by a "more modest" homage in stone, "discreetly" placed among the chestnuts in the place Paul Verlaine.[46] Other works were replaced with monuments to entirely different figures. In the square Vintimille in Paris, a stone statue of Berlioz stands, since 1948, where a bronze of Alfred Lenoir stood before 1941.[47]

Inevitably, not all melted bronzes were replaced, and some of the empty bases were left where they stood. The pedestal of *Ledru-Rollin* was left intact in place Léon Blum in Paris. The pedestal of the statue of Charles Fourier stands to this day in place Clichy. The empty base of the bust of Charles Lenoir in Rennes is a melancholy sight, despite being tucked away in a pretty corner of the municipal gardens. Sometimes the remaining bases were used for other statues. The monument to Jehan de Chelles in Paris surmounts the base previously occupied by *Baudin*.[48] In Toulouse, the monument to Armand Silvestre was transformed into a decorative piece with a marble nude taking the place of the bronze effigy. Only a trace of the original subject of the monument remains. "Silvestre" is inconspicuously inscribed on the bottom of the large stone structure.

Many bases, after long sitting empty, were simply removed. Étienne Dolet's vacant pedestal sat in place Maubert in Paris until 1979. Nothing remains of Bartoldi's famous monument to the siege of Paris, popularly known as the *Ballon des Ternes*. The same is true of hundreds of other monuments across the country. In the Grand Rond gardens in Toulouse an enormous monument—*The Glory of Toulouse*—once stood in the center of a large fountain. After 1945, the municipal council debated replacing it with other sculptures and water features, but the center of the fountain remains empty.[49]

Dealing with the legacy of the lost statues in the postwar period—replacing statues or removing plinths—was part of a more general process of reconstruction and redefinition and of dealing with—or erasing—painful memories of the war. This process also reflected wider debates surrounding how France should be rebuilt physically and symbolically and how four years of occupation should be incorporated into French history and memory. These debates have since been taken up by scholars in a number of fields. Much work has been done on questions surrounding the symbolic reconstruction of France, hinging in particular on issues of memory and the Occupation, Vichy, and the Holocaust.[50] Philippe Burrin describes the paradoxical nature of Vichy as a "site" of memory: "[Vichy] was a deliberate, persistent, and futile effort to organize reality around...memory, to reconstruct a national spirit in which the memory of a mythologized past would shape the perception of the present to create a unified way of feeling, thinking and acting."[51] But Vichy failed in its attempt to construct a unifying and inspiring memory of France, and instead left in its wake a divisive and traumatic memory. "Instead of 'reconstructing the national soul' by filling it with memories of an invented 'France' it left vivid memories of a very real France that continues to be a source of embarrassment and outrage."[52] Julia Kristeva asserted that the symbolic legacy of the Second World War can only be absence. The scale and nature of the cataclysm undermined the symbols through which any meaning could be attached to this conflict. Faced with the horrors of the Holocaust, our systems of perception and representation are incapacitated:

> As if overtaxed or destroyed by too powerful a breaker, our symbolic means find themselves hollowed out, nearly wiped out, paralyzed. On the edge of silence the word "nothing" emerges, a discreet defense in the face of so much disorder, both internal and external, incommensurable. Never has a cataclysm been more apocalyptically outrageous; never has its representations been assumed by so few symbolic means.[53]

Jay Winter describes the inappropriateness and insufficiency of traditional commemorative forms in the face of the Second World War and, primarily, the Holocaust: "both the political character of the [war] and some of its horrific consequences made it impossible for many survivors to return to the languages of mourning which grew out of the 1914–1918 war when they tried to express their sense of loss after 1945."[54] Sarah Farmer argues that when it came to commemorating the Second World War, Europeans "shifted away from their traditional emphasis on the war dead to an unprecedented effort to mark and preserve sites and traces of destruction. Indicating the places of important events took priority over erecting traditional monuments."[55] This preservation of sites was due in part to changes in the character of war. Occupation, exodus, deportation, and bombings all erased the distinction between home and battle front. "In 1945 the war did not have to be brought home with a statue or a stele, it had already been there."[56]

Another contentious issue that arose in the aftermath of the Second World War, one that is closely related to postwar commemoration, is the question of the physical reconstruction of the country. The primary debate surrounding the reconstruction of French cities was how faithfully they should be rebuilt. On one hand, reconstructing cities exactly as they had been meant recreating their distinctive character and the features that made them familiar and unique. On the other hand, rebuilding provided the opportunity to incorporate modern innovations in architecture, to improve infrastructure and sanitation, to adopt more efficient urban planning, and to adapt cities for public transportation and motorized vehicles. In the wake of the bronze mobilization campaign, commemoration and reconstruction overlapped in interesting ways. This intersection of space and memory ensured that reconstruction—or not—of bronze statues in the postwar period was contentious.

Issues surrounding postwar commemoration and reconstruction in France are reflected literally in the replacement of bronze statues. The reasons to rebuild a city faithfully are the same as those that inspired the recreation of an exact replica of a bronze statue that had been lost—to turn back the clock, to undo the damage of four terrible years, to trick posterity. For practical reasons it was often not possible to rebuild exactly, but in many municipalities the impetus was to be as faithful to the old city as possible. A stone copy of a lost bronze was a popular solution. But many French cities in the process of rebuilding turned toward the future at the expense of the places and events of their past. The opportunity to modernize often had consequences for the relics of Vichy, including the naked bases that dotted French urban landscapes.

France in the 1950s and 1960s embraced the automobile. Cars became the central symbols of the new nation, and reconstruction was shaped by traffic.[57] "Paul Delouvrier, prefect of the Seine and Haussman of his day, continued in the line of thinking dominant in Parisian planning since Haussman, according to which the needs of street circulation take precedence over all other urban considerations. 'If Paris,' he wrote, 'wants to espouse her century, it is high time that urbanists espouse the automobile.' Delouvrier found a great enthusiast for his ideas in Georges Pompidou, who affirmed, 'Paris must adapt itself to the automobile. We must renounce an outmoded aesthetic.'"[58]

A tirade by Aragon, inspired by the display of a Ford automobile on the former site of the statue of Victor Hugo in Paris, reveals the tensions created by this "renunciation of an outmoded aesthetic" so prevalent in Parisian postwar reconstruction:

> A Ford automobile, the civilization of Detroit, the assembly line...the atomic danger, encircled by napalm...here is the symbol of this subjugation to the dollar applauded even in the land of Molière; here is the white lacquered god of foreign industry, the Atlantic totem that chases away French glories with Marshall Plan stocks....The Yankee, more arrogant than the Nazi iconoclast, substitutes the machine for the poet...the Ford for Victor Hugo.[59]

Projects to straighten streets and improve infrastructure to make the capital more car-friendly also had consequences for some of Paris' surviving monuments. Place de la Nation, for example, lost its fountain during the expansion of the metro, and *Diderot* was moved several meters to free up the intersection at Saint-Germain-des-Prés. Traffic concerns appear to have been *the* major consideration in the discussion of public statuary in the postwar period. An article in *Le Monde* in March 1953 exclaimed that "the demands of traffic circulation must prevent public squares and avenues [from harboring] effigies of famous people."[60] The art critic Claude Roger-Marx saw two reasons for forgoing the restitution of lost bronzes: they were too cumbersome on thoroughfares now thronging with vehicles, and "the second-hand recreation of past glories always appears somewhat awkward."[61]

In Paris there was also still evidence after the war of widespread hostility to the statues that had been the product of Third Republic "statuemania." Albert Mousset asked in 1949 if the city would "perpetuate the simplistic tradition which, over the last century, has led to *n'importe quel* monument being put *n'importe où*."[62] For Mousset, "a square...is a place of leisure and relaxation. Nobody goes there to contemplate Diderot, Michel Servet or Déroulède."[63] The Parisian press voiced strong opposition to the replacement statuary: "One hesitates to pull down monuments raised by our fathers. But, at least, let us not restore...those eyesores which have been swept away by events."[64]

Those absences that remained—the empty pedestal of Arago in Paris or the empty fountain in Toulouse's Grand Rond, for example—appear to have remained absences by default rather than by design. Debate in Toulouse over what should be done with the fountain in the Grand Rond reveals that bureaucratic indecision and shifting priorities were responsible for the persistence of emptiness rather than any particular gesture to memory. It is also interesting that the absence created by an empty pedestal—emptiness that had a memorial character during the war—has lost this meaning since. Andreas Huyssen concedes that some monuments "stand simply as figures of forgetting, their meaning and original purpose eroded by the passage of time."[65] In the case of our empty bases, this is evidence of what Winter and Sivan have called the "shelf-life" of memory:

> Remembrance is by its very nature vulnerable to decay, and hence has a shelf-life. Even under the delayed impact of the extreme conditions of war, memories do not necessarily endure....Other tasks take precedence; other issues crowd out the ones leading to public work. And ageing takes its toll....This fading away is inevitable.[66]

Commemorative practices are deeply tied to ritual. The ritual associated with monuments perpetuates memory, though the nature of that memory clearly shifts and changes with time.[67] In the case of the empty bases of removed statues, to which no ritual was tied, the memories they invoked became subordinated to other concerns, which were shaped by shifting social and political contexts.

The ambivalence of the French regarding their roles in the Second World War has provided fertile ground for mythmaking.[68] Henry Rousso's study of the place of Vichy in French memory and history since 1944, *The Vichy Syndrome*, examines the political, ideological, geographic, and generational fault lines that have shaped how French society has chosen to remember the war years. According to Rousso, de Gaulle's primary concern during his time in office was national reconciliation, national unity, and national honor. Toward this aim he constructed a "resistancialist myth," which celebrated not The Resistance but the French "people *in resistance*" in an attempt to superimpose a "smoothly polished image of complex events" on a contentious and divisive period.[69] As Antoine Prost observes, the commemorative practices that resulted from this "smoothly polished image" were distorted: "The work of remembrance was made of lies, pious lies, well-meaning lies, but lies nonetheless. [It] has tried to hide the fact that Frenchmen murdered other Frenchmen. Remembering was a way of preserving this kind of secret."[70]

An example of the perpetuation of the resistancialist myth in French monuments is an inscription on the Saint Michel fountain in Paris that reads: "In the year 1944, from the 19th to the 25th of August, after 50 months of German occupation, the people of Paris, at the approach of the liberating armies, rose up against oppression." Similarly, the city of Paris decided in April 1949 that when the monument to the *franc-tireurs* of the Franco-Prussian War was reconstructed it would include an inscription relating the fighters of 1870–71 to the Parisian Resistance of 1940–44.[71]

The resistancialist myth is not built solely on the selective remembering of the resistance and the forgetting of collaboration. It also focuses on acts of German barbarism in order to highlight French victimhood. Thus, the base of the statue of General Mangin long held a sign proclaiming "here stood the statue of General Mangin, destroyed by Hitler, June 1940."[72] In this case, where a stone statue had been destroyed by German troops in the early weeks of the occupation, the empty base easily became a touchstone of outrage and indignation.

The urge to sublimate the contentious and divisive aspects of the recent past coincided with French determination to modernize in the 1950s and 1960s. The chief prosecutor at Pétain's trial spoke of 'four years to be stricken from our history,' and the amnesties of 1951–53 testified to the government's willingness to forget. This focus on the future meant that the absences created by empty pedestals were no longer poignant but, rather, unsightly and embarrassing. The city of Rouen, for example, intended eventually to replace its statue of Flaubert, but decided to remove the base temporarily because an empty pedestal left in place was unattractive and disrespectful to Flaubert's memory.[73] The grandiosely named place de la Concorde—a small square in a working class neighborhood in Toulouse—featured a monument called La Poésie Romane, donated by a pharmacist to a local cultural organization. A figure of a woman in traditional Occitan dress topped a column and basin that served as a public fountain. Although the statue was removed during the bronze campaign, it was not melted because the monument was both a bequest and the property of a private association. After the war, it was not replaced. A local community group wrote to the municipality to complain that the fountain and the basin were in a deplorable state.[74] They asked that the statue be replaced and that the entire monument be restored, or that the whole thing be razed. This request stemmed from two considerations: the deplorable state of the monument and the function of the space it occupied. The original purpose of the fountain—a source of potable water for the community—was no longer relevant in the new era of indoor plumbing. Its role now purely aesthetic, either the fountain had to become a proper monument of which the community could be proud, or else the nature of that space had to be redefined.[75]

In the 1970s and 1980s, the nature of French memory of the war years shifted. In response to the myth of a nation of victims and resisters, a counter-myth was launched by the generation of May '68. This version painted wartime France as a country of cowards and collaborators. This counter-myth revealed a reactionary, xenophobic, anti-Semitic, and authoritarian France and ushered in a period of "obsession" with Vichy and the Occupation. This obsession took the form of an "urge to expose the ugly side of collaboration."[76] In November 2002, the inscription on the Saint Michel fountain was defaced. The original inscription was altered so that the word "German" was replaced by "Nazi," and "50 months of German occupation" was followed by "and good French collaborators like the illustrious Papon."[77]

Despite this drive for reckoning, the symbolic potential of using lost statues to paper over the cracks created by Vichy persists. The case of La Savoyarde in Chambéry is an interesting example. The statue was discovered in a foundry in Germany in 1950. It was returned to Chambéry where it was kept, largely forgotten, in a municipal depot.[78] It was not until the late 1970s that a project to restore and replace the monument was approved by the municipal council. This belated decision coincided with an historical moment in which the community's relationship with the occupant and the Vichy regime had come under the microscope. Paul Touvier's wartime actions and the support that he had received in the region during and following the war brought the comportment of the local population under scrutiny.[79] The occasion of *La Savoyarde*'s re-inauguration in April 1982 provided a welcome demonstration of Chambéry's victimization. In 1985, the original cast of the statue *Leperdit* was discovered in an antique shop in Rennes by the curator of the fine arts museum. With the support of city hall, the statue was recast and re-inaugurated amid much celebration. The mayor of Rennes proclaimed at the re-inauguration: "we are doing the work of memory and of justice, erasing...the ignominiousness of Vichy, which decided to eradicate this statue by melting it down."[80] The symbolism of Leperdit—a man who embodied justice and tolerance, an administrator who protected Rennes from repression, a personality who was a beacon of fairness and moderation in a time of terror, hysteria, and excess—was unifying and healing and, like the re-inauguration of *Pierre l'Hermite* in Amiens, symbolized the return of noble values and virtues that had once defined the city.

The Promise of Community

Public art is enveloped by the promise of community.[81] Monuments not only brought prestige to French communities through an evocation of great deeds of the past, they symbolized a spirit of unity that was desperately sought. The destruction of statuary meant not only a loss of prestige, but an attack on memory, the undermining of community, and the erosion of local identity.

The mourning that the loss of bronze statues elicited during the war—mourning that was made palpable by the profusion of desecrated pedestals—and the uneasiness that their absence created in the postwar period, both testify to the symbolic impact of the bronze mobilization campaign on French communities. While the campaign was motivated by pragmatic economic considerations, the implications of the destruction of national patrimony are clearly symbolic.

At the core of the legacy of the statue episode is absence, and the most interesting feature of this legacy is how forcefully *absence* and *emptiness* are tied to memory. In its evocation of uncertainty and its demand for resolution, the dealing with absence works to erase, or at least minimize, the complexities of the past. On one hand, reconstituting commemorative space by reconstructing statuary serves to fill the absence, while redefining the function of the empty space by eradicating of all traces of a statue's existence eliminates it. Here we see evidence of the central dialectic of remembering and forgetting in the service of contemporary concerns that is characteristic of memory.

In investigating the legacy of Vichy's destruction of bronze statuary for French communities, the broader symbolic significance and impact of the bronze mobilization campaign comes to light, yet another instance of the difficult process of coming to terms with the Vichy past.

Notes

1. In Yvon Bizardel, "Les Statues parisiennes fondues sous l'Occupation, 1940–1944," *Gazette des Beaux Arts* (1974): 129–56.
2. Musée d'Orsay, *À Nos Grands hommes*, CD ROM.
3. Centre Historique des Archives Nationales (hereafter CHAN), F 21 7071. "Sur l'enlèvement des statues et monuments métalliques en vue de la refonte," March 1, 1942.

4. This nonferrous metal fed the German armaments industry.

5. 1,527 is my estimate. 1,750 is the estimate provided by Elizabeth Karlsgodt in "National Treasures: Cultural Heritage and the French State during the Second World War," (Ph.D. dissertation, New York University, 2002). My estimate is more conservative than Karlsgodt's because it is limited to those statues removed between autumn 1941 and spring 1942. Karlsgodt includes monuments removed in this period as well as in a second wave of removals from summer 1942 to autumn 1944. The difficulty in establishing the precise number of statues destroyed lies in the fact that documents compiled by the Ministries of Industrial Production and National Education list only statues that were dismantled, but give no indication of whether they were melted down. A number of monuments, particularly in the later phase of the campaign, were removed from their pedestals but escaped destruction. Many that were destroyed in the second wave were removed during the first and appear on earlier inventories. These factors make calculating a precise number difficult.

6. See Michèle Cone, *Artists under Vichy: A Case of Prejudice and Persecution* (Princeton, NJ: Princeton University Press, 1992); Denis Peschanski, ed., *Collaboration and Resistance: Images of Life in Vichy France, 1940–1944* (New York: Harry N. Abrams, 2000); Robert Gildea, *Marianne in Chains* (London: Macmillan, 2002). For more detailed investigation see Maurice Agulhon, "La Statuomanie et l'histoire," *Ethnologie Française* 8 (1978); Yvon Bizardel, "Les Statues parisiennes fondues sous l'Occupation, 1940–1944," *Gazette des Beaux Arts* (1974); June Hargrove, *The Statues of Paris: An Open-Air Pantheon* (New York: Vendôme Press, 1989); Jacques Lanfranchi, *Les Statues des grands hommes à Paris: cœurs de bronze, têtes de pierre* (Paris: L'Harmattan, 2004); and Georges Poisson, "Le Sort des statues de bronze parisiennes sous l'Occupation," *Paris et Ile-de-France: Memoires*, 47:2 (1996): 165–297. Images of the destruction of bronzes have also become well known; notably Pierre Jahan's photographs in Pierre Jahan and Jean Cocteau, *La Mort et les statues* (Paris: Seghers, 1977).

7. See Kirrily Freeman, "Incident in Arles: Regionalism, Resistance and the Case of the Statue of Frédéric Mistral," *Contemporary European History*, 16:1 (2007): 37–50; "The Battle for Bronze: Conflict and Contradiction in Vichy Cultural Policy," *Nottingham French Studies*, 44:1 (2005): 50–65; *The Battle for Bronzes: The Destruction of French Public Statuary, 1941–1944* (Ph.D. dissertation, University of Waterloo, 2004), as well as Elizabeth Karlsgodt, "Recycling French Heroes: The Destruction of Bronze Statues under the Vichy Regime," *French Historical Studies*, 29:1 (2006), and chapters 8 and 9 of Karlsgodt, "National Treasures."

8 The exception to this is Karlsgodt's article.

9. In "Recycling French Heroes," Karlsgodt found that in Chambéry the strongest criticism of the destruction of statuary came from the left. What has emerged from my research is the role of Vichy's own administration and local elites in this protest.

10. Chamber of Commerce to Prefect, November 13, 1943. Archives Départementales Bouches-du-Rhône, 7 T 3/3.

11. William Cohen, "Symbols of Power: Statues in Nineteenth-Century Provincial France," *Comparative Studies in Society and History*, 31:3 (1989): 491, 495.

12. Cohen, 496.

13. Cohen, 509.

14. Rosalind Krauss, "Sculpture in the Expanded Field," in *The Originality of the Avant-Garde and Other Modernist Myths* (Cambridge, MA: MIT Press, 1985), 279.

15. Timothy Walsh, *The Dark Matter of Words: Absence, Unknowing and Emptiness in Literature* (Carbondale: Southern Illinois University Press, 1998), 25, 104.

16. Municipal Archives Toulouse, 5M 318.

17. Note, Prefecture Bouches-du-Rhône, April 21, 1942. Archives Départementales de Bouches-du-Rhône, 7 T 3/3.

18. Archives de Paris, VM 92 1.
19. Ministry of Industrial Production to prefects, April 1943. Archives Départementales de Bouches-du-Rhône, 7 T 3/3.
20. Archives de Paris, Déliberations du Conseil Municipal D4K3 64.
21. "Les Piedestaux font regretter les statues," *Aujourd'hui*, 12.2.42. CHAN, F 21 7071.
22. Ibid.
23. Anonymous article, n.d. *Université Libre*, CHAN, F 21 7071.
24. Michael Taussig, *Defacement: Public Secrecy and the Labor of the Negative* (Stanford, CA: Stanford University Press, 1999), 33.
25. Ministry of Industrial Production to prefects, April 29, 1943. Archives Départementales de Côte-dOr, W 24725, Archives de Paris VM 92 6.
26. Note, Prefecture Bouches-du-Rhône, April 21, 1942. Archives Départementales de Bouches-du-Rhône, 7 T 3/3.
27. Elaine Scarry, *The Body in Pain* (New York: Oxford University Press, 1985).
28. Charles Maurras, "L'Affaire des statues" n.d. CHAN, F 21 7071.
29. See Andreas Huyssen, *Twilight Memories: Marking Time in a Culture of Amnesia* (New York: Routledge, 1995), also Daniel Sherman, *The Construction of Memory in Interwar France* (Chicago: University of Chicago Press, 1999) and Allyson Booth, *Postcards from the Trenches: Negotiating the Space between Modernism and the First World War* (New York: Oxford University Press, 1996).
30. Elizabeth Hallam and Jenny Hockey, *Death, Memory and Material Culture* (New York: Berg, 2001), 124.
31. Ibid.
32. Booth, 36.
33. Sherman, 73.
34. Antoine Prost, "The Algerian War in French Collective Memory" in Pierre Nora, *Realms of Memory: Rethinking the French Past* (New York: Columbia University Press, 1984), 2:319.
35. Report, Gendarmerie Nationale, Archives Départementales de Cantal, 1 w 65.
36. CHAN F 21 7071 and F 21 7072.
37. Taussig, 1.
38. Taussig, 2–3.
39. Regional Delegate, Laon to Commissariat for the Mobilization of Non-Ferrous Metal, March 7, 1945. CHAN, 68 AJ 313.
40. *L'Œuvre de relèvement dans le Département de la Seine depuis la Liberation. Exposé presenté au Conseil Municipal de Paris et au Conseil Général de la Seine lors de leur session de mars 1950 par M. Roger Verlomme, Prefet de la Seine*, 381. Bibliothèque Historique de la Ville de Paris, 134 933.
41. Yvon Pailhès, *Rouen: un passé toujours présent...rues, monuments, jardins, personnages* (Luneray: Editions Bertout, 1994), 266.
42. *Dictionnaire des monuments de Paris*, Bibliothèque Historique de la Ville de Paris, 149 577, 206. *À nos grands hommes*, Musée d'Orsay (2004).
43. Jacques Ducos, *La Patrimoine des communes de la Haute-Garonne* (Toulouse: Flohic, 2000), 1663.
44. *Dictionnaire des monuments de Paris*, 261.
45. Ibid., 273.
46. Ibid., 577.
47. Pierre Kjellberg, *Le Guide des statues de Paris* (Paris: Bibliothèque des Arts, 1973), 95.
48. Prefecture de la Seine, Service technique d'Architecture, rapport de l'Architecte adjoint, May 20, 1942. Archives de Paris, VM 92 8.
49. Archives Municipales de Toulouse, *Fontaines toulousaines* (2003).

50. "In 1946 the French parliament passed a special law classifying Oradour [Oradour-sur-Glane, site of an SS massacre of French civilians] as an historic monument and mandated that the vestiges of the old town be preserved for eternity." The government intended the ruins as "an image of 'France which had been ravaged.'" The new town built beside the ruins would provide "an image of 'France being reborn.'" Farmer, "Oradour-sur-Glane: Memory in a Preserved Landscape," *French Historical Studies*, 19:1 (1995): 30.
51. Philippe Burrin, "Vichy" in Nora, 1:183.
52. Burrin, 196.
53. Cited in Jay Winter, *Sites of Memory, Sites of Mourning* (Cambridge: Cambridge University Press, 1995), 229.
54. Winter, 9.
55. Farmer, 27.
56. Farmer, 28.
57. Kristin Ross, *Fast Cars, Clean Bodies: Decolonization and the Reordering of French Culture* (Cambridge, MA: MIT Press, 1996), 22.
58. Ross, 53.
59. R.F. Kuisel, *Seducing the French: The Dilemma of Americanization* (Berkeley: University of California Press, 1993), 41.
60. Ibid.
61. Claude Roger-Marx, "Le Moment est venu de mettre des statues sur les socles dépouillés," *Figaro Litéraire*, January 21, 1950. Roger-Marx's suggestion was to put statues from Parisian museums on the empty pedestals. The city of Toulouse is currently doing the opposite—vandalism of public sculptures has forced the municipal council to put the originals in a museum and resin replicas in public squares and parks.
62. Albert Mousset, "Les Statues abusives," *Le Monde*, March 18, 1949.
63. Ibid.
64. Hargrove, 309.
65. Huyssen, 250.
66. Jay Winter and Emmanuel Sivan, *War and Remembrance in the Twentieth Century* (Cambridge: Cambridge University Press, 1999), 30, 10. See also Hallam and Hockey, 8. Even the sites of atrocities, such as Oradour-sur-Glane, suffer this shift: "the passage of time and the impact of weather have transformed the ruins of Oradour and shifted their possible meanings. The ruins have taken on a romantic quality, and convey a mood of not unpleasant melancholy rather than revulsion at horror." Farmer, 43.
67. Nora, 1:7.
68. See Conley, "The Myth of the dernier poème: Robert Desnos and French Cultural Memory," in Bal, Crewe and Spitzer, eds; *Acts of Memory: Cultural Recall in the Present* (Hanover, NH: University Press of New England, 1999), 134–47.
69. Rousso, *The Vichy Syndrome: History and Memory in France since 1944*, Arthur Goldhammer, trans. (Cambridge, MA: Harvard University Press, 1991), 18.
70. Prost, 174.
71. Archives de Paris, Délibérations du Conseil Municipal ,d4K3 66 (April 14, 1949).
72. Hargrove, 304.
73. Municipal Archives Rouen, Dossier statue Flaubert.
74. Note, Cabinet de l'adjoint délégué, Municipal Archives Toulouse, 5 M 323.
75. The monument was restored.
76. Henry Rousso and Eric Conan, *Vichy: An Ever-present Past* (Hanover, NH: University Press of New England, 1998), ix.
77. Maurice Papon was Secretary General of the Gironde prefecture under Vichy and organized the deportation of Jews from Bordeaux between 1942 and 1944.
78. Karlsgodt, "National Treasures," 350.

79. Touvier, an officer in the Milice who tortured and executed Jews and resisters, was the first Frenchman convicted of crimes against humanity.

80. Ville de Rennes, *Jean Leperdit: une statue place du Champ-Jacquet* (1994), Archives Municipales de Rennes, BP 182.

81. See Karim Benammar, "Absences of Community," in Eleanor Godway and Geraldine Finn, eds., *Who is this 'we'? Absence of Community* (Montreal: Black Rose Books, 1994), 36.

13

Reflections on the Literary Vichy Syndrome since 1990: Contexts, Chronologies, Metamorphoses

Richard Golsan

In September 1994, a scandal erupted in France over troubling new revelations concerning the nature and duration of President François Mitterrand's service to the Vichy regime during World War II. Commenting on the hue and cry produced by these revelations, the historian Robert Paxton wryly observed that "Vichy stirs the French public more than either money or sex."[1] At the time, Paxton was referring specifically to the fact that the French seemed more shocked by what he described as "Mitterrand's politics of fifty years ago" than they were by other recent revelations concerning the president's personal life, including the fact that he had an illegitimate daughter. But given the broader political, cultural, and even juridical contexts of 1990s France, Paxton's observation had a much wider application. It served to underscore the power and pervasiveness of the memory of *les années noires* or the dark years in French public life and to stress as well the capacity of that memory to disturb and unsettle the nation's moral conscience. Indeed, in a decade haunted by the past and burdened by traumatic memories of events including the Algerian war and the crimes of communism, the memory of Vichy had the greatest notoriety. So numerous were the "irruptions" of the memory of Vichy and reminders of French complicity in the Nazi Final Solution, the 1990s as a whole seemed to confirm that what Rousso had famously labeled the Vichy syndrome in his 1987 book had indeed reached its most advanced stage, which Rousso called the "Obsessions" phase.

In addition to the Mitterrand scandal, the weight of the Vichy past made its presence felt in a wide range of other contexts, from controversies over monuments and Resistance heroes to the release of widely-discussed films and the publication of some remarkable fictional works. The latter will be my primary focus here, but before discussing the literary Vichy syndrome of the 1990s (and indeed up to the present), I want to look briefly at the most spectacular and controversial manifestations of the Vichy past in the nineties, namely, the prosecutions and trials for crimes against humanity involving René Bousquet, Paul Touvier, and Maurice Papon. In my view, the lessons of these events, taken together, provide an interesting and helpful backdrop to and frame for the discussion of the literary Vichy syndrome over the last 15 years.

In spring 1994, the trial for crimes against humanity of former *milice* member Paul Touvier took place in Versailles. Touvier was being tried for the murder of seven Jewish hostages at the cemetery of Rillieux-la-Pape in June 1944, although in fact he was certainly guilty of numerous other crimes, including several that had been qualified as crimes against humanity by the examining magistrate, Jean-Pierre Getti, in his 1991 indictment. Touvier was found guilty, sentenced to life, and died in prison in 1996.[2]

In retrospect, two things are particularly significant about the Touvier trial. The first is the extent to which it confirmed what Rousso has called the "judeocentrism" of the memory of Vichy in the 1990s. The trial focused attention virtually exclusively on official Vichy anti-Semitism and the Pétain regime's complicity with the Nazis. Moreover, in order to secure the conviction of the accused, it was necessary to distort the historical record by insisting on a virtual fusion of Vichy with Nazi Germany, at least where organizations like the *milice* were concerned.

The second thing concerns the accused himself. During the trial it became abundantly clear, through readings from Touvier's infamous "green notebooks" from the 1980s as well as in courtroom outbursts by the accused, that Touvier remained completely immersed in the attitudes and hatreds that characterized him during the Occupation. As several observers put it, Touvier seemed to be a prisoner of his past, and indeed was "fossilized." It was as if the long postwar period did not exist, and, to the degree that Touvier can be considered symptomatic of the memory of Vichy in the 1990s, the Dark Years had become disconnected from the continuum of history, erupting into the present in all their immediacy and horror.

When the Papon trial took place two and a half years later in Bordeaux, it underscored the judeocentric nature of the memory of Vichy in the 1990s, in that the former Vichy civil servant was accused of crimes against humanity for his role in organizing the deportations of Jews from the Bordeaux region to the death camps to the east between 1942 and 1944. But while Vichy complicity in implementing the Final Solution was the central issue, early in the proceedings the trial was almost derailed by the controversy surrounding Papon's postwar career, and specifically his role as prefect of Paris police. In that capacity, in October 1961 Papon was responsible for ordering the brutal suppression of Algerian protesters ; as many as 200–300 protestors were killed and many more were beaten and held in detention camps around the city. In the press, both at the time and during the Papon trial, many compared French police brutality in 1961 to Vichy and Nazi crimes during the Occupation, and in fall 1996 Pierre Vidal-Naquet went so far as to suggest these crimes were virtually identical. Vidal-Naquet asserted that Papon should be on trial for crimes against humanity as much for his actions in 1961 as for his actions between 1942 and 1944.[3]

Whatever the merits of these comparisons, and, more generally, whatever both trials accomplished in terms of fulfilling a so-called "duty to the memory" of the victims, the point is that both trials ended up distorting the memory and meaning of "Vichy" itself. As an historical referent at least, the term *Vichy* became at once too narrow and too broad. In focusing attention on Vichy's anti-Semitism and criminal complicity with Nazism, the trials generally ignored the broader historical specificities of the period and of the Pétain regime itself. At the same time, in linking the crimes of the Pétain regime with those of the Fifth Republic in 1961 during the Papon trial, they pointed to an apparently profound *parenté* that tended to suggest that, at least in moral terms and perhaps in historical terms as well, the blight that Vichy signified had overflowed its chronological boundaries to infect the postwar period. In short, Vichy seemed to become both a metaphor for Evil and, as Éric Conan and Rousso put it, *un passé qui ne passe pas* (a past that does not pass) in historical terms.[4] Finally, because the prosecutions dealt ultimately only with so-called small fry and not with the man principally responsible for implementing the Final Solution in France, Vichy Chief of Police René Bousquet, who had been gunned down by a crazed publicity seeker before he could stand trial, the Touvier and Papon trials seemed to many to be simply warm-up acts for a main event that never occurred. These trials were therefore incapable of providing a truly cathartic and effectively symbolic coming-to-terms with the criminality of the Vichy past. The memory malaise, so to speak, remained.

In the introduction to his 2001 book, *Vichy: l'événement, la mémoire, l'histoire*, Rousso offers his own assessment of the changing memory of Vichy in the 1990s, and many of the developments he identifies coincide with those I have just discussed here. As the trials and prosecutions for crimes against humanity suggest, the memory of Vichy focused increasingly on "la réparation…et la judiciarisation,"[5] a process that, Rousso notes, exposed it to the paradox of trying to repair or compensate for a crime that had long been considered irreparable. At the same time, the memory of Vichy increasingly lost both its national and historical specificity as it became subsumed in the now global and hegemonic memory of Nazism and the genocide of the Jews. That memory itself became paradigmatic of other mass crimes and genocides from Bosnia to Rwanda. By association, Vichy became linked to these contemporary crimes as well. As a result, Vichy and its memory became almost if not completely impossible to define in strictly national and even European terms. In its fluidity it seemed to elude disciplinary boundaries, becoming as much a moral construct as an historical moment or event. It is the fluidity and ambiguity of the memory of Vichy in the 1990s, which one critic colorfully described as a living corpse growing in the basement,[6] that the works belonging to what I describe as the literary Vichy syndrome of the 1990s set out to evoke and explore.

My attempt here to sketch out the parameters of the literary Vichy syndrome of the 1990s is of course not the first effort to characterize and chronologize the literature dealing with *les années noires* and their memory. In fact, more ambitious efforts than my own have sought to cover virtually the entire postwar period. It is helpful to discuss briefly some of these efforts, not only to contextualize my own observations but to highlight difficulties and limitations that any effort to characterize or chronologize the literary memory of Vichy will encounter.

In *The Vichy Syndrome*, Rousso does address the memory of Vichy in postwar literature, and the novel in particular, as he defines and fleshes out the stages of memory of the syndrome. But his approach is both intentionally and necessarily selective. Rousso is primarily interested in a younger group of writers associated with the so-called *mode rétro* whose works influenced and helped define the "broken mirror phase" of the early 1970s. During this period, the heroic myths of Gaullism and Resistance were shattered, and collaborationism and Nazism itself were often cast in a dubiously ambivalent light. Rousso criticizes the writers in question, and Patrick Modiano in particular, for "playing with the ambiguity of commitment and reacting strongly, almost too strongly, against any notion of ideological determinism."[7] Apart from these writers and their works, Rousso leaves literature largely out of the discussion.

In an effort to fill in Rousso's blanks and to situate the works of Modiano and others in a properly literary historical context, in 1992 Alan Morris published *Collaboration and Resistance Revisited: Writers and the "Mode Rétro."*[8] In his study, Morris not only discusses the novelistic antecedents of the *mode rétro*, he offers a kind of preliminary chronology of a literary Vichy syndrome that coincides roughly with Rousso's scheme, although it collapses Rousso's so-called "Unfinished Mourning" and "Repressions" phases covering the years 1944 to roughly 1970 into one "phase," which Morris places under the heading of "Heritage." Morris characterizes the Heritage period as essentially a pitched and long-running battle between, on the one hand, provocative texts that sought to rehabilitate Pétainism and collaboration by challenging and debunking the orthodoxies and myths of the Resistance and Gaullism, and, on the other, texts that sought to quash these provocations and restore the aura of sanctity to the myths themselves. Given the general political climate of the period, the literary champions of the Gaullist and Resistance mythology largely won out.

According to Morris, the transition from the period he labels the Heritage to the *mode rétro* occurred around 1970, with the disappearance from the scene of Charles de Gaulle and the enormous shadow he cast. After 1970, in the face of new provocations by *mode rétro* writers including Modiano, Pascal Jardin, Marie Chaix, Evelyne Le Garrec, and Michel Tournier in *Le Roi des aulnes*, there was little or no literary "resistentialist" response. As a result, novelistic efforts to deny profound political and moral distinctions between collaboration and Resistance, and to emphasize and exaggerate the role played by contingency and ignorance in individual political choices, encountered little or no resistance, so to speak. They therefore fell on fertile literary, cultural, and psychological ground. Hence their impact.

Morris's approach has the virtue of underscoring the fact that the political and moral provocations evident in the works of the *mode rétro* writers were not as original as is often assumed, nor as groundbreaking as Rousso's analysis implies. In fact, from early in the postwar period, works by writers as diverse as Jean Genet, Roger Vailland, Marcel Aymé, Roger Nimier, and Jacques Laurent, to mention only a few, anticipated the sacrilegious views of the next generation, while not, of course, robbing the latter of their thunder.

Morris also does a good job of challenging from a literary standpoint Rousso's strict periodization of the postwar memory of Vichy, and of demonstrating that the *mode rétro* writers were not as original in their provocative evocations of the Dark Years as is generally assumed. But his approach (as well as Rousso's) is limited because it fails to take into account a number of considerations and difficulties with which a more comprehensive and exhaustive effort to assess the postwar literary memory of Vichy would need to contend. The first of these is essentially generational. In dividing literary works chronologically and thematically into two broad categories centered on 1970, Morris ignores important generational differences that undoubtedly shaped the writer's work more than these categories allow. For example, where the *mode rétro* writers are concerned, Michel Tournier and Pascal Jardin experienced the war directly as adolescents, whereas Patrick Modiano was born after the war. It is hard to imagine that direct experience of the conflict as opposed to only indirect experience would not have a significant impact on the writer's outlook and work.

Generational considerations, then, point to some shortcomings of chronological and thematic models. But neither generational nor thematic and chronological models can accommodate the cases of important writers who have continued to write about the Occupation over many years and indeed throughout their entire careers. Often, the works of these writers become increasingly self-revelatory and self-referential, following internal, personal trajectories generally impervious to the stimulus of external events. As a consequence, any effort to situate them in particular "generations" of writers or broad chronologies risks distorting their work. As Susan Suleiman writes in her recent book *Crises of Memory and the Second World War*, this is especially the case with such central figures as Jorge Semprun. Semprun's many works—fictional and autobiographical—dealing with his deportation and incarceration at Buchenwald span much of the postwar period and articulate a very personal ongoing struggle on the writer's part to come to terms over time with the trauma of his wartime experience. To abstract any of these works from the internal logic of Semprun's writing and experience and to read them primarily in relation to contemporary external events or the works of a specific generation of writers would be, Suleiman implies, not only to distort each work's meaning but to risk betraying the integrity of Semprun's literary project as a whole.[9]

Apart from these difficulties, there are other obstacles as well that hinder efforts to construct a comprehensive and inclusive chronology and narrative of the literary memory of Vichy. The political climate as well as shifting aesthetic tastes can profoundly affect a work's impact and reception, and the work may therefore be overvalued or underestimated at a given time. Consider the postwar novels of Louis-Ferdinand Céline. In a perceptive article on Céline's postwar trajectory, Nicholas Hewitt points out that the writer's literary eclipse immediately after the war had less to do with the scandal of his anti-Semitism and collaborationism than with "changing attitudes to his writing," and specifically "the reading public's inability to recognize and deal with the formal revolution being undertaken"[10] in such works as *Féerie pour une autre fois* and *Guignol's Band*. By contrast, Céline's popular comeback in the late fifties and early sixties had precisely to do with his willingness to politicize his work by concentrating on his wartime reminiscences and offering scandalously revisionist views of the conflict. According to Hewitt, such works as *D'un château l'autre* and *Nord* achieved notoriety precisely because they catered to "the public's voyeuristic fascination with collaboration."[11] Moreover, in Hewitt's view, these works anticipated the *mode rétro* in their provocative evocations of the ambiguities of the war itself.

In historical terms, what the timing of Céline's postwar success suggests is, first, that where public opinion and certainly literary tastes were concerned, there was not really a "Repressions" phase, as Rousso avers. Moreover, the timing of *D'un château l'autre*, *Nord*, and *Rigodon* underscores fluctuations in Morris's heritage period not yet taken into account. Finally, for Hewitt at least, Céline's trilogy constitutes a direct antecedent to the *mode rétro*, suggesting that the passing of De Gaulle was not as important in this context as is commonly assumed.In turning, finally, to the literary Vichy syndrome since 1990, the cases and issues I've just discussed should serve as so many warning signs that any effort to offer a thorough and comprehensive literary history of the memory of *les années noires*, embracing thematics, generational issues, and a host of other factors, is doomed to frustration. I will therefore not attempt such an exercise here. Just the same, many of the most important and provocative works dealing with the Vichy past

written since 1990 do reveal profound similarities of vision, inspiration, and even structure. These similarities are, I believe, intimately linked to the changing perception and
understanding of the Vichy past during the same period. Here I want to recall the vision
of Vichy that emerged during the trials of Touvier and Papon: primarily judeocentric,
to be sure, but fragmented and subject to conflation with other historical traumas. At
the same time the memory of Vichy, above all in its status as a metaphor for moral and
political evil, becomes more immediate, part of the here and now, an eruption of the
past into the present. In the intensity and even abruptness of its manifestations, moreover, it does not appear to form part of the continuum of the history, and it is not safely
removed from the present by intervening events. The works I want to consider reveal
the Vichy past in this light. They are haunted by it and "occupied" by it.

I will discuss four works by four writers: Marguerite Duras's *La Douleur*, Patrick
Modiano's *Dora Bruder*, Lydie Salvayre's *La Compagnie des spectres*, and Amélie
Nothomb's *Acide sulfurique*. Of the four writers, only Duras experienced the war
directly, serving in the Resistance. But the text I wish to discuss by her, *La Douleur*,
is hardly typical of works written by those who experienced the war and wrote pro-
Resistance texts in the early postwar years. Nor does *La Douleur* mimic the provocative
texts of the later *mode rétro*, although through some of its characters, situations, and
attitudes recall many of these works. In fact *La Douleur*, published in 1985, marks a
transition between the *mode rétro* and the works by Modiano, Salvayre, and Nothomb
that typify the literary memory of Vichy since 1990.

Novel, memoir, or perhaps *autofiction*, *La Douleur* offers a fragmentary and profoundly disturbing account of the writer's experiences in the Resistance during the
war and in awaiting her husband's return from Buchenwald after the Liberation. As
Duras explains in a brief prefatory note to the work, the texts recounting these experiences were discovered in a cupboard by chance some 40 years after they were written.
Duras tells us that she has no memory of having written them, and she has no idea why
she abandoned them in a country house where they risked being destroyed by annual
floods. She acknowledges as well that the effect of re-reading them is overwhelming:
"Comment ai-je pu écrire cette chose que je ne sais encore nommer et qui m'épouvante
quand je la relis?"[12] (How was I able to write this thing that I still don't know what to
call and that terrifies me when I reread it?)

In the same way that the texts making up *La Douleur* force the writer to re-experience,
indeed, to relive the war in the present, so the reader encounters the Occupation and
the writer's experiences and emotions in all their immediacy and urgency. The writer's
insistent use of the present tense almost entirely throughout helps accomplish this aim.
Similarly, just as Duras encounters the past almost as a ghost, detached from her daily
life and the flow of history but that suddenly invades and traumatizes her present existence, so the reader is presented abruptly with the horrors and brutality of the war with
no historical context and no soothing distancing effects in the text. Vichy is a haunted
and haunting past, thrust upon writer and reader alike.

If, as I suggested earlier, *La Douleur* marks a transition in the way the Vichy
past is presented textually, in several of the stories recounted it remains close to
the thematics and sensibility of the *mode rétro*, at least as characterized by Rousso.
The Resistance's torture of an informer in the story of "Albert de Capitales," and the
narrator's reaction to it, cast the Resistance in a far from heroic light. In their brutality in torturing and humiliating the German informant, the Resistance fighters

appear no better than the collaborators. They are in fact perhaps worse, since part of their motivation for being in the Resistance is ostensibly to oppose such practices. Similarly, "Ter le Milicien" is essentially a double of the quintessential *mode rétro* antihero, Louis Malle's Lucien Lacombe, the wily Quercy peasant boy and namesake of Malle's 1974 film. Ter is equally politically naive, equally amoral, and yet equally—and disturbingly—sympathetic. Perhaps more so than even Lucien Lacombe, he typifies the *mode rétro* antiheroes Rousso condemns as "stooges without conscience or morality."[13]

If *La Douleur*, then, remains essentially divided between the *mode rétro* and the literary works of the 1990s, that is, textually allied to the more recent phase in the way in which the Vichy past emerges, but in several instances at least thematically closer to the *mode rétro*, the text that marks the full transition to the latter phase is Patrick Modiano's 1997 text, *Dora Bruder*. Like *La Douleur*, *Dora Bruder* is a hybrid text— part autobiography, part historical document, part philosophical meditation, and part *autofiction*. In simplest terms, the book recounts the writer's effort to reconstruct the wartime destiny of the adolescent Jewish girl Dora Bruder, her flight from a boarding school where she was apparently safely hidden away, and her subsequent arrest and deportation to Auschwitz. Modiano's investigation leads him to an exploration of the monstrous machinery of the Holocaust and especially French complicity in it. In this sense, the book echoes the judeocentric focus of the trials and the broader Vichy syndrome in the 1990s.

But *Dora Bruder* is not only the story of an adolescent girl and the horrific crime to which she fell victim, along with millions of others. The book is also a moving account of the writer's troubled relationship with a father who abused and abandoned him. And it is also a meditation on history, not just the Vichy past but, through allusions to the writer's youth, to the Algerian War of Independence, and, through Dora's father's past, to World War I. What emerges in the work is a vision of history, and especially what began with World War I and passed through Vichy, as a crushing and dehumanizing process which, as Modiano puts it at the end of the book, "sullies and destroys you." To the degree that this history is that of the twentieth century itself, it is, as Alain Finkielkraut has recently described it, "a monster which cannot be incorporated in the continuum of human time."[14] It is a terrifying anomaly that still hovers over the present, and, as Modiano insists, its crimes are as real, immediate, and traumatic as they were at the time of their commission.

To the extent that in *Dora Bruder* "Vichy" comes to designate and encompass this nightmarish history haunting the present, while serving as well as the ineluctable point of origin of Modiano's family trauma, it becomes a metaphor for something much larger and at the same time more intimate than the actual historical moment of France's Occupation by Nazi Germany. In that it embraces these larger connotations, moreover, Vichy cannot be consigned to the past because it exists in and haunts the present. Textually, this presence of Vichy is signaled in a variety of ways and circumstances. For example, the narrator states that on certain evenings, occupied Paris literally emerges in "furtive reflections" in the Paris of today. In another episode the narrator watches a film dating from the Occupation and senses the presence of the gaze of spectators from that time, a gaze that "impregnates" the film and allows the writer directly to experience their feelings and their state of mind. As these and other passages reveal, what began as a straightforward, "historical" inquiry into the fate of a particular person during a specific moment in history turns into something much more complex, and very definitely of the present moment.

Published in the same year as *Dora Bruder,* Lydie Salvayre's 1997 work *La Compagine des spectres* differs from both Modiano's text and *La Douleur* in that it is not a hybrid text but explicitly a novel. Moreover, the author did not experience the Occupation directly, as did Duras, nor as an integral part of a fractured and troubled family past, as did Modiano. In fact, Salvayre's personal links to historical trauma are less French-specific than Spanish, since her parents were Spanish Republican refugees.

Despite the fact that Salvayre comes to Vichy by a very different route from that of Modiano and Duras, the textual evocation of that past in the novel strongly resembles its representation in *La Douleur* and *Dora Bruder.* Vichy emerges and indeed erupts into the present in the reminiscences and outbursts of the crazed mother Rose, who experienced the Occupation as a child and who now regales her daughter Louisiane and the process server, *Maître* Échinard (there to evict her and her daughter from their apartment), with horrific accounts of her family's suffering during the war. Rose also hurls invective at *Maître* Échinard, accusing him of being Joseph Darnand, head of Vichy's *milice.* She rails at Pétain, whom she refers to as "Putain." These are not, however, simply the ravings of a madwoman. As Louisiane explains, her mother literally lives during the Occupation, or, more accurately, the Dark Years are for her completely intertwined with the present. Rose's "atemporal" mind is constantly shuttling back and forth between the year 1943 and the 1990s. Nineteen forty-three, we learn, is the year when her brother was murdered by the *milice.*

Toward the end of the novel, we also learn that Rose has not always been obsessed with Vichy. In fact, the Dark Years began to overwhelm and haunt her present at precisely the moment she became aware of the full extent of French criminality and specifically Vichy's complicity in the Nazi Final Solution. Louisiane reveals that in 1978 Rose read the famous article in *L'Express* that exposed René Bousquet's responsibility for the infamous 1942 *Vel d'Hiv* round-ups, when French police rounded up some 12,000 Jews in Paris. The Jews were incarcerated in brutal conditions, and most were deported to their deaths at Auschwitz. On learning of Bousquet's crime, Rose decided that in order to suppress "*la gangrène du monde*" and the stink it propogates,[15] it was necessary to kill Bousquet, and it was her failure to accomplish this aim that triggered her current delusional state.

Despite Rose's apparent insanity, the power, lucidity, and brutal eloquence of her recollections and outbursts against Vichy criminality are such that they come to dominate the novel completely. They literally overwhelm Rose's life and then her daughter's. More than that, they serve as a kind of mirror or echo chamber for the present, which reflects and amplifies the sinister, cold, and inhuman bureaucracy *Maître* Échinard represents, as well as the modern crimes and genocides that devastate the world of the present, from Algeria to Rwanda. For her part Rose believes in ghosts, but ghosts of a special kind. Explaining her theory of ghosts to her daughter, Rose affirms that the ghosts that haunt the present "sont les morts assassinés par Putain qui ressuscitent et viennent nous regarder vivre" (the dead murdered by Putain who return to life and come watch us live). They also go "là où la mort pue, et la mort pue en maints endroits de la planète" (where death stinks, and death stinks on the planet in many places).[16] Through its ghosts, then, Vichy comes not only to haunt the present, but to define or designate it in all its horror. In the final analysis, it would appear, Vichy is synonymous with Evil and death itself.

If, through their modes of representation as well as their evocations of a Vichy past very much implicated in the present, *Dora Bruder* and *La Compagnie des spectres* define a particular form of literary response to the memory of Vichy that begins to emerge in the 1980s and becomes quite visible in the 1990s, what of the new century and indeed the present? The last work I would like to discuss in this context is Amélie Nothomb's highly provocative 2005 novel, *Acide sulfurique*. A bitter satire of global culture's fascination with television "reality shows," *Acide sulfurique* describes in harrowing detail a new show, called simply *Concentration*. The participants, rounded up randomly in the streets of Paris, are thrust into a televised concentration camp, authentic in all its horrific details, and from which prisoners emerge only after their deaths. The show is wildly successful, and the sufferings and even the deaths of the inmates are all part of the public's viewing experience. Television viewers are in fact allowed to vote on who lives and who dies.

In *Acide sulfurique* there are few explicit references to Vichy or, for that matter, to the Nazi camps themselves. But the fact that the novel takes place in France and that the inmates are French leaves little doubt as to the actual historical reference. In fact, no reference is really necessary. *Acide sulfurique* operates a perfect fusion of the nightmare of the Vichy past, already increasingly subsumed in the Nazi past and the horror of the Genocide of the Jews, and what is for Nothomb at least the horror of our cultural present. In the process, democracy and totalitarianism become indistinguishable, reality and virtual reality collapse into one another, and death itself is reduced to pure spectacle.

If *Acide sulférique* is indeed the most recent manifestation of the literary Vichy syndrome since 1990 I have been discussing here, then both thematically and structurally it takes us in important ways beyond works like *Dora Bruder* and *La Compagnie des spectres*. In Nothomb's novel, Vichy and the nightmare it represents do not erupt in the present because they are the present, merely dressed up in modern clothes. For Nothomb, Vichy has become our destination now.

Conclusion

By way of concluding, I would like to return to the Papon trial and a provocative essay written after its conclusion by Jean de Maillard entitled "À quoi sert le procès Papon?"[17] In his essay Maillard makes two claims that help illuminate the works characterizing the literary Vichy syndrome since 1990, and that allow us to draw some conclusions here. First, Maillard asserts that Vichy and even Nazism itself were, in the final analysis, not the real stakes of the trial. Rather, they were conveniently diabolical symbols for the real culprit, the real accused, which is the nation-state and political power itself. In the brave new world of genocides, globalization, the death of reassuring progressive ideologies, and the isolation and devaluation of the individual, a new "collective unconscious" is being formed, according to which the nation-state incarnates nothing less than an omnipresent and ever-menacing evil.

While the works by Modiano, Salvayre, and Nothomb discussed here are obviously also and primarily about Vichy and Nazism, the taint and the blight that Vichy and Nazism represent bleed over into the present, and, in the process, the present is revealed in many ways to be no less menacing than the past to which it is intimately linked. In Nothomb's novel, the game show *Concentration*, whose so-called organizers clearly enjoy absolute political power, is as dehumanizing and murderous as the Nazi death camps themselves. Reference to the past in *Acide sulfurique*, in fact, simply helps to understand better the horrors of the present. In *La Compagnie des spectres*, the delusional Rose likens the process server *Maître* Échinard to Joseph Darnand, thereby linking Échinard to the brutalities of the Putain regime. But the comparison itself also allows Rose to put a face on the bureaucratic inhumanity that threatens and humiliates her in the present. Even in *Dora Bruder*, access to the horrors of Vichy and the Holocaust itself are filtered through references to the postwar world and to postwar political and institutional power. For Modiano, the latter are on occasion almost equally menacing in their capacity for inhumanity and oppression as their Vichy and Nazi predecessors.

Maillard's second observation is essentially a corollary of the first. One of the cherished assumptions of the Papon trial was that understanding the past, and the factors that made Nazism possible, would help prevent crimes like the Holocaust from repeating themselves in the present and future. Failing that, at least remembering the crimes and their victims would accomplish the same purpose. But, Maillard writes, both of these assumptions were sadly and indeed tragically misguided. Six months of endless legal proceedings did not and could not make it possible to foresee or prevent similar crimes. Maillard concludes therefore that humanity is doomed to re-experience "all those horrific adventures against which we sought to fortify ourselves."[18]

It is precisely this intuition, I believe, that lies at the heart of the works discussed here. The anguished invocation and interrogation of the Vichy and Nazi crimes, the haunting sense of their *actualité*, and their linkage to crimes, real or imaginary, in the present point to the fragile hope that the past can perhaps illuminate, educate, and hopefully prevent future crimes. But the works also testify to the simultaneous recognition that this hope is delusional, futile. It is this paradoxical intuition that gives these works their tragic power, and makes them perhaps as much about the present and future as about the Vichy past.

Notes

1. Robert Paxton, "Symposium on Mitterrand's Past," *French Politics and Society* 13:1 (1995): 19.
2. For a detailed account of Touvier's life and the legal case against him, see the introduction to my *Memory, the Holocaust, and French Justice: The Bousquet and Touvier Affairs* (Hanover, NH: University Press of New England, 1996), 1–49.
3. For an account of Papon's career, his trial, and the controversy surrounding his role as Prefect of Paris Police in suppressing the Algerian demonstrations in October 1961, see the introduction to my *The Papon Affair: Memory and Justice on Trial* (New York: Routledge, 2000), 1–34.
4. Conan and Rousso entitled their book on the scandals surrounding the Vichy past in the 1990s *Vichy, un passé qui ne passe pas*. The book has been translated into English as *Vichy: An Everpresent Past*, Nathan Bracher, trans. (Hanover, NH: University Press of New England, 1998).
5. Henry Rousso, *Vichy: l'événement, la mémoire, l'histoire* (Paris: Gallimard, 2001), 43.

6. Éric Conan, "Enquête sur le retour d'une idéologie," *L'Express,* July 17–23, 1992, 27.
7. Henry Rousso, *The Vichy Syndrome: History and Memory in France since 1944,* Arthur Goldhammer, trans. (Cambridge, MA: Harvard University Press, 1991), 129.
8. Alan Morris, *Collaboration and Resistance Revisited: Writers and the 'Mode Rétro' in France* (New York: Berg, 1992).
9. Susan Suleiman, *Crises of Memory and the Second World War* (Cambridge, MA: Harvard University Press, 2006).
10. Nicholas Hewitt, "Céline: The Success of the *monstre sacré* in Postwar France," *Substance* 102 (2003): 34.
11. Hewitt, 35.
12. Marguerite Duras, *La Douleur* (Paris: P.O.L., 1985), 10.
13. Henry Rousso, *The Vichy Syndrome,* 129.
14. Alain Finkielkraut, *Nous autres, modernes* (Paris: Ellipses, 2005), 192.
15. Lydie Salvayre, *La Compagnie des spectres* (Paris: Seuil 1997), 157.
16. Salvayre, 150.
17. Jean de Maillard, "À quoi sert le procès Papon?" *Le Débat* 101 (September–October 1998), 32–42.
18. Maillard, 38.

Part III

War Imagined

Newspapers, Novels, and the Comic Book War

Libby Murphy

Henri Barbusse, like other journalists and publicists, was critical of his colleagues in the press corps during World War I. At the same time, like so many soldier-novelists, Barbusse was an avid reader of newspapers. His letters to his wife often included annotated newspaper clippings, with images and articles from the daily papers serving as points of departure for his own descriptions of life at the front.[1] In one letter, Barbusse took this dialogue with the press one step farther, describing a visit to the trenches by a group of journalists and dignitaries made, as always, under the watchful eye of a commanding officer. "Yesterday, in the trenches," he wrote to his wife, "we had a visit from some journalists led by officers of the general staff...it is with a certain irony and even contempt that we [soldiers] look upon these tourists of the trenches."[2]

This anecdote, first documented in his letters home, would continue to preoccupy Barbusse. In his 1916 serial novel *Le Feu*, he would take his revenge on the "tourists of the trenches," having his soldier-protagonists react to a similar frontline visit. In this often-cited scene, as the group of visitors approach, one soldier, Barque, mutters under his breath, "It's the tourists of the trenches," before hamming it up for his comrades crowded around him: "This way, ladies and gentleman!" One of the visitors, an artist, approaches the group "rather timidly, as one would at the zoo in the *Jardin d'Acclimatation*," after exclaiming with a hint of disbelief, "Oh! Oh!...*poilus*...[Real] live *poilus*." As the "tourists" move out of sight, the soldiers overhear the officer turned tour guide address some of the visitors as "messieurs les journalistes."[3] After Marthereau asks if the journalists are the ones who "stuff [our] skulls," Barque returns to his antics, parodying a far-fetched newspaper article in a pastiche that delights his fellow soldiers.

In this short scene, Barbusse posits a disconnect between journalists and soldiers with respect to representations of the war. But he also points to a sense of alienation felt by his soldier-protagonists from their own lived experience. Barbusse's soldier-protagonists are emptied of their humanity by visitors who treat them like animals in a zoo. Indeed, they see their very reality called into question by civilians more familiar with caricatures of them in print media than with them as human beings and as historical actors.

A similarly ironic moment of recognition occurs toward the end of *Le Feu*, when the few remaining members of the original squad go on leave. The soldiers stand in dismay before a shop window where a pitifully childish war scene has been staged using mannequins and toy guns and featuring a kneeling German soldier surrendering to a baby-faced French officer. An elegant lady, having taken a minute to appreciate the shop window, approaches the group, asking for confirmation. "Tell me, you *messieurs* who are real soldiers from the front lines, you have seen that kind of thing in the trenches, right?"[4] The delight the small crowd of civilians derives from the stunned but affirmative responses made by the soldiers is matched in the following scene by that of a café-goer who, spying the soldiers ordering drinks, exclaims, "Real live *poilus!*"[5] These encounters between the front and the home front, Barbusse suggests, can only confirm for civilians the resemblance of the real, historical soldier with his ubiquitous comic book copy, and not the other way around. Next to his comic book counterpart, the real poilu fades into a state of invisibility and complete irrelevance.

This fading into fiction described by Barbusse can be understood as a kind of Debordian "recession" of soldiers "into representations" of themselves over which they have no control.[6] Of course these soldier-protagonists are themselves already representations, products of Barbusse's observations and imagination. But Barbusse's soldier-protagonists achieve the consistence and authority of real historical actors precisely by virtue of this staged encounter between two different levels of fictionality (figure 14.1).

Tropes like Barbusse's "tourists of the trenches" and "soldiers in the shop window" point to an epistemological rift between the front and the home front. By problematizing the authenticity of noncombatant war stories, these texts establish a hierarchy of legitimacy and of truth-telling media, placing themselves firmly on top.

The "trench tourist" and "soldiers in the shop window" scenes likewise point to the increasingly gendered topography of knowledge construction articulated in texts—written by soldiers but also by civilians—that attempt to showcase their own authenticity.[7] Those who fundamentally misapprehend the "true" war experience—that of the frontline soldiers—are gendered as feminine. In Barbusse's scenes, one of the trench tourists, for example, is an "artist," effeminated by his "floppy hat" and "floating tie."[8] The artist, the poet, the neurasthenic, and the dandy are all subcategories of the *embusqué* or shirker, a character type circulating in Great War culture and routinely marked as feminine: frail, timid, overly concerned about his appearance. Alongside the shirker, the pot-bellied civil or civilian is likewise routinely characterized as emasculated or hen-pecked. Meanwhile the elegant lady in the "soldiers in the shop window scene" is just one of the many ignorant or indifferent female characters used in Great War texts by soldiers and civilians alike to represent civilian incomprehension.[9] The rift between front and home front thus maps neatly with the fact-fiction divide characterizing so much of Great War writing, with fact being imagined as resolutely masculine and fiction as feminine.

This gendered conceptual framework reappears in Barbusse's war diary, which like his letters to his wife, contains notes and observations for *Le Feu*. In one fragment, Barbusse lays out his intention to offer an antidote to the phony war peddled in mass culture: "Go after sappy descriptions.... Déroulède bullshit and Detaille

Figure 14.1   "Journalists at the French Front Listening to Explanations of a French Staff Officer," from the British weekly magazine *The Times History and Encyclopedia of the War.*

paintings... shirking camouflage artist. He paints socialite portraits with cold creams, rouge, pomade Rosa and face powder."[10] The naturalistic, photo-realistic aesthetic Barbusse would employ in *Le Feu* would remasculinize a war story co-opted by the make-up artists of the comic book war. Of course this aesthetic would also go a long way toward camouflaging the traces of *Le Feu*'s own literariness and indeed artificiality.

Barbusse was well-placed to appreciate the irony of the "tourists of the trenches" and to critique the Déroulèdian "bullshit" of the mass press, since he was himself a huge name in the world of French journalism. Barbusse's journalistic career had ridden the wave of the prewar press boom in France that had begun in the 1880s and that was still going strong on the eve of the First World War. The biggest innovation of this prewar press boom had been the precipitous rise of the mass marketed and mass distributed *journal d'information*, dominated by four huge Parisian dailies—*Le Petit Parisien, Le Petit Journal, Le Journal,* and *Le Matin*—selling an average of over a million copies apiece.[11] Of these four, *Le Petit Parisien* and *Le Petit Journal* had a truly national distribution, while *Le Journal* and *Le Matin* catered to a more Parisian and more uniformly middle class readership.[12]

Despite the mobilization of personnel, the interruption of distribution networks, paper shortages, and rising prices, the four big Parisian dailies continued to thrive during the war years with sales figures near their prewar averages.[13] These four papers worked closely with the War Ministry's Press Bureau, which tightly controlled the flow of information to and from the front. In their coverage of the war, the big dailies largely followed the tone set by the government and the army, echoing the nationalistic high diction of the *union sacrée*.[14] In the absence of war correspondents with direct, uncensored access to the frontline soldiers, most papers kept at least one former army officer on the payroll to provide "expert" commentary on the laconic daily *communiqués*.[15] The War Ministry authorized the four big dailies and the *Écho de Paris*, the preferred paper of the general staff, for distribution at the front.[16] Other papers, particularly smaller *journaux d'opinion*, and later the nonconformist or oppositional press, were not banned, but they could reach the front only if mailed directly to a soldier by his family or friends.[17]

Barbusse's decision to publish *Le Feu* at *L'Œuvre* is thus highly significant. As we will see, the publishing of serial novels was a time-tested marketing strategy of the French press, and one that served the papers well during the war years. What is remarkable about Barbusse's serial is that it positions a critique of the mass press precisely within the very medium being critiqued: the newspaper. Equally significant is Barbusse's decision to publish his bombshell, not in the paper with which he was under contract, the huge and highly innovative daily *Le Matin*, but at *L'Œuvre*, a left-leaning, anticonformist opinion paper with daily wartime sales figures one-tenth of those of *Le Matin*.

L'Œuvre did not enjoy the huge following—or resources—of mainstream papers like *Le Matin*, but it did see its sales figures rise from roughly 55,000 copies when it was relaunched as a daily in September 1915 to 125,000 copies by 1917.[18] While a precise breakdown of *L'Œuvre*'s readership is difficult to establish, there is reason to believe that it reached a similar, though wider, audience to that of *Le Canard enchaîné* and other papers considered by contemporaries as belonging to the same family of publications.[19] *Le Canard enchaîné* was printed on *L'Œuvre*'s presses and acknowledged its alliance to its sister publication by including it in its list of "patriotic papers."[20] The two papers employed many of the same writers, including Georges de la Fouchardière and Victor Snell, and even worked in tandem to get certain articles past the censors.[21]

L'Œuvre's influence and notoriety far exceeded its sales figures. Its editor-in-chief, Gustave Téry, had worked for *Le Journal* and *Le Matin* before founding *L'Œuvre* as a weekly in 1904. Like most papers, *L'Œuvre* shut down operations in August of 1914. When it resumed publication as a daily on September 10, 1915, it set an aggressive new tone and positioned itself in opposition to the mainstream national press. "Imbeciles," *L'Œuvre*'s famous tag line insisted, "don't read *L'Œuvre*."[22] *L'Œuvre*'s campaign against the censors, its almost daily attacks on *L'Action française*, and its unique serial novels would contribute to its growing success and growing reputation.

Early in the war *L'Œuvre*, *Le Canard enchaîné*, and *Le Crapouillot* ran satirical spoofs of the serial novels published in the big Parisian dailies. In the mainstream press, stars of the prewar serial novel continued to churn out far-flung fictions from behind the lines, reworking time-tested plot lines and enlisting prewar heroes for the epic struggle against the dreaded *boche*. Arsène Lupin, Maurice Leblanc's gentleman burglar, Joseph Rouletabille, Gaston Leroux's globe-trotting reporter, and the various officers of Jules Mary's patriotic military novels were all heroes from prewar popular fiction mobilized during the war in serial novels designed to sell readers fantasies of revenge and retribution. The frustrations of industrial warfare—anonymity, alienation, and emasculation—are kept at bay in these texts and many others like them that recast a global war of men and material as a duel between France and Germany and that depict the war as a series of adventures in which human agency, energy, and will dominate.

In Leblanc's *L'Éclat d'obus*, for example, the world conflict is reduced to a battle of the wills between the selfless and intrepid French Sergeant Paul Delroze and Hermine d'Andeville (aka Major Hermann), a diabolical cross-dressing female spy and handmaiden to the Kaiser. In this text, as in Leblanc's later *Le Triangle d'or*, the energy and intelligence of the novels' soldier-heroes are backed up by the interventions of the omnipotent Arsène Lupin, who returns to serve his country "in [his] own way."[23] Around the same time that Arsène Lupin was unraveling the mystery of the *Triangle d'or* in *Le Journal*, Gaston Leroux's star reporter and amateur detective Joseph Rouletabille had been called back from the trenches to save Paris from the menace of a super canon fabricated in the German factory town of Essen.[24] Lupin and Rouletabille thus fed a larger fantasy cultivated in such novels of a benevolent, God-like force working behind the scenes for the good of *la patrie*.

L'Œuvre's first wartime serial novel, written by journalist and humorist Georges de la Fouchardière, would be decidedly different in tone and content from these heroic fictions. *L'Araignée du Kaiser* pokes fun at the spy novel and the science fiction novel. The text features a den of German spies on the *Côte d'Azur*, mad scientists, fanciful inventions, biological warfare, and an insane Kaiser.[25] The far-fetched adventure plot is led by the unwitting hero Boulot, who would become a stock figure in La Fouchardière's repertoire. Boulot, the Parisian *mécano*, loudmouth, and drunkard-turned-hero, saves France from the weapons of mass destruction *L'Araignée* and *La Guêpe*. He then heads off for the trenches, where the role of poilu fits him like a glove. The story ends with Boulot, the *boute-en-train* (entertainer) of his company, telling tall tales of his adventures to his fellow soldiers.

Boulot is an early iteration of La Fouchardière's Alfred Bicard, aka "Le Bouif" (cobbler), a character who would grace the pages of *Le Canard enchaîné* in a series of editorials, "Les Propos du Bouif," that would continue well into the postwar period. The same year that Alfred Bicard made his debut in *Le Canard enchaîné*, he also renounced his prewar antimilitarism, dragged himself away from his beloved bistro, and marched off to war in the dime novel *Bicard dit le Bouif, poilu de 2e classe*.[26] Like Boulot before him, Bicard is a picaresque figure whose primary concern is not with saving his country, but with saving his own hide.[27]

The tone and substance of press coverage of the war would be in a constant state of evolution during the war years, with editors, journalists, and serial novelists revising their strategies as public opinion changed and as competing representations of the war took hold. The poilu type embodied in Bicard and Boulot, for example, originated not in the satirical press but in *Le Journal*, the same paper that brought readers the debonair and omnipotent Arsène Lupin and his heroic poilu partners. In the spring of 1915, from his hospital bed in Tours, soldier novelist René Benjamin sent his series of sketches *Les Parisiens à la guerre*, later published under the name of its title character *Gaspard*, to *Le Journal*'s literary director Henri de Régnier. Benjamin was among the first soldier-writers recruited by the mass press to provide stories about the war, but he would certainly not be the last. Newspapers, such as *L'Œuvre* in the case of Henri Barbusse, and even some book publishers, such as Hachette in the case of Maurice Genevoix, began making deals with soldier-writers almost as soon as the war began.

Gaspard is important because, with the creation of his plucky Parisian snail merchant turned poilu, René Benjamin launched the first significant literary volley in what would be one of the most heated cultural battles of the war years. Debate over poilu iconography continued throughout the war in newspaper articles, pamphlets, novels, plays, posters, cartoons, postcards, and many other media. Perhaps no single subject was more widely discussed than the "légende du poilu," and no problem considered more vexing than the appropriate way to represent the French infantry soldier: as larger than life Superman or as down and out Everyman, as hero or hot-head, warrior or quick wit, patriot or *pícaro*. (See figure 14.2.)

While such writers as Georges de la Fouchardière and René Benjamin constructed a new poilu-type to replace the traditional one offered by Leblanc or Leroux, others attempted to steer public debate away from typologies altogether. In his 1916 study *L'Armée de la guerre*, Louis Thomas, writing under the pseudonym Capitaine Z..., argues, for example, "today there are fifty different kinds of soldiers, and it would be silly to try to reduce them all to the type of 'poilu' so dear to reporters short of copy."[28] Thomas goes on to provide a catalog of what are, nonetheless, a series of soldier types. (See figure. 14.3.)

If studies of soldier character types achieved such prominence during the war, it is in part because of the extraordinary column space accorded in newspapers, and most notably via the serial novel, to the articulation of these types. It might come as a surprise to readers of modern newspapers to find an entire fourth of a page, the *rez-de-chaussée*, of the best-selling French wartime newspapers devoted to serial novels, especially considering these papers were forced to cut back to a four- and sometimes two-page format during the war. In some cases, two serial novels were run at once, increasing even more the percentage of column space dedicated to fiction in a medium that aspired to be the nation's number one source of facts about the war.[29]

Combating the comic book war was no simple task, and more often than not, it turns out, the terrain separating the types articulated by soldier-writers at the front

Figure 14.2   Cartoon from the article "Civilians in the Trenches" in the August 1, 1916, issue of the French humor magazine *Fantasio*.

Figure 14.3 "Glory to our brave poilus," *Image d'Épinal* published by Pellerin & Cie and featuring poilu types from all walks of life.

and those co-constructed by novelists and newspaper publicists behind the lines was no more substantial than a rhetorical *mise en garde*.[30] The French literary imagination's obsession with typologies, its go-to strategy for making sense of—or at least for attenuating—the anxieties of cataclysmic social change, had followed the newspaper and the novel into the First World War. The poilu would be just one type in a whole cast of Great War characters collectively constructed in popular media. Together, newspapers and novels isolated a series of types and a set of situations that could be talked about by groups having diverse experiences of the war. The same characters described verbally in newspaper articles and novels were described visually in cartoons, postcards, ads, illustrations, and films circulating in war culture.

The poilu (French infantry soldier), the *civil* (civilian), the *embusqué* (shirker), *la Parisienne* (frivolous Parisian woman), the *tirailleur sénégalais* (colonial soldier), the *boche* (German), the *marraine* and *filleul* (war godmother and godson) were among the many character types constructed in the mass press, in novels, and in visual culture. If we take a step back and look at this "printed digest" (to borrow Vanessa Schwartz's term) of the war in the aggregate, we can better appreciate the anxieties surrounding the comic book war: a version of the war drawn in bold, repetitive, recognizable strokes, whose aim is to make the evermore complex war more manageable, by imposing structure on the war story and isolating a set of common denominators.

The Great War comic book aesthetic functioned both verbally and visually. Advancements in print technology—including photomechanical reproduction techniques, innovations in typography, and improvements in color lithography—all contributed to the intensification of image-based or image-heavy print media immediately before the war. Prewar transformations in the illustrated press would find new impetus during the war, with newspapers using images as they had traditionally used serial novels, as marketing strategies to draw in and keep readers, and, during the war, as fillers for the wartime information gap.

Some of the best-known commercial artists and illustrators of the time were regular contributors to illustrated magazines,[31] including humor magazines like *Le Rire rouge* and *La Baïonnette,* which also contained short stories and commentary from mobilized writers and artists. The trench paper *Le Crapouillot* took some of these illustrators to task, echoing in many ways Barbusse's critique of the "trench tourists" and their armchair war stories. An article entitled "Les Dessins de guerre" in the April 1917 issue of *Le Crapouillot,* for example, lambastes civilian caricaturists for their treatment of the French soldier and promises that *Le Crapouillot* will publish only authentic drawings created by frontline soldiers. In a description that recalls Barbusse's attack on Déroulèdian "bullshit," the article's author Le Rousseur rails against "the idiotic caricature of the Poilu: bearded, shaggy-haired, jovial, who treats war like a sport and plays cards under a deluge of 420 millimeter shells, the blissful Poilu who never worries, sings while he goes over the top and dies with a satisfied grin on his face."[32]

The same kinds of critiques applied to the new medium supplanting illustrations in print media: photography. *L'Illustration,* a longtime leader in the French illustrated press and a common target for soldier satire, found itself in competition during the war with newly created photographic magazines such as *J'ai vu, Sur le vif,* and *Le Miroir,* whose famous tag line was "*Le Miroir* pays whatever the price for photographic documents relating to the war that are of special interest."[33] (See figure 14.4.)

Cinquième année. — N° 58.   Le Numéro : **25** centimes.   DIMANCHE 3 Janvier 1915.

LE MIROIR

PUBLICATION HEBDOMADAIRE, 18, Rue d'Enghien, PARIS

LE MIROIR paie n'importe quel prix les documents photographiques relatifs à la guerre, présentant un intérêt particulier.

BRANCARDIERS TRANSPORTANT UN BLESSÉ SOUS LE FEU DE L'ENNEMI

Bien que non-combattants, les brancardiers sont loin d'être des embusqués. C'est le plus souvent sous la mitraille qu'ils vont effectuer la " relève ", et beaucoup tombent au champ d'honneur.

Figure 14.4   "Stretcher-Bearers Transporting a Wounded Soldier under Enemy Fire," from the January 3, 1915, issue of *Le Miroir.*

Picture taking, like trench tourism, became yet another theme soldier-writers could exploit in their attempts to reestablish the fading boundaries between fact and fiction. In a scene in his 1919 novel *Clavel soldat*, for example, Léon Werth calls attention to the artificiality and manipulability of the photographic medium. In this scene, a stretcher bearer who has made a cottage industry out of selling photos of poilus and their camps decides to expand his offerings to include "a more warlike scene: a convoy of wounded soldiers at Xivray." The trouble is that the wounded are generally evacuated at night. So the stretcher bearer brings in stretchers and enlists loitering soldiers to play the parts: "And the fake casualties and their fake stretcher bearers look straight into the camera."[34]

Staged photographs from the front, postcards, maps, toys, board games, commemorative plates, dioramas, military-inspired clothing patterns, advertisements—all of these visual representations of the war were available at affordable prices for consumption alongside the countless sketches, drawings, illustrations, and caricatures that filled newspaper pages, novels, magazines, and albums. Soldier-writers of the First World War like Léon Werth understood that people wanted to see the war as much as they wanted to read about it, and they were well aware of the ways in which readers' imaginations were affected by images of all sorts.

Soldier-writers understood the inherent dangers of the verbal and visual comic book aesthetic—the Great War as fiction or as spectacle. They expressed most keenly the anxieties surfacing in many different texts and contexts over not just the war itself but also the production of knowledge about the war. The comic book war, soldier-writers increasingly worried, would keep public debate from confronting some of the harshest realities of the war. A great deal of energy and column space was being spent talking about how the relatively limited set of types and experiences identified in war culture should be represented and understood. The messiness, confusion, and excess of processable detail of the real, historical war and its mass of individual participants were being replaced in the media by a streamlined and exaggerated set of types and situations.

For soldier-writers, newspaper publicists, and readers alike, coming to terms with the war would mean coming to terms with its means of representation, coming to terms with a modern, media situation in which the categories of fact and fiction threatened to collapse. The Great War would reveal the limits not just of literary typology, but of the literary in general. What many readers and writers of the Great War wanted was information, not imagination, facts and not fiction. The time-honored French typologizing impulse would ultimately buckle under the dizzying complexity and horrifying realities of the war, realities that budding practices of censorship and propaganda could not even begin to keep under wraps, given, among other things, the ubiquity in city streets and village squares of widows, orphans, amputees, mutilated and blinded soldiers and *gueules cassées* (the disfigured). These historical actors would be flesh and blood reminders of the limits and indeed of the dangers of the comic book war.

For Barbusse, the "tourists of the trenches" and civilians were to blame for the breakdown of factuality as a conceptual category during the war. They had robbed the soldier of his voice and of his powers as storyteller. Barbusse would be just one of many wartime novelists to call into question the possibility of writing history and of authenticating experience under the regime of the comic book war. In a chapter of *Le Feu* entitled "La Grande Colère," the soldier Volpatte recounts his convalescent leave behind the frontlines. The poilu's story, Volpatte argues, is being co-opted by civilians and shirkers behind the lines. So well do civilians profess to know the reality of this war, Volpatte maintains, that after the war the soldier's story of his experiences will count for nothing. "[When] you come back, if you come back," he insists, "this bunch of jokers will say you're the one who's got it all wrong, with your little truth about the war."[35] Representations circulating in the comic book war, Volpatte predicts, will trump lived experience in the historical war.

In *Clavel soldat*, Léon Werth pushes the analysis of the breakdown of factuality a step farther. Like Barbusse, he insists upon the rift between the front and the home front and decries civilian incomprehension and insensitivity. Werth's bitter and disillusioned main character André Clavel compares civilians to spectators and the war to a giant circus. The only difference is that circus goers are far less bloodthirsty than civilian spectators. While the former will sometimes cry out "Enough," Clavel bitterly muses, the latter cry out "More."[36] Clavel echoes Barbusse's "tourists of the trenches" comparison and Volpatte's complaints that civilians claim to know the war better than soldiers: "They are informed by the newspapers and by films, with their special effects. And they go there, too, to the war. People organize excursions to take them there."[37] Civilians at the circus and on excursions to the trenches fit into an elaborate system of metaphors of spectacle, theatricality, and artificiality used by Werth throughout *Clavel soldat*, including hundreds of references to popular imagery and songs, history painting, postcards, and toy soldiers. These metaphors serve to mark Clavel's growing alienation from the war and from himself and to describe his losing battle against the comic book war.[38]

"How is it," the narrator asks as the novel opens, "that André Clavel has ideas so different from those of *Le Petit Parisien* or the *Revue des Deux Mondes*?" The answer, the narrator suggests, lies in Clavel's "sensibility," which tends toward "pity," and in his robust critical mind, which resists any attempt to create an "artificial order" by recycling a previous one.[39] For a time, Clavel is lucid enough to observe his fellow soldiers as they are slowly absorbed into the shadowy irreality of the comic book war. Their capacity for independent thinking and their very mastery of language are eroded, finishing in "this kind of torpor," out of which the only distinguishable sounds are those of "moans, snippets from newspapers, and song lyrics."[40] So contaminated are the men's minds by the media they've absorbed that they tell personal stories based on "what they've read" or on "stories from the newspapers."[41] The war experience so confounds any separation between fact and fiction that soldiers often can't even distinguish, Clavel observes, "between what they've seen and what they've been told."[42]

Over time, the very mythologizing impulse that Clavel so decries begins to creep into his own imagination. He begins to see reality through eyes trained by the comic book war. One rainy night, for example, he pauses to reflect on the sight of no-man's land. While the truth of this image is clear in Clavel's mind, he can also understand the impulse to translate this image for the comic book war. "[The] truth is that this is as boring as a missed bus," Clavel muses, but it would make a "great *tableau* for one of those war diaries they publish in right-thinking magazines."[43] Throughout the novel Clavel fights against the brain-numbing seduction of the comic book war, with its ready-made myths and legends. By the end of the novel, however, he has been completely worn down. Giving up on any attempt to tell the truth about the war, he finds himself "ready to repeat any old story from the newspapers."[44]

Toward the end of *Clavel soldat*, André Clavel, reflecting on the clichés and epithets bandied about in the comic book war, poses a question that newspaper publicists like Gustave Téry at *L'Œuvre* had already taken to heart: "À quand le réel?"[45] When and how would there be a return to reality? How would the conceptual framework of factuality ever be restored? What was the exit strategy for the comic book war?

Soldier-narratives like those of Barbusse, Dorgelès, or Werth, with their newspaper-reading scenes, their tourists of the trenches, and their questioning of the status of story-telling, went a long way toward theorizing the inadequacies of French wartime journalism. The initial gap between the papers' understanding of the war and the horrific realities of the war was so traumatic that writers like André Maurois and Léon Werth near the end of the war, and Céline and Jules Romains well after it, would still use the press and Barbusse's tourists of the trenches as shorthand markers for all that was ironic but also uniquely modern about the war. The comic book war revealed the limits of literary typology, and indeed of the literary, in making sense of cataclysmic change. In order for public confidence in journalism to be restored, there would have to be a new division of labor. The novel and the newspaper, two media with a long tradition of interdependence and a proven track record of lucrative symbiosis, would eventually have to part company.

The reform of the serial novel undertaken during the war at such papers as *L'Œuvre* would be an important gesture toward significantly changing French journalistic practice. What *L'Œuvre*'s editor-in-chief Gustave Téry had in mind was a masculinization of the serial novel that would be an important step toward recalibrating the fact to fiction ratio in the modern French newspaper. On April 9, 1916, *L'Œuvre* ran an article entitled "Littérature" critiquing the serial novel. The article took the form of a conversation about literature between two cultural authorities, the author, a humble noncommissioned officer, and a wounded officer, *le général blessé*. In this article, the Wounded General laments the fact that French letters have yet to find the right tone for talking about the war, and he points to the inadequacy of prewar narrative patterns for dealing with the war as a literary subject. For the Wounded General, French letters have been emasculated by the dominance of sentimental novels and are thus grossly incapable of rising to the challenge that the hyper-masculine *culture de guerre* poses. "Today," he insists, "it's all about tobacco, cheap wine, face smashing and hell's bells. So, your literature can't hack it, and it's shitting itself, if I dare say it like that."[46] By the time this article was published, *L'Œuvre* had already taken steps to remedy the sad state of the serial novel. The Wounded General's comments can be seen, then, as just one part of a general campaign to prepare readers for the reforms *L'Œuvre*'s editor-in-chief Gustave Téry had in mind, that is, to prepare them for the bombshell of Barbusse's *Le Feu*.

In late December, 1916, just after the awarding of the Goncourt Prize, *L'Œuvre* would congratulate itself on the success of *Le Feu* and lay out more explicitly the program for serial fiction reform that it had, in fact, begun in the early months of the war. In an article introducing the paper's next short story, "Le Nécessaire et le superflu," Téry gives his readers credit for recognizing all along the paper's self-proclaimed mission to "hone and test the critical thinking skills of its readers." As a part of this larger mission, Téry concedes, *L'Œuvre* has undertaken to "renovate, restore, and rehabilitate the *feuilleton* genre."[47]

For the first year after its relaunching, *L'Œuvre* had continued with a policy of alternating between satiric or comical *feuilletons* and short hybrid narratives like the February 1916 *La Captivité de Grand-Père*—billed as the "simple narration of a brave man who saw his village invaded by the Kaiser's soldiers."[48] Even as its lighthearted summer serial

Comment je suis allée me marier sur le front was running, ads began to appear for the next novel, Henri Barbusse's *Le Feu*, which promised to be a full-length docu-*feuilleton* resembling *Captivité*, but by a renowned journalist, publicist, and novelist whose service at the front made him well-placed to give an eyewitness account of trench life in all its details.

Early ads worked *Le Feu* into the ongoing discussion running in *L'Œuvre*'s pages on the correct iconography of the poilu, promising that *Le Feu* would show "French soldiers in the picturesque variety of their lives, in their grandiose simplicity, and the rough majesty of their long and precious sacrifice."[49] Grandiose simplicity, rough majesty: these early ads for *Le Feu* were designed to prepare readers for the new kind of novel they were about to read. A public accustomed to Manichean binaries would need to learn to see grandeur and simplicity, heroism and humility, as compatible. The generic hybridity of *Le Feu* likewise had to be explained: "It's simultaneously a real-life epic [*l'Épopée réelle*] and the gripping story [*Roman empoignant*] of these children of the people torn from their work to face the invader."[50]

While the tone of *L'Œuvre*'s docu-*feuilletons* had been predominantly serio-heroic and that of its earlier *romans-feuilletons* had been mostly comic, it wasn't until *Le Feu* and later Pierre Chaine's *Les Mémoires d'un rat* that *L'Œuvre* really managed to combine the two tones and the two tactics in one text. Téry's strategy for writing the war was to blend the comic and the heroic and to showcase variety, experimentation, and innovation. Barbusse's text, while predominantly serious in tone and grim in its depictions of mass death, is far from being completely original in its iconography and characterization. It draws upon many of the same strategies, including the use of humor, of soldier *gouaille* (irreverent humor), and *débrouillardise* (resourcefulness) used in other, less "serious" and less celebrated *feuilleton* novels of the time.

A text like *Le Feu*, hyperrealist in its abundance of minute details, went a step farther, offering a photographic aesthetic to counter the comic book aesthetic in war culture at large. While such writers as Werth and Cendrars would challenge the truth-telling potential of the technologies of photography and cinema, the older Barbusse appears to have bought into the prevailing fantasy of photography as the ultimate objective medium.[51]

Le Feu was marketed by *L'Œuvre* as a serial novel capable of providing unmediated access to reality, and ads described Barbusse much as one would describe a star reporter, chasing down the war story "sur le vif," with bombs falling all around him. *L'Œuvre*'s own review of *Le Feu*, written by Henri Bataille and published at the time of the Goncourt Prize announcement, sets up a familiar comparison between Barbusse's factuality and the fictitiousness of the comic book war. With all the "[Fake] poilus…cardboard lyricism…all the shameless, sugary-sweet adulterations…of those scribes charged with keeping up the gilded lie of the war," Bataille writes, "oh what a relief to see reappear so suddenly the great, simple and august face of the Truth!"[52]

Every literary trick of the trade is used by Barbusse precisely to construct the truth-status of his text. In addition to the rhetorical warnings about newspapers and tourists of the trenches, the novel uses a self-effacing narrator who serves as an almost mechanical set of eyes and ears through which the sights and sounds of war and the dialogues of its participants can be captured. The text engages with, comments upon, and sometimes even reproduces the very same types articulated in the comic book war at large.

Le Feu is less important to the history of the French novel than to the history of French journalism. The novel won Barbusse the Goncourt Prize he had been yearning for ever since his prewar disappointments with *Les Suppliants* and *L'Enfer*. More important though, it marked a decisive step in the evolution of Barbusse's career-long reflections on mass media and journalism. In his postwar works *Les Enchaînements* (1925) and *Faits divers* (1928), Barbusse would continue his critique of the "society of the spectacle" begun during the war. In *Les Enchaînements*, Barbusse describes his era as one marked by "collective insanity," a time that allows "realities" to be replaced by "words."[53] Later, in his dedication to *Faits divers*, a collection of short stories, many of which deal with the war, he expresses his hope that "these notations, taken at random, here and there, in our alarming contemporary civilization, [will] teach readers to recognize the implausibility of the truth, and awaken public opinion, lulled by blissful legends, to the real face of our twentieth century."[54]

In the interwar period, the newspapers would take up Barbusse's challenge to recognize the true face of the twentieth century. The new journalistic genre of *le grand reportage*, a kind of literary nonfiction, would replace the worn-out serial novel as a primary marketing strategy of the national press. The great postwar heroes of *le grand reportage*, a genre that had existed before the war but only really developed in France after it, had lived through the war, many of them as reporters. They had experienced the frustrations of wartime journalism and had seen the curiously high fiction-to-fact ratio in the papers. They had seen the prewar staples of the *faits divers* and the serial novel lose their urgency and credibility under the weight of world-historical events. In giving pride of place to works of *grand reportage*, the press would answer the call made by novels such as Barbusse's and the countless works of *témoignage* or eye-witness accounts that had flooded the literary market and, with time, fleshed out the newspaper's columns during the war.

Le Feu and similar novels had given shape to the chaos of the war. They had allowed readers to be guided through the war by a narrator whose legitimacy and authority were established by various textual means. In their critiques of the press and in the dialogues and newspaper reading scenes we have seen, these novelists carved out a very important place for narrative fiction in the division of intellectual labor in wartime society, placing novels higher than newspapers in a hierarchy of media able to transmit the "truth" about the war. The press would react to this unique hierarchy in the postwar period, eventually eliminating the novel from the newspaper in favor of this new, properly journalistic form of writing, *le grand reportage*, imported from the United States but given a unique French twist by French wartime docu-*feuilletons* like *Le Feu*. One of the most famous *grands reporters*, Albert Londres, had made his big break covering the destruction of Reims for *Le Matin* during the early days of the war. His reports, unlike the false or euphemistic ones so many people decried, were widely respected. Good reporting, like good works of *témoignage*, provided a balance between documentary accuracy and compelling storytelling. The general perception throughout the war had been that neither good reporting nor an adequate amount of space for nonfiction, literary or otherwise, had been provided by the newspapers and that readers had had to seek extra-journalistic correctives in certain kinds of novels and first-hand accounts.

The evacuation of fiction from newspapers didn't come overnight. Journalism and literature continued to overlap well after the war. Many of the postwar's star novelists—Joseph Kessel, Colette, Jean Cocteau, Blaise Cendrars, Saint Exupéry, Jean Giraudoux—were also *grands reporters*, and papers would often publish *grands reportages* in several parts, like a serial novel, to keep readers coming back for more.[55] What seems to have developed, however, was a new division of labor between newspapers and novels that was unique in the history of the French press, which had always relied heavily on literature for its success. Critiques of the press by soldier-novelists and hybrid genres, many of them developed during World War I, paved the way for this new kind of journalism in the postwar period, with papers selling themselves as providers of facts and leaving fiction for the literary market.

Notes

All four figures of this chapter are from the private collection of Anthony Langley.

1. Henri Barbusse, *Lettres de Henri Barbusse à sa femme, 1914–1917* (Paris: Flammarion, 1937). Among the papers cited in Barbusse's letters are *Le Matin, Le Journal, L'Écho de Paris, L'Intransigeant, L'Illustration, Le Bulletin des Armées*, and the *Bulletin des Écrivains*.
2. Barbusse, *Lettres*, 136. All translations from this and subsequent texts are my own.
3. Henri Barbusse, *Le Feu: Journal d'une escouade* (Paris: Flammarion, 1965), 57–58.
4. Barbusse, *Le Feu*, 324.
5. Barbusse, *Le Feu*, 325.
6. Guy Debord, *Society of the Spectacle*, Ken Knabb, trans. (London: Rebel Press, 2006), 7.
7. Margaret R. Higonnet, "Not So Quiet in No-Woman's-Land," in *Gendering War Talk*, Mariam Cooke and Angela Woollacott, eds. (Princeton: Princeton University Press, 1993), 205–226.
8. Barbusse, *Le Feu*, 57.
9. See Paul Géraldy's extremely successful *La Guerre, Madame* (Paris: G. Crès & cie, 1916).
10. Henri Barbusse, *Carnet de guerre, présenté et annoté par Pierre Paraf*, published in *Le Feu*, 387. Poet-playwright and Franco-Prussian war veteran Paul Déroulède is perhaps best known for his 1872 collection of poems and lyrics, *Chants du soldat*.
11. Raymond Manevy, *La Presse de la IIIe République* (Paris: J. Foret, 1955), 123. According to Manevy (144) traditional *journaux d'opinion*, had sold on the order of 50 to 100,000 copies a piece.
12. Manevy, 124.
13. Claude Bellanger, Jacques Godechot, Pierre Guiral, and Fernand Terrou, eds., *Histoire générale de la presse française*, vol. 3 (Paris: Presses Universitaires de France, 1972), 428. Bellanger et al. cite the following figures for sales of the four big Parisian dailies, *L'Écho de Paris*, and *L'Œuvre* in 1917: *Le Petit Parisien*, 1,683,000; *Le Matin*, 999,000; *Le Journal*, 885,000; *Le Petit Journal*, 515,000; *L'Écho de Paris*, 433,000; *L'Œuvre*, 108,000.
14. Michael Nolan, "'The Eagle Soars over the Nightingale': Press and Propaganda in France in the Opening Months of the Great War," in *A Call to Arms: Propaganda, Public Opinion, and Newspapers in the Great War*, Troy R. E. Paddock, ed. (Westport, CT: Praeger Publishers, 2004), 53, 57.
15. Gabriel Perreux, *La Vie quotidienne des civils en France pendant la grande guerre* (Paris: Hachette, 1966), 255.
16. The right-wing, pro-Catholic, pro-military opinion paper *L'Écho de Paris*, the paper that published Maurice Barrès's daily wartime articles, was openly promoted by the General Staff. Manevy, 150.
17. Bellanger et al., 420.
18. Bellanger et al., 438. Sales figures rose steadily between 1915 and 1919. The paper sold 55,000 copies in September 1915, 63,000 in December 1915, 89,000 in 1916, 125,000 in 1917, 116,000 in 1918, and 135,000 in 1919.
19. Laurent Martin, *'Le Canard enchaîné' ou les fortunes de la vertu: histoire d'un journal satirique, 1915–2000* (Paris: Flammarion, 2001), 71. *Le Crapouillot* was published once a

month and sold 1,500 copies at the height of its wartime success. Bellanger et al., 439. The weekly *Canard enchaîné* reached 40,000 readers by 1918, with 40 percent of its readers residing in Paris and 20 percent fighting at the front. Martin, 76–77.

20. Martin, 73. This list was published in the February 27, 1918, edition.

21. Martin, 65, 71.

22. According to René de Livois, this slogan first appeared on the walls of Paris on September 8, 1915, two days before the paper's relaunching. René de Livois, *De 1881 à nos jours*, vol. 2 of *Histoire de la presse française* (Lausanne: Éditions Spes, 1965), 402.

23. Maurice Leblanc, *Le Triangle d'or* in *Arsène Lupin*, vol. 3, Francis Lacassin, ed. (Paris: Éditions Robert Laffont, 1986), 781. *L'Éclat d'obus* (September 1915) and *Le Triangle d'or* (May 1917) were both published in one of the biggest Parisian dailies, *Le Journal*. Jules Mary's *Sur les routes sanglantes, récit de la grande guerre* was published in *Le Petit Parisien* in 1915.

24. *Rouletabille chez Krupp*, part of the series *Les Aventures extraordinaires de Rouletabille reporter*, was published in seven installments in the monthly magazine *Je sais tout* from September 1917 to March 1918.

25. *L'Araignée du Kaiser* ran from September 10 to October 24, 1915.

26. Georges de La Fouchardière, *Bicard dit le Bouif, poilu de 2e classe* (Paris: E. Mignot, 1917). For more on Bicard, see Allen Douglas, *War, Memory, and the Politics of Humor: The Canard Enchaîné and World War I* (Berkeley: University of California Press, 2002).

27. The soldier-type reactivated by René Benjamin, the *poilu-picaro* is, of course, not entirely new. The great social typologies of the mid-nineteenth century, *les physiologies*, had already codified the figure of the *titi* or *gamin de Paris*, veteran of the Napoleonic campaigns and, later, with Hugo's Gavroche, of the barricades. Gaspard and Bicard also resemble the pre-war *comique-troupier*.

28. Louis Thomas, *L'Armée de la Guerre*, published in 1916 by Payot, reproduced at http://www.greatwardifferent.com/Great_War/French_Artillery/Armee_01.htm (accessed January 29, 2008).

29. For a survey of wartime serial novels, see Dominique Kalifa, "Guerre, Feuilleton, Presse 1913–1920," *14/18. Aujourd'hui. Today. Heute*, vol. 2 (Paris: Éditions Noêsis/Péronne, Centre de Recherche de l'Historial de la Grande Guerre, 1998).

30. See J. P. Daughton, "Sketches of the Poilu's World: Trench Cartoons from the Great War," in *World War I and the Cultures of Modernity*, Douglas Mackaman and Michael Mays, eds. (Jackson: University Press of Mississippi, 2000).

31. Hautot, Hansi, Poulbot, Willette, Weber, Faivre, Steinlen, Hermann-Paul, Bofa, Capy, Manfredini, Forain, Morin, Florès, Sem, Valloton, Zislin. For more on wartime cartoons and illustrations, see Arsène Alexandre, *L'Esprit satirique en France* (Paris: Berger-Levrault, 1916).

32. Le Rousseur, "Les Dessins de guerre," *Le Crapouillot*, April 1917. Cited in Philippe Dagen, *Le Silence des peintres: les artistes face à la Grande Guerre* (Paris: Fayard, 1996), 63.

33. *Le Miroir*, January 3, 1915. Legionnaire poet Blaise Cendrars was a regular contributor to *Le Miroir*. In his post-war account of the war, *La Main coupée*, Cendrars recreates a scene in which he attempts to calm an angry general who has just discovered he has been sending photographs to *Le Miroir* and other papers behind the lines. "Oh, it's no big deal, general. It's to make a little extra money. *Le Miroir* pays me a *louis* and I clink glasses with my buddies. I send them picturesque stuff, nothing but safe stuff, old news…nothing to worry about." Blaise Cendrars, *La Main coupée* (Paris: Denoël, 1946), 266.

34. Léon Werth, *Clavel soldat* (Paris: Éditions Viviane Hamy, 1990), 336.

35. Barbusse, *Le Feu*, 145. Dorgelès makes a very similar argument in *Les Croix de bois*.

36. Werth, 116.

37. Werth, 373.

38. In the *entre-deux-guerres*, the metaphor of war as theater will be developed at length in novels and stories that push even farther the theory of the pseudo- or comic book war. See in particular Cocteau's *Thomas l'imposteur*, Drieu la Rochelle's *La Comédie de Charleroi*, and Céline's *Voyage au bout de la nuit*.

39. Werth, 12.

40. Werth, 178.

41. Werth, 142.

42. Werth, 143. Clavel's theory of the breakdown between fact and fiction is seconded by the narrator in Alexandre Arnoux's 1919 *Le Cabaret*. "I got all the details from my buddies... I've gotten so used to it that I have to think hard to remember that I didn't see anything. But my buddies explained it to me, what do they know themselves?" Alexandre Arnoux, *Le Cabaret* (Paris: Fayard, 1919), 64.

43. Werth, 357.

44. Werth, 375. In much the same way, Dorgelès's main character Sulphart will end the war telling tall tales of his adventures in the trenches that will draw the ire of his fellow soldiers.

45. Werth, 269.

46. Jacques Baudier, "Littérature," *L'Œuvre*, April 9, 1916.

47. Gustave Téry, "Le Nécessaire et le superflu," *L'Œuvre*, December 23, 1916.

48. From an ad for *La Captivité* from January 5, 1916 in *L'Œuvre*.

49. June 10, 1916: first ad for *Le Feu*.

50. Ad from mid-July 1916. The adjective "empoignant," along with "poignant" and "palpitant," was a stock descriptor in serial fiction ads. The syntax of the description, "le Roman empoignant de ces enfants," points, however, to the many semi-fictional autobiographies appearing in newspapers, commonly described as "le roman de," meaning lived adventures of.

51. An avid amateur photographer, Barbusse had his trusty Kodak *Pige-Tout* with him in the trenches. A good deal of his correspondence has to do with the pictures he has taken and with the camera equipment he wants his wife to send him.

52. Henri Bataille, "Le Feu," *L'Œuvre*, December 16, 1916.

53. Henri Barbusse, "Préface," *Les Enchaînements* (Paris: Flammarion, 1925), ix.

54. Henri Barbusse, *Faits divers* (Paris: Flammarion, 1928), vii–viii.

55. Fabrice Schlosser, *Les Légendes de la presse* (Lyon: Aléas, 2002), 157.

15

The Impact of the Great War on the Writing of Fernand Léger

Béatrice Vernier-Larochette

It's perhaps cruel to say, but the war of 1914–1918 offered me an opportunity; it allowed me to discover the People and to renew myself completely.

(Fernand Léger)[1]

Fernand Léger, the twentieth-century painter recognized for his innovative work that stressed the importance of the object and of pure color, also expressed himself extensively in writing.[2] In fact, the public came to understand his plastic principles all the better through his esthetic meditations on the evolution of art and his work. Prior to World War I the artist already enjoyed a certain reputation in the Cubist movement, but four years in the trenches would change his esthetic conception. In letters to Louis Poughon, a friend from youth, he shared this experience as a soldier, which he expressed in drawings that would not be exhibited for some 30 years.[3] Yet he did not follow the path of André Masson or such writers as Genevoix, Cendrars, Barbusse, and Dorgelès, who did publish their war stories.[4] The silence concerning the war's atrocity that we find in Léger, but also in numerous soldiers on leave or after the war, has been attributed to self-censorship,[5] to the inability of language to make this hell believable,[6] and to society's own implicit censorship of the conflict.[7]

Yet Léger's postwar art and writing—letters, travel diaries, lecture notes, articles—implicitly refer to this lived experience, this space of war, in the way he views the world, lays claims for his plastic esthetics, and describes his relation to others in the public sphere or close to him. He imagined getting his work accepted as a struggle, the public and the art world being presumed enemies whom he had to conquer and win over to accepting the pertinence of his goals. He associated them in his plastic search, creating a fraternal relation resembling the one developed on the front with his brothers in arms. However, this cooperation that he often extolled is quite particular in his writing, where the expression of an intimate self is notoriously absent, resulting from the disindividualization he underwent when he enlisted in 1914 and from his anxiety at having survived his friends. After the war, the simple soldiers whom he continued to admire deeply were often represented in his paintings (workers busy at work or at rest, circus people), and in his artistic reflections he often referred to individuals of modest means.[8]

Witness to the slaughter produced by the use of more efficient arms, after the war Léger became fascinated with the modern machine, which shaped his plastic art during his so-called mechanical period (1919–22).[9] Léger's writing and art are unquestionably influenced by the Great War. They represent the tribute he pays to comrades who perished, as well as a way to justify his own survival.

Armed Struggle and Plastic Struggle: Similar Forms of Combat

Although Léger always remained aware of his work's avant-garde aspect, which called for "a conceptual realism achieved by the dynamism of plastic contrasts,"[10] in all his reflections he persisted in seeing the world and society as a place of tensions, involving only victims and aggressors. "It's a bewitchment, a skillfully regulated fascination, the stores want a victim, and often they get one" (*FP*, 113). For Léger, this state of affairs stemmed from the conflict of 1914–18: "Once he returns to normal life, the man of 1921 keeps within himself the physical and moral tension of the harsh years of war. He has changed; economic struggles replace front-line battles" (*FP*, 207). Léger's writing often displays a military thematics, as when he defines his attraction for the intense activity of life: "If I consider life head-on, I love what is commonly called the state of war, which is nothing other than *life speeded up*. Peacetime being life slowed down" (*FP*, 107). He characterized his times as "resulting from an instructive war in which all values were laid bare" (*FP*, 115). Presenting an overview of the new idea of pictorial art, he defines the innovative artist's action thus: "Each artist possesses an offensive weapon enabling him to brutalize tradition" (*FP*, 103). The entire progression of art even resembles a combat for Léger, a delicate confrontation between the notion of subject and of object: "Today's artistic evolution can be considered as a battle that has been waged for fifty years between the conception of *subject* that emerges during the Italian Renaissance and the interest in the *object* and pure tone that is increasingly being asserted in our modern ideas" (*FP*, 187). Léger explains this esthetic assertion once more through his experience in the trenches: "I left Paris fully in an abstract manner, in a period of pictorial liberation. I immediately found myself once again surrounded by the French people...; once I had become part of that reality, the object never left me."[11] In a 1937 lecture given in Anvers and later in Helsinki, although Léger announces the importance of color in modern life, especially in what he calls minor arts (advertising, display art, fashion), he stresses how these new entities must struggle to establish themselves with respect to painting: "The minor arts are patiently awaiting the outcome of this closely-fought battle before taking off behind the winner" (*FP*, 219). In one of his last texts, which Henri Laugier presents as a true "chef-d'œuvre,"[12] Léger sums up once more the difficult course of artists in terms of belligerence, beginning with the work's creation, and moving from its realization to its acceptance by the public. The doubts and certainties accompanying this process he likens to a soldier moving along on a battlefield: "You go forward, you go back, you keep your head down....You have to slip through the eye of a needle, hide, play dead, crawl carefully without making a sound." Later when the work takes form, the artist tries to judge his work. But there too nothing is simple, and Léger refers to a situation he experienced during the war: "You start to lift your head up, after all you want to see it, everything that's crushing you and rejecting you, like the war up above the trenches." Similarly, when he states his view of the artist's work, an "action in freedom" but that underlies a marginal situation, he expresses it thus: "It was upright and in a state of war with society that these lively works were conceived and forged."[13] As Sylvie Forestier notes, in Léger "the artist, observed by those around him, is always in danger."[14] The notions of "freedom" and "truth" are inseparable for Léger from his artistic combat, for they are his faithful companions, just as his friends the soldiers were: "These works were achieved with freedom and truth as companions in arms."[15]

As a simple soldier in the trenches for three years, Léger had to fight the German enemy, the cold, rats, and the unfairness of some superior officers, observing sometimes bitterly the more favorable situation of particular friends and always brushing up against imminent death.[16] Unquestionably, these extreme conditions marked him, leading him later to view the world as being in a situation of "insurrection," which he considered necessary nonetheless, since it drove the artist constantly to surpass himself and to acquire indisputable authenticity. "Dispassionately you view this entire life before you, and you feel that you fully possess the truth of your times."[17]

Ever-present, the war in which Léger participated underlies his comments and his art. According to Dieter Koepplin,

> in his own way the painter transposed the events he experienced during the war. He related them to the modern world, a world dominated by mechanisms invented by man that were as fantastic as they were dangerous: machines, industrial products, cars, film, advertising, etc.[18]

However, the position of "artistic combatant" confronting potential enemies—the public and art critics—is by no means a solitary one. As when he was at the front, Léger waged a collective battle, always including others in an adventure associated with formulating a new conception of art.

Writing a Collective Plastic Engagement

Although the individual to whom his work is directed is taken initially for an enemy, Léger tries to tame him and make him into a friend or assistant. Writing is thus not only a form of instruction but also a sharing of esthetic reflections and observations concerning surrounding reality. Léger clearly wishes for his work to be well understood in order to create a relation of harmony with others, those close to him and the general public. With this in mind he encourages listeners, readers, and friends to adhere to his approach in a kind of shared engagement.

It should be noted that Léger rejects the idea of individualism, constantly stressing how he values the beauty of any collective act, "these admirable modern machines, masterworks of perfection, collective works, such as those that created the Roman and Gothic cathedrals of anonymous Beauty."[19] This same will is found in the attitude he adopts in his relations with others, above all with his art dealer Léonce Rosenberg, with whom he stressed the benefits of a cooperative relationship: "We will never do anything well unless we work together" (*CA*, 27). Moreover, when Léger refers to his plastic objectives in his lectures, he often uses the pronoun "we" and the corresponding imperative, forms referring to association and the group. "We live surrounded by beautiful objects that slowly reveal themselves and that people perceive" (*FP*, 127), he stated in 1924. When Léger addresses architects, he reveals that although the artist is defined by a struggle with the support of art, he visibly shares in the cause of these spatial ideas men: "The painter-artist, who hates dead surfaces, can get along with you" (*FP*, 181), he states, adding that "we must *all* struggle *together* and join the battle" (*FP*, 182). In a 1936 article that once again called for the esthetics of a "new realism,"[20] he placed himself within a group of artists involved in the renewal of art: "We have freed color and geometric form" (*FP*, 196).

The tone of his numerous articles and lectures resembles that of a friendly guide when he uses the first-person plural imperative form to encourage his listeners to open themselves to the world around them. "Let us become aware of everyday deeds and actions, the anxiety of the Beautiful fills three-quarters of daily life" (*FP*, 28), he declared in a 1924 Sorbonne lecture. "Let's stroll about and look around," he suggests in a 1934 speech.[21] "Let us introduce color…and mix it skillfully," he proposes elsewhere (*FP*, 131). Léger encouraged his contemporaries to join him in the new artistic transposition of expressing an ordinary reality. Using instructive presentations—with the rhetoric of demonstration (eulogy and praise) and debate (the example)—he attempted to make his work accessible to others, for, as Michel Meyer observes, "argumentation aims at suppressing distance by adhering to a thesis one seeks to have shared by all."[22] Persuaded that a collective process is useful, Léger found that the first-person plural and its corresponding imperative form offered a way to defend his work and his status as an artist by maintaining a partnership with other creators as much as with the public and art critics.

This wish to advance collectively does not mean avoiding personal convictions, however, for the "I" remains and displays a personal independence that is important to Léger. "I apply the law of contrasts.…In that way I believe I can control the situation because I reach a 'multiplicative' state that any constructed object can attain only with difficulty" (*FP*, 63).[23] Léger was quick to reaffirm his independence in responding to attacks: "Our 'dear enemies' who wish to take every advantage must remember that I remain free to evolve as I wish."[24] Concerning cooperative efforts with his art dealer, he readily pointed out his refusal to adapt to the constraints of the art market. "I intend to make very complete paintings, but in my own way."[25] In his book on the artist's work, Pierre Descargues acknowledges his whole character: "he has a certain roughness and a clear way of seeing things. He is not troubled by doubts."[26] This attitude that includes the wish for co-creativity without denying his personal creative responsibilities resembles what Émile Benveniste observes concerning the pronoun "we": "the use of 'we' blurs the too sharp assertion of 'I' into a broader and more diffuse expression."[27] Thus, despite the often didactic character of Léger's texts and his conviction that the basis of his esthetics is solid, he made it clear that he does not want to assume the position of leader. In his writing, he repeats his wish to share his goals and not impose them, as Kosinski notes: "Until the end, Léger remained faithful to the ideal of collaboration, as we see in his writings on the role of color and mural painting in the urban environment, and especially in his own mural paintings and window decorations, his mosaics and his tapestries."[28]

A Relation of Egalitarian and Peaceful Alterity

In this shared plastic adventure, Léger was careful to nuance his statements so as to avoid possible misunderstandings and remain on good terms with his correspondents, even though he viewed the world as being in a state of permanent struggle. He wrote to his art dealer, "I say this, dear friend, in the friendliest of spirits as always, but yet with a slight desire for the kind of clarification that only the solid friendship that joins men for life can produce" (CA, 60). Léger extended this idea of a partnership between artist and public when, in his speeches or writing, he used the indefinite pronoun "one" (on) next to "we." Instead of being an indeterminate expression, according to Sartre "one" displays the desire to eliminate the conception of oneself and the other in order to guarantee a total cohesion: "I am not opposed to the Other, for I am not 'me'; instead we have the social unity of the *they*."[29] "One cuts the flowers in the garden to have them next to one in the apartment, with paintings," Léger observed when explaining how he chooses his colors.[30] In one of his late texts, referring to the path artists take, he chose to assess the situation globally: "But still, a sense of self and a bit of pride go a ways. This life is coolly seen up against one, one feels oneself as existing fully in the truth of the period and that one should use as quickly as possible this speed that creates emptiness and solitude.[31] Valorizing "one" here, Léger steps back from his own words, asserting his wish to be involved in a collective artistic enterprise, like the mission of the soldier participating in an army's advance. However, Mildred Szymkowiak notes that Heidegger observes that the "one" in no way implies a passivity at the heart of its process, but rather an evermore significant presence, which is that of the person involved in the process: "The *one* is thus less one of the figures of the other ('the others') than an indeterminate community with all and no one, which I must not exclude (because I could not) but rather recover in the first person, in a being that is properly my own."[32]

This attitude that encourages partnership without neglecting individual engagement evokes an association in which Léger does not wish to be overbearing in his relation to another, an association found in his letters. Thus, although he begins his letters to his mistress, his wife or his friend André Mare with "Dear Pet," "My Daughter," "My Dear Old Friend," the closing signature[33] ("Your Bear," "Your Father," "Your Friend") suggests he considers himself as being over by the other and hence exposed and vulnerable.[34] Establishing a balanced relation to another seems vital for Léger, a reminder of a war where extreme conditions of existence favored and even required mutual support, as he wrote to Poughon in the fall of 1914: "The squadron, this little family of ten guys, is the framework of the whole machine.... The feeling of sociability is finally what saves everything.... I only know ten guys, but I know them well. There are some great ones. All of humanity is present here" (CG, 11).

More often than not, the differences between social classes faded, and noncommissioned officers sometimes felt compassion for their soldiers, something Léger experienced himself: "I was lucky enough to encounter a great major who understood that I couldn't hold up against that kind of life. The day after the tragic event when I almost cashed in my chips...he assigned me to duty under him as a stretcher-bearer. In fact, I'm his secretary" (CG, 16). Deeply affected by this support, Léger could not forget this generation of Frenchman of simple means who died for their country. Afterward he strove to direct art toward the masses who, he believed, had been treated unjustly.

"It's the people's time. Their turn has come now to play their role. It's enough that they're sacrificed in every war—and we artists must do everything we can in our area to help them" (*GB*, 248). He took part in left-wing political movements, published articles in the journal *Plans* (a union journal), joined in 1932 the Association of Revolutionary Writers and Artists, which was made up of communist sympathizers, and was very active during the rise of the Popular Front by giving lectures more frequently in working-class settings.[35]

In his professional relations with others, when Léger collaborated it was without reserve, but in his personal relations he hesitated to share his more intimate feelings, a result of having participated in the conflict of World War I. It was probably not easy to reaffirm his identity after four years of anonymity during which the self was condemned to silence so that self-sacrifice to the national cause was possible, a period the artist characterized as "four years without color" (*FP*, 206).

Self-abnegation in Postwar Writing

In his war memoir, the painter André Masson recalls the words spoken by one of his superiors in 1914: "Forget those whom you left behind, forget even your name. You're no more than a myriad of men." Masson added, "It's hard for a 'myriad' to become a man again."[36] The self-negation that occurred upon joining the military did not affect the artistic position Léger adopted after the war, but it did persist in the way he laid bare the self.

During the high points of the 1914–18 battles, Léger gives extensive descriptions of things around him, which fill his letters to his friend Poughon. In attempting to have Poughon share in this experience as much as possible, he fails to express his own view of the situation. "In his reporting, the writer who is overwhelmed here clearly circumscribes the range of his writing since he mentions only the events he takes to be significant, facts that most often are outside him."[37] If Léger thus effaces himself from his wartime letters, later he continues to eliminate all personal expression from his writing. The copious notes taken during his travels favor an observational style of writing that focuses on what is different rather than on personal reflections, a "chronicle of the world and of others rather than of oneself," observes Georges Gusdorf in what he calls "the exterior journal," as opposed to "the personal journal."[38] "Naturally, if I stop to reflect, if I close my eyes, I perceive dramas lurking around this exaggerated dynamism, but I came to *watch*, and I'll go on," Léger remarks in notes he took in New York in 1931.[39] Similarly, the letters sent to his mistress are quite unusual in their form, for even if this affair was supposed to remain hidden, one can sense the difficulty of considering the other in their complete intimacy based on their own person.[40] Léger avoids naming this woman, and he refers to her through a special play of pronouns in which he sometimes mixes together in the same sentence "I," "you," and "we": "I'm really working a lot—she too, you'll tell me";[41] "Dear Pet my love, how adorable you are in this long letter—where she is acting and fulminating—where she is working" (*CR*, 140); "Where are you? What is she doing?" (*CR*, 46). If we accept Émile Benveniste's theory, this intimacy is clearly one that cannot easily be affirmed. "The form that is called the third person really does contain an indication of a statement about someone or something, but not related to a specific 'person.'"[42] Implicitly here,

Léger disapproves of intimacy in correspondence since he steps back from the situation and considers it from the outside: "She's going to write to tell me about all the violence and everything unexpected in dear pet's life, who is so precious to her bear who is always thinking so much about her" (CR, 100). He is no longer the speaker of this language, which is not addressed to his mistress anymore; rather, it seems to refer to her as if she were a nameless woman he might encounter in the world, with the result of avoiding an epistolary dialogue. As Benveniste explains:

> Consciousness of self is only possible if it is experienced by contrast. I use I only when I am speaking to someone who will be a *you* in my address. It is this condition of dialogue that is constitutive of *person*, for it implies that reciprocally I become *you* in the address of the one who in his turn designates himself as *I*.[43]

Elsewhere, the self-portrait that Léger performs in a letter to his mistress in which he takes her place to describe himself reveals once again his resistance to self-representation:

> The color in the raging world towards 1918—I was so little that I couldn't pee alone yet!...So this bear, I must tell you, paints once in a while when I give him his freedom— and if you saw his painting—it's terrible—you don't just see it—you hear it and listen to it, and it even has a smell, a bear smell that sometimes can't be stood (CR, 166).

The doubling that Léger engages in parallels what Bernard Beugnot notes when he takes the theatricalization of the letter frequently to be the way writer displays modesty: "The letter is indeed a staging of the self..., which is just as much the development of an imaginary persona as the reserve, the modesty, and the distance that mask the individual."[44] The game that Léger goes in for here reveals a resistance to express what he is, to communicate to his correspondent the depth of his character.

The way Léger signs his letters to his various correspondents is important in this respect as well, since the final signature never reflects the private man. In addition to the kindly names with which he closes his letters ("Your sailor bear," "Your old friend," "Your angry father," "Your old guy"), the artist always adds F. Léger, whether in letters to his mistress ("Your bear F. Léger"), to his friend Poughon ("Your friend F. Léger") or to his wife ("Your Father very sad F. Léger"). The person named Fernand—lover, husband, friend—is only an F., and Léger, the patronym of the painter, alone persists. The closing word of his letters thus reaffirms his identity as an artist, a public man but in no way a husband, lover or friend. This effacement is noted by the poet Guillevic, with whom Léger contributed to a work on his art entitled *Coordonnées*: "Socially of modest means, Léger did not wish to stand out: very polite, very urbane, and yet timid with a deep-seated silence. Neither weighty nor completely present, meditative in his entire being."[45] For Pierre Arnaud, this artist's reticence to insert himself personally within a reality can be seen on a pictorial level since he painted only a single self-portrait in 1905 (one whose shapes are difficult to make out moreover) and only a few sketches of himself throughout his life.[46]

Even if Léger was extensively engaged in promoting his work, clearly this investment in no way can be taken as the expression of his private person, a consequence arguably of a war in which personalization itself had to be erased. Léger's uneasiness in affirming the private person also stems from the difficult situation of having outlived his comrades in arms.

Homage to the Lost Soldiers and the Writing of the Survivor

The esthetics that Léger argues for after the war is aimed at plain citizens. He maintains that the ability for artistic appreciation should no longer belong only to an elite but also to the people because, as he states in a text of 1936, art must adapt itself to the social realities of its times. "The working class has a right to all that. It has a right to have mural paintings on its walls signed by the best modern artists, and if it has enough time and leisure it too will be able to settle there to live and enjoy these paintings" (FP, 200). To accomplish this mission, Léger made plans to provide art instruction for the disadvantaged, and he began working in working-class public spaces. At issue implicitly is the wish to do justice to this social class by including it in the art world that previously was reserved for privileged intellectuals, a decision Léger made during the conflict of 1914–18:

> During the war I discovered this great social body of the people, and the more I pursued this discovery, the more the idea of human generosity grew in me, and the more I was convinced that it was necessary to struggle to prevent the people after the war from once again becoming victims in the hands of the powerful, which so often happens for simply economic interests.[47]

This engagement is witnessed in his partnership with Le Corbusier on murals, mosaics, and stained glass creations located on ordinary religious sites (the churches of Audincourt and Assy). His activities, writing, and speeches, which constantly stress that "between today's artist and the people a path must be established" (FP, 203), were a way of paying homage to his departed soldier friends, prompted by the desire to privilege their memory tacitly as d'Aubigné does, according to Agnès Conacher, in writing Les Tragiques: "A work of mourning, the survivor's debt, Les Tragiques were written to recall the past, to rescue from the oblivion of death the victims of wars, martyrdom, and the Saint-Barthomew's Day massacre in particular."[48] One can even see Léger's attempt to justify his own survival for, as Brice Parain stresses concerning his own return from war, "as survivors of so many deaths, death was for us the true cost of war, a cost that had not been paid by those of us who had not died."[49]

One can sense the attitude of the survivor in Léger when he marvels at life unfolding before his eyes and encourages his readers to take pleasure in the simplest of landscapes. Léger had become aware of the rapid rhythm of twentieth-century life and that adaptation was necessary in order to capture this simultaneity. "If you like design and objects, then December is when you must observe the countryside.... But for that, you have to walk slowly, look around, really see, and not hurry everywhere."[50] In persisting in making the public sensitive to the charms of the surrounding world, he also displays a personal awareness of the world's beauty and the ephemeral aspect of human existence. "One must live in intensity, not day by day but hour by hour. Each new event must be grasped just at the moment when the projector sweeps it away" (FP, 237), he advises in the 1946 article "L'Œil du peintre." This reflection strikingly parallels the writing of Pascal de Duve, an AIDS survivor who, in his travel diary written on a cargo ship sailing between France and the Antilles, also urges his "clan" to remain sensitive to the beauty of life. "Brothers and sisters of misfortune..., open your eyes to the marvels of things great and small especially, all those things in which those whom Death has not yet courted, those for whom Death is distant and abstract, cannot truly take pleasure, as we can."[51]

In this attitude of taking pleasure in life, Léger attempted to exorcise the death that indiscriminately had struck human beings like him, a death he always sensed to be imminent. To counterbalance the fear of this moment that destroys human beings and actions, he involved himself in all the activities he could, and he experienced a certain unease if he felt he was not participating in life's unfolding. On one occasion he expressed to his friend Poughon how frustrated he was, upon returning to his birthplace after the Great War, to see how much it had changed during his absence. "How our beautiful city has changed. It's nothing like what it was. We're really like foreigners, my friend" (CG, 92). Later, in 1945 after his return from exile in the United States,[52] he expressed to his student Nadia Khodassievitch his disappointment at withdrawing from a society that had continued to develop during his time away: "how I regret having left France and not having been with you.... This work will have to be begun again with French workers, who will have to be brought to understand our art. It's fascinating work" (GB, 247). This is the Léger who realizes he has momentarily failed to pursue the collective artistic mission in which he was so invested. We also find what Jules Vallès rightly notes concerning the state of exile: "[it is] being absent, and in addition not having a name, being dead."[53]

In Léger's correspondence the idea of calm and of things not energized appears as a negative notion, which he sometimes associated with death. "I find it silly that the mountain is immobile. It should move or have animals that animate it. It's dead" (CR, 17), he wrote to his mistress in 1931 during a stay in Austria. The same year, while crossing the Atlantic to New York, the ocean provoked in him the same feeling of inertia: "The ocean is very pretty this morning, it is absolutely like English anthracite in the form of flowing lava, and so supple. Like in the mountains, what's missing there are giant animals; there's not even a hand, the head of a drowning victim, nothing" (CR, 28). This observation recalls the battlefields he described in his correspondence from the front. "Human heads that are almost mummified are sticking up in the mud. They're tiny in this sea of earth" (CG, 66). The absence of linguistic referents in his letters, as well as the absence of spatial reference points that he noted to his correspondent during his time in the trenches, reveal the apocalyptic aspect of this lived experience that later marked his psyche:

> Nothing left. Not a stone or a piece of wood, holes, mud, water in the holes and bits of human flesh. Whole bodies that are better preserved. From that and that alone I estimated that I must be in the lines where waves of assaults came from a couple of days ago after the bombings. Now I wanted to find the Kraut lines. There was no sign except bits of human flesh and things. The shoes were really what guided me. When I saw boots and green helmets, I knew that I had gone beyond our lines.... I simply can't describe all that to you (CG, 66).

This terrible ordeal in which the individual must certainly be destabilized by the loss of reference points continued to haunt Léger. According to Dorothy Kosinski, it directed his plastic choices: "For Léger and those of his generation, the First World War was the great break, not only because it led to chaos, tragedy, and destruction, but also because it brought out aspects that can be seen as specifically modern and that challenge notions of temporal and spatial continuity."[54]

These images of a war long over that assailed Léger once again are one symptom of post-traumatic stress disorder (PTSD), which John Wilson and Terence Keane describe as follows: "The first PTSD reexperiencing criterion...involves 'recurrent and intrusive distressing recollections' of the event....Note, that includes raw images, such as a visual memory of parts of the experience, perceptual processes...and organized or disorganized forms of thought."[55] Léger perceived these moments of frozen immobility, absence, and solitude as suffering and weakness, reminders of the calm experienced after the attacks when the dead were being counted. Afterward, he repeatedly attempted to make up for the quiet moments. "Always a need for movement around me" (CR, 192), he notes to his mistress. Wilson and Keane associate hyperactivity with symptoms of a previous trauma.[56] Thus all his writing, as well as the numerous lectures and trips that clearly reveal his desire to find support for his work, also reveal Léger's fear of stillness. Synonymous for him with death, this stillness had already been experienced on the front where he had been struck by the atmosphere of immobility after the attacks. "It was the complete opposite of the previous day's pandemonium. I was truly in the desert with no living being around me" (CG, 66). The act of writing that Léger performs during his travels—in notes and letters—to communicate his remarks to his friends resembles what Amélie Schweiger observes in her study of Flaubert's travel correspondence. "Letters and travel notes seem to be a way of warding off the danger of the dissolution of the self."[57] Moments of calm and pause represent for Léger a non-renewal, even the possible disappearance of his being, the ghost of a death that will follow him for years later.

Léger's writing constantly attempts to reaffirm his position as an artist involved in the cause of the people, a goal that manifestly justifies his own survival by offering him the illusion of always being in combat alongside his departed companions.

Conclusion

The conflict of the Great War in which Léger was reluctantly involved shaped his actions and thoughts throughout his life. The military terms chosen to express his perception of the world reveal a being deeply marked by a frontline experience. Similarly, his plastic conception of an art he attempts to renew and the choice of a collective artistic way of working are imbued with the idea of trench warfare in which life and army units move forward only as a group. The relation of equality that Léger strives to develop also reflects an underlying will to recreate the atmosphere of fraternity that often prevailed among soldiers. Moreover, joining with someone is a way to avoid solitude, introspection, and calm, which for Léger were synonymous with a death he constantly feared, years afterward this terrible conflict.

This painter, whom critics of plastic arts characterize as a witness to his times,[58] leads us to consider all his writing and artistic work as an implicit testimony of his wartime experience, following the point of view of Conacher who likens the act of testimony to a necessary moral duty, "that of giving or giving back a story (histoire) to those who made History (*Histoire*)."[59] Gijs Van Tuyl describes this situation in Léger's case thus:

> The impressions and experiences of the front added two new dimensions to his life and his work, conferring upon him the status of a triple virtuoso of modern art. His ability to identify with modern, mechanized life, and his involvement on the side of the man in the street, gave his 'modernity' more impact and depth.[60]

Léger's writing pays homage to his departed companions, but it also serves to validate his existence since he attempts to exorcize a death that he always feels to be imminent, even years after the war. His written expression thus represents a space where the trauma of the Great War can certainly be seen in the form of persistent forms of anguish, but that Léger successfully employs in an innovative plastic engagement. The trenches, this space of war that Léger occupies in 1914 and that for four years constituted his lived experience, continued to inhabit him, and despite himself it directed his actions throughout his life, in both his artistic and his written reflections.

Notes

Chapter translated by Daniel Brewer.

1. Fernand Léger, *Fonctions de la peinture* (Paris: Éditions Gallimard, 1997), 247 (henceforth *FP*). All French quotations have been translated.
2. "This *object* that was enclosed in the *subject* becomes free, this pure color that could not assert itself will emerge" (*FP*, 188).
3. Fernand Léger, *Une correspondance de guerre à Louis Poughon, 1914–1918*, Christian Derouet, ed. (Paris: Éditions du Centre Pompidou, 1990) (henceforth *CG*). Léger's war accounts are also found in his correspondence with his wife, Jeanne Lohy-Léger, and in the correspondence sent to his student Nadia Khodassievitch, in *Fernand Léger: vivre dans le vrai*, Georges Bauquier, ed. (Paris: Adrien Maeght, 1984) (henceforth *GB*). Léger enlisted as a sapper in the engineering corps and later was a stretcher-bearer.
4. Henri Barbusse, *Le Feu: journal d'une escouade* (Paris: Flammarion, 1916); Blaise Cendrars, *La Main coupée* (Paris, Éditions Gallimard, 1975); Roland Dorgelès, *Les Croix de bois* (Paris: Albin Michel, 1931); Maurice Genevoix, *Ceux de 14* (Paris: Flammarion, 1950); André Masson, *La Mémoire du monde* (Geneva: Éditions Skira, 1974).
5. From the front, Léger writes the following to his friend Poughon: "Duchêne was killed. Above all, if it's not too late *yet, not a word to my mother*" (*CG*, 18) (Léger's emphasis).
6. Carine Trévisan stresses that "it is as if regular language could not confront this experience," "Le Silence du permissionnaire," in *Écrire la guerre*, Catherine Milkovitch-Rioux and Robert Pickering, eds., (Clermont-Ferrand: Presses Universitaires Blaise Pascal, 2000), 202.
7. See Frédéric Rousseau, "Écho d'une correspondance de guerre," who notes concerning Henri Barbusse: "Hence quite naturally the novelist transcribes a certain number of his terrifying visions, which not everyone will like. Cru in particular criticized him for using too many of them, for adding so much that is macabre and bloody that it hurts the pacifist cause." Rousseau also reveals how the editorial staff of the newspaper *L'Œuvre* censored Barbusse's writing of *Le Feu*, which he first published in serial form. Henri Barbusse, *Lettres à sa femme, 1914–1917* (Paris: Buchet-Chastel, 2006), 23.
8. "Le Mécanicien" (1920), "L'Hommage à Louis David" (1948), "Les Constructeurs" (1950), "Les Deux Cyclistes" (1951), "Les Trapézistes" (1954).

9. The paintings "Les Hélices" and "La Ville" (1919) reveal Léger's plastic reflection on the modern machine's impact on society and the world of work.

10. Dorothy Kosinski, ed., "Un langage pour le monde moderne: l'œuvre de Léger de 1911 à 1924," *Fernand Léger: le rythme de la vie moderne 1911–1924* (Paris: Flammarion, 1994), 17. Léger describes his search as follows: "I oppose curves to straight lines, flat surfaces to sculpted forms, pure tones in places to shades of grey" (*FP*, 63).

11. Fernand Léger, quoted in Christopher Green, "La Peinture de Léger pendant et après la guerre," *Fernand Léger: le rythme de la vie moderne 1911–1924*, 45.

12. Henri Laugier, introduction to Fernand Léger, "C'est comme ça que cela commence," *Europe*, 508–509 (1971), 58.

13. Léger, "C'est comme ça," 58, 59.

14. Sylvie Forestier, "Présentation" (*FP*, 9).

15. Léger, "C'est comme ça," 60.

16. His friend Poughon, for instance, succeeded in escaping from the front and had a position as prefecture adviser, which caused Léger to remark angrily in one of his letters, "You know, Louis my friend, I know how much each thing costs. I know what bread is, wood, socks, etc. What about *you*, do you know? No, you can't know that, you weren't in the war" (*CG*, 5) (Léger's emphasis).

17 Léger, "C'est comme ça," 58.

18. Dieter Koepplin, "Les Éléments mécaniques," in *Fernand Léger: le rythme de la vie moderne 1911–1924*, 220.

19. Fernand Léger, "De l'Acropole à la Tour Eiffel," lecture given at the Sorbonne in 1934, in *Pour un réalisme du XXe siècle: dialogue posthume avec Fernand Léger*, Roger Garaudy, ed. (Paris: Éditions Grasset, 1968), 242.

20. "Realisms vary in that the artist lives in different periods, in a new milieu, and in a general way of thinking that shapes and influences his mind" (*FP*, 192).

21. Fernand Léger, "De l'Acropole à la Tour Eiffel," 238.

22. Michel Meyer, "Aristote et les principes de la rhétorique contemporaine," Introduction, Aristote, *Rhétorique* (Paris: Librairie générale française, 1991), 52.

23. Consider the following statements: "I am asserting then something that has already been said" (*FP*, 25); "To my way of thinking I'll say that pictorial realism…" (*FP*, 26); "I ask the architects present here kindly to remember that I'm their friend, that I deeply admire their social and architectural work.… But I don't forget that I'm a painter, and so I must speak certain truths to them, or what I take to be truths" (*FP*, 176).

24. Fernand Léger, "Fernand Léger précise," *Fernand Léger: la poésie de l'objet, 1928–1934*, Christian Derouet and Marie-Odile Caussin-Peynet, eds. (Paris: Éditions du Centre Georges Pompidou, 1981), 19.

25. Fernand Léger, *Fernand Léger, une correspondance d'affaires* (Paris: Éditions du Centre Georges Pompidou, 1996), 27 (henceforth *CA*).

26. Pierre Descargues, *Fernand Léger* (Paris: Éditions du Cercle d'Art, 1955), 170.

27. Émile Benveniste, *Problems in General Linguistics*, Mary Elisabeth Meek, trans. (Coral Gables: University of Miami Press, 1971), 203.

28. Dorothy Kosinski, ed., "Avant garde et tradition," 165.

29. Jean-Paul Sartre, *Being and Nothingness*, Hazel E. Barnes, trans. (New York: Philosophical Library, 1956), 203.

30. Fernand Léger, "Réponse de Fernand Léger," *Fernand Léger: la poésie de l'objet, 1928–1934*, 22.

31. Fernand Léger, "C'est comme ça," 59.

32. Mildred Szymkowiak, *Autrui* (Paris: Flammarion, 1999), 227.

33. "Which certifies the letter's origin and meaning," Christian Bank Pedersen, "Noms spectraux, littérature en correspondance: sur les signatures de Franz Kafka dans les lettres à Milena Jesenská," *Poétique* 32 (2001): 109.

34. It should be noted that this kind of signature is sometimes interpreted differently: "the few letters signed 'your mother' are those where the sender seeks to exercise the entire weight of her maternal authority," Anne McCall Saint Saëns, "Nom, mais alors? Les signatures dans la correspondance de George Sand," *George Sand: une œuvre multiforme*, Françoise van Rossum-Guyon, ed. (Amsterdam: Rodopi, 1991), 102. The condensed French grammar book by Maurice Grévisse gives another interpretation: "The possessive adjective can have an expressive value and indicate, relative to the person or thing in question, the speaker's interest, affection, scorn, submission or irony," *Précis de grammaire française* (Louvain-la-Neuve: Éditions Duculot, 1995), 109.

35. Forestier, "Présentation" (*FP*, 17).

36. Masson, *La Mémoire du monde*, 100.

37. Béatrice Vernier-Larochette, "L'Écriture de guerre de Fernand Léger: un autotémoignage de vie," *Revue de l'Aire* 32 (2006): 250.

38. Georges Gusdorf, *La Découverte de soi* (Paris: Presses Universitaires de France, 1948), 41.

39. Fernand Léger, *Mes Voyages* (Paris: L'École des loisirs, 1997), 14.

40. Simone Herman was a young Belgian artist who entered the artist's academy in October 1930. This affair lasted ten years.

41. Fernand Léger, *Fernand Léger: une correspondance restante*, Christian Derouet, ed. (Paris: Éditions du Centre Georges Pompidou, 1997), 160 (henceforth *CR*).

42. Benveniste, *Problems in General Linguistics*, 197.

43. Benveniste, *Problems in General Linguistics*, 224.

44. Bernard Beugnot, "De l'invention épistolaire: à la manière de soi," in *L'Épistolarité à travers les siècles*, Mireille Bossis and Christian Porter, eds. (Stuttgart: Franz Steiner Verlag, 1990), 35.

45. Eugène Guillevic, "Le Grand Coordonnateur," *Europe*, 818–819 (1997), 9.

46. Pierre Arnauld, ed., *Fernand Léger: peindre la vie moderne* (Paris: Gallimard, 1997), 11.

47. Fernand Léger, *Entretiens: notes et écrits sur la peinture: Braque, Léger, Matisse, Picasso*, André Verdet, ed. (Paris: Galilée, 1978), 75.

48. Agnès Conacher, "*Les Tragiques* d'Agrippa d'Aubigné: les qualités d'un témoignage ou écho d'une histoire qui est arrivée et d'une histoire qui aurait pu être," *French Studies*, 57:1 (2003), 25.

49. Brice Parain, *Recherches sur la nature et les fonctions du langage* (Paris: Gallimard, 1942), 8.

50. Fernand Léger, "Notre Paysage," *Fernand Léger: la poésie de l'objet*, 21.

51. Pascal de Duve, *Cargo Vie* (Paris: Éditions Lattès, 1993), 149.

52. When Léger became aware of the German occupation, he took refuge in the south of France, then in fall 1939 he traveled to the United States where he had several friends and where his work was more appreciated than in France.

53. Jules Vallès was in exil in London from 1877 to 1879. See Dolorès Djidzek-Lyotard, "La Consigne de proscription: sur la correspondance d'exil de Jules Vallès," *Expériences limites de l'épistolaire: lettres d'exil, d'enfermement, de folie*, André Magnan, ed. (Paris: Champion, 1993), 39.

54. Dorothy Kosinski, "Un langage pour le monde moderne," *Fernand Léger: le rythme de la vie moderne 1911–1924*, 19.

55. John P. Wilson and Terence M. Keane, *Assessing Psychological Trauma and PTSD* (New York: Guilford Press, 2004), 20.

56. "Inability to relax, discontent with self-comfort activities," in Wilson and Keane, *Assessing Psychological Trauma and PTSD*, 37.

57. Amélie Schweiger, quoted in Bernard Beugnot, "De l'invention épistolaire: à la manière de soi," in *L'Épistolarité à travers les siècles*, 36.

58. "Léger was first of all a witness," Guy Marester, "Fernand Léger et 'Les Constructeurs'," in Pierre Arnauld, ed., *Fernand Léger: peindre la vie moderne*, 112.
59. Conacher, "*Les Tragiques* d'Agrippa d'Aubigné," 19.
60. Gijs Van Tuyl, "Palette mécanique dans un espace cybernétique: la peinture moderne de Léger à la lumière de la réalité virtuelle," in *Fernand Léger: le rythme de la vie moderne 1911–1924*, 195.

16

Artists, Commemorations, and Political Culture: The Example of the Great War

Annette Becker

Museums, cemeteries!...Truly identical in the sinister way they mingle bodies unknown to one another....The daily frequenting of museums, libraries, and academies (these cemeteries of wasted efforts, these Calvaries of crucified dreams, these registers of halted bursts of energy)...truly is for artists what the extended guardianship of parents is for intelligent young people, intoxicated with their talent and their ambitious....Welcome, then, the good arsonists with sooty fingers! Here they come! Here they come! Spread fire along the library shelves! Divert the canals to flood the museum's vaults!

So exclaimed Marinetti in the "Manifeste initial du futurisme" appearing in *Le Figaro* on February 20, 1909. Earlier in the article he announced, "We want to glorify the war—the only way to cleanse the world—militarism, patriotism, the destructive acts of anarchists." Marinetti's "no future" was transformed into war's future, in a twentieth century marked first by mass warfare before becoming, almost simultaneously, the century of mass murders and genocides.

Artists always have to be taken seriously. A museum is what cemeteries in fact have become, especially the non-cemeteries made up of all the traces of war and massacres of the two world conflicts. Marinetti both won and lost. Today's Italians, his distant compatriots, have crafted the concept of *Museo al aperto* to describe spaces where the traces of wars and mass death have become places of commemoration, mourning, and the mooring of despair.[1] These "open" museums are numerous today in France, from the site of Verdun to that of the Somme, from the village of Oradour to the buildings of Drancy and the remnants of the Struthof concentration camp. There are also the numerous plaques that recall the murder of Jewish primary school children and the murder of Algerians drowned in the Seine in 1961. Does the impossibility of projecting oneself historically into a too painful past explain this intensive *patrimonilisation*, this reappropriation of places that have become either real or substitute cemeteries and thus museums? Through a museum-like present are we attempting to find the past once again?[2]

Contemporary artists also inhabit these places of death. By situating their works outside of traditional museums, they offer their messages to passersby who have—or have not—chosen suddenly to confront this art and the past thus revisited. In commissioning public art, the state has taken over the double market of memory and of art, entrusting its ministry of culture and its regional directors of cultural affairs to work out the details. When events judged to be particularly important occur, projects are overseen directly by the office of the president of the Republic (François Mitterrand in 1984 and Jacques Chirac in 1996 and 2006 at Verdun) or the prime minister (Lionel Jospin at Craonne in 1998). Commemoration, memory, and forgetting try somehow to get along with contemporary art. In the calculation of political gain and financial cost, will it be the state or local groups who will pay the bill?

In the name of the "duty to remember," and often forgetting the duty to history (since a footloose and fickle memory is so much easier to instrumentalize than an unbending critical history), a spectacular return to the Great War in collective consciousness has occurred since the 1990s in France, as in other Western European countries. Numerous aspects of this commemorative upsurge—historical, memorial, political, editorial, media—were disputed or strengthened. Eighty years after the armistice, it became increasingly apparent that the end of the fighting meant neither the end of conflict nor of war. Although a consensus was reached over suffering and mourning, beginning with the massive erection of commemorative monuments between 1919 and 1922, all the other aspects of understanding the war remained contentious. The Third Republic, by continuing the craze for statues marking the years between 1880 and 1914, had commemorated itself despite and through the public debates in the ceremonies of 1918–20.[3] What could the Fifth Republic do, especially during the period of political cohabitation? Moreover, how could the enduring works of artists be dished up with a political sauce, linked inevitably to temporary circumstances and contexts, in this case the context of the speeches inaugurating these works, which were followed by counter-speeches and controversy? The 1998 debates of the eightieth anniversary of the armistice clearly showed that France continues to be haunted by the long memory of the Great War and its repression or symmetrical forgetting.[4]

In the following pages I discuss the interaction between contemporary art, public commissioning, and political discourse, focusing on the Biron monument commissioned from the artist Jochen Gerz in 1993 and on the commission Lionel Jospin's office made to five artists in November 1998 for the anniversary of the armistice. The speech Lionel Jospin gave at the time led to a huge Franco-French debate that is far from over. When I wrote this article, I had just returned from the Chemin des Dames in France, where I attended the reinauguration of Haïm Kern's statue, which was installed in 1998 and had been wrecked and damaged for a second time.

"Nothing Is Stronger than Traces": What Artists Display[5]

July 14, 1996—the day was not selected by chance—marked the inauguration of a new war memorial in Biron, designed to replace the obelisk erected after the Great War.[6] For its creator Jochen Gerz, a German artist born in 1940 who has lived and worked in France for quite some time, the memorial is a "living monument." When the mayor of the small Dordogne community requested support for the extensively damaged

1914–18 war memorial, Gerz was selected for the project because he had completed several public space monuments during the previous two decades. "Public space is the true gallery of today, the place where art has the most future," he observes. His various "anti-monuments" play on the paradox of dead and living, always aiming to heighten awareness of the difficulties involved in remembering, of this past that won't pass because it won't pass on, and because, for Gerz, it is constantly being turned over without being reappropriated. Gerz has aptly summed up what he has at heart and how he views this project, and his work in general, in the political context:

> Death is always a scandal; it is unacceptable, but it is also a part of life. There is "in addition" the organized, industrial death of the twentieth century. No peace can be made with that or with the ideas that produced those deaths, which are as impossible to explain or to excuse today as they were yesterday. That's what the commemoration of death is. You can't commemorate a genocide the way you commemorate May Day. Memory is a day without a date. There is no grave or art work that could house these deaths. Only the living can bear witness to them and look after them. By dealing with absence, one is not very far from presence and from what you call the celebration of living people. I think that our fragility, whether we are alone or not, at home or elsewhere, becomes accentuated through this absence. Like absence, art too always has a relation to people. If it's a celebration, then it must contribute, despite everything, to making life beautiful.[7]

With this in mind, Gerz asked all the inhabitants of the small village a "secret" question, which only they and he knew. If we judge from their answers, they must have been asked to reflect on what they thought would be important enough today for them to risk their life. Their answers were transcribed on plaques of red enamel affixed to a new obelisk that was identical to the previous one. The "living" obelisk is in Dordogne stone from the region. Jochen Gerz also reattached the two previous plaques to the monument: "To the dead of the Great War, perpetual memory" (nine names) and "Deported 1944" (two names).

Gerz wanted to return their monument to the villagers themselves who expressed themselves anonymously. It is a kind of interactive monument, dynamic in a double sense since the villagers said what they wished there about their own life and about the death of their loved ones in war throughout the century, and since space remained for new plaques to be added. The monument is no more static than the dead. Thus it is a political work in the primary sense of the term, a work of citizens for their city. The 127 plaques express a suffering, a manner of seeing today:

> All throughout my youth, I saw my grandmother cry for her sons. She lost two of them, the third one had only one foot, and the fourth one had to leave. She almost lost her mind, the poor woman. She cried out of sadness, anger, and fear. During the Second War, I was expecting a baby, and we slept in the forest for fear the Germans would find us. But later the Germans were so good to my son who worked for Bayer and who had kidney trouble. They did everything for him before he died.

All the villagers, along with the artist, unthinkingly combined the two world wars, like an echo of the Thirty Years War that one of the Barrès-like fighters of the Great War spoke about (the war of land and the dead), a fighter who became a politician passionately interested in rhetoric—General de Gaulle.

In late 1997, Lionel Jospin's office decided to commemorate November 11, 1998, in striking fashion by soliciting five artists: Haïm Kern, Alain Fleischer, Christine Canetti, Ernest Pignon-Ernest, and Michel Quinejure. In viewing the various artists' works, what is striking is how acutely they represent the totalization of war. For them, not only men but women, children, and countrysides were transformed or destroyed. One way or another, in whatever medium they worked, they all showed that the First World War was a catastrophe and a tragedy, and that the work of mourning and the work of memory must never be separated from a reflection on their opposites, namely, forgetting and repression.

In 1998, Bosnia was on everyone's mind, and the Sarajevo of 1914 was more present than ever, as the artist Christine Canetti puts it:

> I kidnapped this heavy, bitter subject, which evoked loss, loneliness, abandonment, and death. This commission allowed me to focus on 1914–18 and to see in these violent stakes the same atrocities and ravages as in current wars, two hours away from us.... Working on the themes of memory, the passage from life to death, I like to dig around in history.... Sometimes an artist feels useless. This work of transcription is what repositions the artist in his or her role as an intercessor with a social mission.[8]

By letting the women of the occupied territories speak in her beautiful work of glass, a symbol of fragility and tenacity, Canetti succeeded in reestablishing the balance between the war front and behind the lines, expressing the "home front" and the specific nature of the dramas of those who were occupied, especially occupied women.

Alain Fleischer felt that he was "being watched by the dead"[9] while he was surveying the battlefields, hence the title of his work. He photographed hundreds of soldiers' gazes in snapshots from the period, but he did not complete the development process. On the day of the exhibition at Arras, hundreds of developing pans awaited the spectators. In the red light of a photography laboratory spread over a battlefield, eyes slowly became visible, those of men whose look seemed to have existed only for that death. The forever open eyes of Alain Fleischer's soldier are, as he says, "practically immaterial...the opposite of a war memorial. They didn't know that this would be the last photo taken of them."[10]

One way or another, the artists reveal the enlightening blindness of war, perhaps precisely because they make no claim either to do history or to exercise a duty to effective memory, but simply, as Paul Klee put it, to exercise their profession: "art does not reproduce the visible, but it makes visible." Kern: "War is an abomination. Are those who practice it abominable beings? Those who endured it are surely victims. What about the warmongers?"[11]

The videographer Michel Quinejure conveyed very effectively both the ties to the First World War—he filmed mud and the fire of the bronze foundry juxtaposed with images of mud and fire brought back from the front by filmmakers, his colleagues of 1914–18—as well as the unending mass death of the twentieth century, from the Holocaust to contemporary conflicts. "Never that again, yet twenty years later everything started over.... The Great War, *la der des ders*, perhaps was only a rehearsal. One must be wary of final 'dress rehearsals.'"[12]

Ernest Pignon-Ernest's bronze trees broken forever are an exemplary metaphor for the conflict. The artist who drew so many human bodies that have been tortured, raped, and excluded decided that for this work he could not show the soldiers' bodies. Hence he chose to cast stumps of 20-year-old trees in bronze. "Nature won all that, life wins out. The trees grow on the bodies of those who died there. Bodies, flesh, blood, land."[13] (The works of Paul Nash comes to mind here, a British artist of the Great War whose tree stumps are metaphors of mutilated men.) As early as 1915, Edith Wharton had certainly best described war's amazing transformation of nature and men: "Our road ran through a bit of woodland exposed to constant shelling. Half the poor spindling trees were down, and patches of blackened undergrowth and ragged hollows marked the paths of the shell...there was something humanly pitiful in the frail trunks of the *Bois triangulaire*, lying there like slaughtered rows of immature troops."[14]

All the artists were especially struck by the site of Craonne along the Chemin des Dames, this forest that had grown up again on a vanished village that had had to be rebuilt elsewhere because it was nothing but a vast cemetery. Christine Canetti spoke of the antinomy between the beauty of teeming nature and the horror of suffering. Haïm Kern said that he wanted to return to nature through his work, which was not supposed to remain a foreign body in the forest.[15] (See figures 16.1–16.3.)

On the plain of Craonne, in the heart of the Chemin des Dames, a tall, thin bronze net rises up, its title highly evocative: "they did not choose their sepulcher." Enmeshed in it are heads, always the same yet always different. Placed at various heights and angles, they never attract light, shadows, or the spectator's gaze the same way. The artist was moved by the suffering and dying experienced by the soldiers of the Great War, and especially by the number of anonymous soldiers, mutilated by unimaginably powerful artillery. "I want this sculpture to be in close physical proximity to these men so that, enmeshed in history, they return to us from earth to light."[16] On the plain, almost 90 years later, farmers continue to plow up bits of metal, bone, and rot. But they never find faces. Paralleling the philosophy of Emmanuel Levinas, Haïm Kern gives back a face and a life to those hundreds of thousands who disappeared, swallowed up by earth and fire. The sculpture's net becomes a chain metaphor of mourning, of an infinite and inextinguishable suffering.

The Politicians Arrive: "The Echo of This Cry Must Not Fade"[17]

As soon as Haïm Kern's sculpture had been commissioned, in other words before the artist himself knew what he was going to express, the prime minister's office decided that this work would provide the backdrop for an important speech by Lionel Jospin. The sculpture was designated for "a political use," and it was going to become "the monuments of monuments," put to a national purpose.

Lionel Jospin began his remarks in front of Haïm Kern's sculpture by referring to a "sacred place." How could such a comment be questioned when one is speaking on such a symbolic level? On Jospin's right as well as on his left, however, the "sacred" is interpreted differently, which explains the divergent consequences of the ceremony of November 5, 1998. All that was remembered of the speech was that it was time to "reintegrate the mutineers" in national memory. The headline of Le Monde read, "The Republic honors the 1917 mutineers." Throughout his term in office, Lionel Jospin insisted on a memorial consistency, which he summed up in a speech delivered on April 26, 2001, at the National War Veterans Office on the occasion of the inauguration

Figure 16.1   Haïm Kern, *Ils n'ont pas choisi leur sepulture* (detail), Chemin des Dames, Aisne, France. Photo Annette Becker.

Figure 16.2 Haïm Kern, *Ils n'ont pas choisi leur sepulture*, Chemin des Dames, Aisne, France. Photo Annette Becker.

of a plaque to honor Georges Morin, a veteran of the Great War, a resistance fighter, and who died in deportation (once again the Thirty Years War).

> The NWVO ensures that the rights of the generations of those who saw battle are protected. Through you, the nation expresses its obligation to all those who gave their life in its name. For almost forty years, the government has ensured that this gratitude has been confirmed, both by solidarity and by the labor of memory.... This labor of memory requires determination and clarity. We must not fear confronting our past, with its shadows and its lights. We must struggle against forgetting and against the distortion of facts. We must be able to view our history head-on. The government has taken pains repeatedly to do that. On the occasion of the eightieth anniversary of the armistice of 1918, I first affirmed the need for truth. During my visit to Craonne, on the Chemin des Dames, I expressed the wish that the memory of soldiers who, after a hard fight, had refused to be sacrificed irresponsibly, be fully reintegrated in collective memory.

In 2001, Lionel Jospin revisited the first political-memorial commotion he caused in Craonne. Why did he choose the Chemin des Dames, the symbol of the 1917 cases of disobedience—the so-called mutinies—to commemorate the end of the Great War of 1918? Why then did he evoke Nivelle and Pétain, the disastrous attack, the catastrophic health conditions, the repression? In a period of political cohabitation, the site of Verdun had become "presidential" because it had come to symbolize the primary

Figure 16.3 Haïm Kern, *Ils n'ont pas choisi leur sepulture* (detail), Chemin des Dames, Aisne, France. Photo Annette Becker.

site of the Great War in national memory and the commemorations of François Mitterrand in 1984 and Jacques Chirac in 1996. As such, Verdun had to be set aside. Moreover, the socialist prime minister wished to show his faithfulness to the memory of his father, who had come from the occupied territories, who with his red armband had been deported at the age of 16 to work in Germany during the Great War,[18] and who had remained a fundamentalist pacifist even into World War II.

Standing before the sculpture, with the one exception of the dated concept of "shot as an example" (isn't every punishment exemplary[19]), Lionel Jospin courageously attempted to speak of yesterday's war with the words of today, without forgetting the work he had come to inaugurate. Not only did the speech take the place of Haïm Kern's sculpture, but some seized on the speech as a way of viewing the work, which inflated the debate. Jospin was vilified by the democratic right, insulted by the extreme right, and rejected by "the left of the left," the former finding that he had said too much and the others that he had said too little.

The sculpture became the "monument to the mutineers" for the anti-war left, and it was attacked and "harmed" twice by partisans of military order and/or the extreme right. In 1998, the latter tried to break the sculpture, leaving behind a *vive Pétain* and a swastika. Each time the plaque bearing the name of Lionel Jospin was replaced, it was stolen or marked up with graffiti. In 2006, the sculpture was damaged once again. Repaired now, since November 11 it bears a notice from which Jospin's name has disappeared.

Thus, this period of commemoration has pushed to feverish heights the process of instrumentalizing the soldiers of 1914–18, with everyone seeking to use Jospin's comments for or against their own political view of the war, forgetting both history and the artists' works, reduced to pretexts for this intense Franco-French debate. For 80 years, one side has seen only heroes and has said nothing of the suffering, and the other has seen only victims and has said nothing about assent.

In memorial consciousness, it is better to be a victim than an agent of suffering and death. The latter, always received and always anonymous, is never given. One is always its victim, unless one is the victim of its leaders who are then promoted to organizers of the massacre. In the twentieth century, the time of tragedy par excellence when one must identify with victims and thus identify the executioners, the dialectic of mutineers and murderous generals was essential. But by making fighters into those who were sacrificed to military butchery, the process of victimizing the soldiers prevented reflecting on the case of civilians, the true new victims of the conflict, but also on the soldiers themselves, betrayed once again concerning the essential question of their assent and their suffering. What was betrayed was the connection between accepting violence and the level of suffering undergone. Contrary to belief, the Great War was a past that didn't pass on. Historiography spoke of cases of violence, cruelty, mutual accusations of barbarism, suffering, and assent at the very moment when people, hand in hand, were building Europe. Thus, displaced onto intense Franco-French debates was what could not be said in the French-German context.

Everyone is so fearful of not having suffered enough in this century of total war that "the victims' competition"[20] takes on a dimension of urgency revealed by an immoderate and anachronistic vocabulary. For militant pacifists, not only were the combatants merely nonconsenting victims, the rebels were the only true heroes. Weren't the "mutineers" of 1917, because of their very act of rebellion, the precursors of European unity? And wasn't the Nivelle offensive "the first crime against humanity,"[21] as the mayor of Craonne himself said? We know now that the blindness concerning the violence expressed by the indicted and traumatized combatants continued in the interwar generations, who did not perceive that the culture of violence resulting from the Great War radiated throughout the world. Yet brutalization caught up with them irreparably, for it came to lodge at the heart of Western societies. Although it took almost 80 years to reflect on this phenomenon, the weight of mourning and of pacifism still won out. More has been written—and dreamed—about the Christmas truces (as in the 2005 film *Joyeux Noël*) and about fraternization than about hatred of the enemy. More thought has been given to a few "mutineers" and to those "shot as an example" than to the millions of consenting and suffering soldiers.[22] As painful as it might be, that one's grandfather or father was killed in the anonymity of artillery battles and gas attacks is easier to admit than that he could have killed.

All the artists commissioned by the state showed in various ways the horrors of death and suffering, hence Lionel Jospin's remark recalling that Haïm Kern's sculpture is "dedicated to all those without a name who lost both their youth and their future." These artists' work transfigured mass death and gave violence to violence. Emotion broke out of the museum, then, perhaps nearer to the experience of war; outside there is never dust or forgetting disguised in museography. Marinetti himself would have been surprised.

Nevertheless, politicians generally don't look much—if at all—at the works whose realization they nonetheless think holds promise. They favor works of art before forgetting them. What the political stripes of the minister of culture are has no bearing on this. If the director of historical monuments who oversaw Biron had not tragically died, the Gerz monument, to which he was pathologically opposed, would never have been erected. The minister of culture was strikingly absent during the November 1998 ceremonies, having decided in part to yield ground to the prime minister, but also since he had been outstripped by the artists he himself had selected. What actually accompanied the work of Pignon-Ernest were the voices of little schoolchildren singing the Marseillaise, as well as the bouquets that nameless people hung on the Haïm Kern's work, thus making it "living" as was the monument of Jochen Gerz. And November 11, 2006, children and pacifists singing the Song of Craonne along with everyone, the absurdity of war.

We read on one of the plaques at Biron:

My cousin left Biron for Dachau. They made him play the violin during the hangings. That saved his life. He never touched a violin since. It's surely madness to give one's life, but if we are plunged once again into war, it must indeed be done, to defend the country, one's loved ones, the soil, as our grandfathers and fathers did. Without politicians, there would be no war. We should listen to the people of the land.

Notes

Chapter translated by Daniel Brewer.

1. Annette Becker, "Musées ouverts, traces des guerres dans le paysage," in *Des musées d'histoire pour l'avenir*, Marie-Hélène Joly and Thomas Compère-Morel, eds. (Paris: Noêsis, 1998).
2. "Les Musées des guerres du XXe siècle: lieux du politique?," Mireille Guessaz and Sophie Wahnich, eds., *Tumultes*, 16 (April 2001).

3. I extend warm thanks here to Maurice Agulhon. Annette Becker, "Du 14 juillet 1919 au 11 novembre 1920; mort, où est ta victoire?", *Vingtième Siècle, Revue d'Histoire* 49 (1996): 31–44.

4. Not only the memory of the Great War, but the entire twentieth century is questioned at present in this way. Paul Ricœur, *Memory, History, Forgetting*, Kathleen Blamey and David Pellauer, trans. (Chicago: University of Chicago Press, 2004).

5. Alain Fleischer in the film by Michel Quinejure, *Quatre artistes sur les traces de la Grande Guerre* (DAP, 1998).

6. Jochen Gerz, *La Question secrète, le monument vivant de Biron* (Arles: Actes Sud, 1996), and "Un artiste allemand redonne vie au monument aux morts d'un village de Dordogne," *Le Monde*, July 20, 1996. "Jochen Gerz, le monument vivant de Biron," *14–18 Aujourd'hui, Today, Heute*, 1 (1998): 154–57.

7. Conversation with the students of the École des Beaux Arts de Dijon, August 2000.

8. Christine Canetti, "Cinq artistes pour la commémoration de la Grande Guerre," *14–18 Aujourd'hui, Today, Heute*, 3 (2000), 230–35.

9. *14–18 Aujourd'hui, Today, Heute*.

10. *14–18 Aujourd'hui, Today, Heute*.

11. Haïm Kern, *14–18 Aujourd'hui, Today, Heute*, 236–37.

12. Haïm Kern, *14–18 Aujourd'hui, Today, Heute*.

13. Interview in Michel Quinejure, *Quatre artistes*.

14. Edith Wharton, *Fighting France, from Dunkerque to Belfort* (New York: Charles Scribner's Sons, 1915), 165–66 (June 22, in the north of France).

15. The two artists in the film by Michel Quinejure.

16. Haïm Kern (1998), quoted in *La Lettre du Chemin des Dames, Bulletin d'information édité par le Conseil général de l'Aisne*, special issue, 3 (November 2006), 2.

17. Lionel Jospin. The Prime Minister is quoting Blaise Cendrars here, after Aragon and before Barbusse. At Verdun in 2001 he will cite Genevoix, which illustrates the literary tradition of French-style political rhetoric.

18. Annette Becker, *Oubliés de la Grande Guerre: humanitaire et culture de guerre, 1914–1918: populations occupées, déportés civils, prisonniers de guerre* (Paris: Noêsis, 1998).

19. Perhaps he can be reproached for having forgotten the thousands of soldiers who were sentenced to hard labor and received harsh sentences that were rarely commuted, who should have found defenders except that, like other forgotten ones of war, they weren't of the right political stripe.

20. Jean-Michel Chaumont, *La Concurrence des victimes: génocide, identité et reconnaissance* (Paris: Découverte, 1997).

21. Noël Genteur, the mayor of Craonne, answered Lionel Jospin: "The dignity that you render today to the soldiers is a measure of your humanity. [In this place,] the first crime against humanity went unpunished."

22. Nicolas Offenstadt, *Les Fusillés de la Grande Guerre et la mémoire collective (1914–1999)* (Paris: Odile Jacob, 1999).

17

La Victoire en chantant: The Cinematographic Transfer of World War I in Equatorial Africa, between Colonialism, Patriotism, and Politics

Floréal Jiménez Aguilera

Ideology and the imaginary supplant a lack of knowledge, sometimes commonly with the imagery of "savages," sometimes slightly by inspiring fashions, sometimes deeply by contributing to the transformation of ways of seeing and creating. When the colonized bursts on the scene and regains his freedom, truths suddenly appear; but more than traces exist of what the colonial relation jointly produced. Immediate history constantly dismisses a more distant history.

(Georges Balandier)[1]

*L*a Victoire en chantant (1976, 90 minutes, Jean-Jacques Annaud), retitled *Blancs et Noirs en couleur* (*Black and White in Color*) after it received an Oscar for the best foreign film, is both singular and original because of several historical features, the distorted way they are perceived, and the relations between them that the film presents, together with the exaggerated and satirical way it treats them, all the while without abandoning an apparent seriousness designed to stigmatize colonialism and war. Although this structure and its variations, without the additional parody, can be seen or imagined in most colonial films produced after the 1960s,[2] it is doubtless most apparent in this film, where it is much more effective. The film appears to minimize the French-German military confrontation,[3] focusing instead on the recruitment of Africans and their integration into the military system of the period. It shows them placed in the trenches facing German installations that the French authorities are afraid to attack, in a strategy that was not applied in Africa where recruited soldiers were sent to the European front.[4]

It is interesting to note as well that this film is probably the only non-documentary French film (co-produced with the Ivory Coast and the German Federal Republic) whose theme is World War I in French Equatorial Africa (F.E.A.). The film is set in Oubangui, a place that serves to display and highlight, in calculated and manipulated fashion, the colonial world and the way it works in economic and human terms. The film selects certain events from this context and juxtaposes them, in significant yet limited fashion, with particular forms of the European conflict and its development. The film then projects French-German relations up to their legible contemporary outcome, without predicting or guessing at political forms and issues in Europe beyond the good relations that already existed between France and Germany when the film was made.

The film repetitively presents most of the elements of the historical and colonial narrative,[5] whether real or mythical.[6] These include the hero who, perhaps unwillingly, has lost status and lacks drive, a civilian or a soldier outlawed from society,[7] prey to solitude and seeking consolation or fleeting amusement with local women and in violent adventures where he is the tool of Western expansion, long expeditions through inhospitable lands, relations with the colonizers, priests or missionaries charged with converting the natives and subjecting them psychologically and ideologically, bearers for the whites, paddlers or accompanying workers who sing to ease their efforts. There are also fights with stubborn local communities, the altruism of doctors bringing care, comfort, and healing, bivouacs by day or at night attended to by Africans, and the enslavement, ignorance or animality of natives. Without completely leaving this set of images behind, the film provides its devalorizing version, whose variables no longer correspond to the glorifying and propagandistic expression of colonial films and their spirit of adventure. Consequently, none of the characters appear as traditional heroes possessing a positive personality allowing spectators to identity with them, and doctors, teachers, and other charismatic and enterprising representatives of a tangible "civilizing mission" are absent. In the traditional film, one or more characters attract or kindle the spectator's adherence,[8] contributing to questioning specific forms of meaning shaded by the film's ambiguity. In Annaud's film, the traditional film deteriorates, erasing its almost always present values, whether latent or not. The ambiguity of the film's possible meanings continues in a filmed interview with the author. As so many others have done, he too must mention and praise the positive work of colonialism, following an often-repeated argument recalling the building of infrastructure, medical work, and the economic development of the former colonies.[9] He thus seems to regret the anti-colonialist tinge of his work.

The way the film's treatment of colonialism has been considered theoretically adds further ambiguity to the film's concrete meanings, yet without lessening its pertinence or preventing the public from having a certain freedom to enjoy it. Thus in the booklet accompanying the DVD and its additions,[10] especially in his interview, Jean-Jacques Annaud observes that if a situation can be imagined, it is because it can be lived. And if it can be lived, then necessarily others already have done so. Consequently, events justifiably could be transformed excessively, some forgotten and other invented, with the result placed plausibly in the story of made-up characters and realistic developments. No one knows everything, and if someone knows certain things, they are considered, transformed, and possibly reconstructed in a way determined by how they are thought of, given all the variables produced by culture and surroundings. Thus, one of the plot's actual hidden dimensions is composed of the unmediated daily life of individuals such as they construct, live, and experience it, shaped by material and social contingencies, emotions and impulses, the interaction between multiple elements, and an individual and collective cultural filtering, with all the necessary reservations and clarifications. This is an aspect of history's unknowns, which are absent from the archives, remaining unspoken either on purpose or involuntarily because they are deemed to be unimportant in the march of history. Yet the cinematographic imaginary in particular can construct another history or a complementary and more random history, one that with the filmmaker's involvement reveals how a society functions mentally and socially, over time or in a given period. We have here some of the major features of cinematographic creation, which concern the various relations between film and society, and the evaluation of their importance and of their forms, as well as the conscious or unconscious effects on the cultural meanings of characters that they can reveal.

Jean-Jacques Arnaud's film is based on his strong and determining personal experience, which is intimately connected both to the colonial world immediately following the independence of the African states and to direct cinematographic references to that world made by the work of Jean Renoir and by political cinema of the 1970s, tinged as it was with colonialism or post-colonialism.[11] The film and its scenario stem from the director's extended stay in Cameroon when he was a young National Active Service Volunteer and a recent graduate of the Institute for Advanced Cinematographic Studies. During his stay, he discovered a county that captivated him. He came to know it better through history books consulted in libraries, where his attention was grabbed by the episode of the taking of Mora peak by French and English troops fighting the Germans in 1914–15. He visited the site and the surrounding region, collecting extensive documentation on the event. The DVD booklet provides a detailed description, including facsimile reproductions of the historical and geographical documentary material from the period, from both the French and the German side. Several years later this research resulted in *La Victoire en chantant*.

We do not wish to return to the kind of "impressionistic" analytic criticism inaugurated by André Bazin and linked to the *auteur* perspective and later pursued by the *Cahiers du cinéma*. Yet we cannot exclude this experience in theorizing the relations between cinema and society, in which the main perspective must be a historical and anthropological one that situates filmic expression and its human and social breadth outside the narrow limits of a scientific and strictly formalist framework of dissection that is often governed by semiology, however useful that framework might be. Cinema and society, cinema and history,[12] and the "politics of *auteurs*" converge in the "pluridimensionality of the filmic object" (*pluridimensionalidad del hecho filmico*[13]), with a mutual enrichment that involves especially the first two. The *auteur* becomes a cultural intermediary between personality and the imaginary, between society and the period in which he develops, extending his creation toward an inevitable universality. Marcel Oms defined this phenomenon thus:

> Every cinematographic representation of a historical moment necessarily involves a retroactive imaginary figuration of a form of the present. Or a series of forms. Whether he intends to or not, the artist who attempts to recreate the past projects a set of personal mental images that he structures based on a narrative need that is conditioned by his idea of the period in question.[14]

In addition to Jean-Jacques Annaud's African experience and its cinematographic consequences, one must add that of Georges Conchon, the project's final scriptwriter, whose contribution adds to the film's latent ambiguity. Jean-Jacques Annaud aimed to make a film about his discoveries and explorations, at first a short-length then a full-length film, although he never succeeded in completing a satisfactory script. His meeting with Georges Conchon was decisive. Africa was quite familiar to this upper echelon administrator of the French Republic posted in French Equatorial Africa (*secrétaire des débats* of the French Union from 1947 to 1959). As general secretary of the Central-African Republic Legislative Assembly until 1960, upon the country's independence he wrote its constitution. He later turned to literature and film, taking inspiration from his African experience.[15] Thus the superiority complex of the white man endures following the colonial period, his post-colonialism guaranteeing a natural moral, intellectual, political, and economic continuity.

   La Victoire en chantant opens with a panoramic landscape shot, sweeping from the German installations with the imperial flag floating above them to the native village in front of which stand some tumbledown French houses with a flagpole flying a washedout French flag. The landscape is unchanging, the sparse vegetation of scattered bushes possessing nothing exotically tropical, with no line of demarcation separating the two nationalities. On the German side reign order, discipline, and the training of native troops, whereas a nonchalant and carefree attitude prevails on the French side. A few of the principal players of colonialism are present: the soldier (a sergeant in the colonial infantry), missionaries busy swapping fetishes for holy images and statues of the Virgin, the man of science (a young geographer fresh from the École Normale Supérieure), and traders. This group represents a rather unlikely concentration of activities and functions given the heightened isolation and poverty of the region and its inhabitants. Although organization in the colonies was less than perfect and efficient, it was far from being as weak as the film suggests. French Equatorial Africa was divided into 49 districts (24 civilian and 25 military),[16] each of which was divided into subdivisions overseen by a civilian or military administrator whose authority extended for some 75 kilometers beyond the administrative center where he was located. At least 60 percent of the territory was under such control, whose effectiveness depended upon the success of pacification, which remained incomplete and was constantly challenged.[17] Although they belonged to the past, the region's original political and cultural structures—organized arabized states, sultanates, traditional chieftain structures, a feudal system but in which powers and commercial exchanges were delegated[18]—are completely absent from *La Victoire en chantant*. The region becomes real only through the presence of the white man and his Ubu-like manner of settling in. There is no teacher or school, no businessman, builder or land franchise holder here, no flamboyant chief district administrator in white pith helmet, and no doctor to whom the care of the local population has devolved. The only believable absences are those of a doctor and a teacher, the "hussar of the Republic" according to Jules Ferry who endlessly sang the praises of colonialism. Medical networks and schools had been established in Africa, but they were very rare in French Equatorial Africa. The description of the colony is given by the geographer in a letter full of deception that he sends to one of his professors. He writes it at his desk, reading the sentences out loud, in front of a portrait of Jean Jaurès that suggests his political affinities and his humanist aspirations:

> Africa is far from being the hell that overblown colonial literature presents it as. There where I expected to encounter wild beasts, I meet only dogs, cows, chickens, and the occasional duck. There where I expected ferocious savages brandishing assegais, spears, and poisoned arrows, resolved to make a feast of the travelers, I see only peaceful country folk whose pastoral existence recalls that of our peasants of Indre-et-Loire and Creuse.
>
> The only dangers I have to fight against are boredom and especially the company of the handful of compatriots whom I must frequent. Where are the empire builders, these modern-day knights, these fearless pioneers we dreamed about as we read adventure novels? And it's been more than 6 months since we had a visit or received any mail. Apparently this isn't an unusually long time to wait in this part of Africa. But I'm eager to learn what's happening in Paris. What's at the Théâtre Français, the Opéra, what is Debussy going to present, what's happening with Péguy, what are our socialist friends doing?
>
> Concerning colonial affairs, tell them to be reserved in their opinions about the inferiority of the black race. Seen up close, this inferiority is due less to the shape of the skull or the makeup of blood than to the feeling we have about our own superiority. Although it might surprise you, I'm pleased to confirm that in many ways the natives are quite deserving of the fine name of human.

When the young man, dressed in a tidy white outfit, finishes writing his letter he settles in on the veranda outside the bungalow where he lives, ready for a glass of water served by an African servant. Before pouring the water from a carafe, he wipes the glass with his handkerchief. An initial doubt thus is felt concerning the sincerity of his egalitarian feelings toward Africans.

While the letter is being read, punctuated with images constituting a different tone of colonial propagandistic bombast, the camera shows a black employee asleep behind merchandise who is awakened by the inexplicable blows from one of the Réchampot brothers, the local traders, returning from a sexual siesta. Another colonist, the owner of the local "bar," is seen nodding off on his bed, listless and unresponsive to the desire displayed by the young woman next to him. The geographer's words are illustrated by languid and sultry scenes of everyday life in an African village: natives busy at small tasks, a soldier lying asleep against a hut surrounded by chickens, and a wandering dog. The colonists' activities and preoccupations, with their cruel futility, outweigh those of the colony's inhabitants. Those of the natives are eliminated, and although their sufferings are shown they are secondary and last for only a few short scenes, relegated to the mention made of them by the white man. As in the majority of colonial films, Africans are presented as naïve, simple-minded big children. But in fact, they're playing a role in order to deceive the white men all the better. For example, as a good Muslim the assistant of the small village temple performs his prayers then hurriedly dons a crucifix for the arrival of two missionaries, before whom he makes the sign of the cross. In their proselytizing travels, the two religious men enjoy the songs the bearers sing in their native language. However, the song refers to how fat and heavy the one is, and the smelly feet of the other.

When the mail and a few newspapers arrive, the colonists learn of the death of Jaurès and of the declaration of war between France and Germany. Meeting in the bar, the members of the community are incredulous. For decades though this war had been expected, even hoped for in France, as a form of revenge against the "Jerry" that would make it possible to recover Alsace and Lorraine. One of the traders thinks the more likely enemy would have been England, eliminating the revanchism and the wish to recover the two provinces that drove French nationalist thinking since 1870. He forgets the "Entente cordiale," unaware of the recent and constant bargaining among the great powers to divide up Africa between 1908 and 1911.[19] French Equatorial Africa was directly involved in these deliberations: in exchange for recognizing the French protectorate in Morocco, Germany received 275,000 square kilometers of French Equatorial Africa, enlarging its colony of Cameroon bordering the Oubanqui River. Moreover, by the time the members of the community learn of the declaration of war, military operations had already begun several months earlier in their region. Of course these operations are not as large, in terms of troop movement and numbers, as those in Europe, but they involve several thousand men from several nations, and they are proportionally just as deadly. An article appearing in 1916 in the *Bulletin de la Société de géographie et d'études coloniales de Marseille* bears this out.[20] Although porterage in the film has a picturesque quality to it and is shaped by the colonial imaginary, it was also deadly and dramatic,[21] requiring as many men as the recruitment of potential soldiers, the "Black Force" so dear to lieutenant-colonel and later General Mangin.[22] Although it is shown without suffering and as a normal activity without unpleasant consequences for Africans, porterage designated their subordination and their constant submission. The two kinds of recruitment incited often large-scale insurrections that were already endemic, caused by an over-taxation of the population, forced labor, and the challenging of local powers, none of which is expressed in the film. Moreover, the conquest of French Equatorial Africa was one of the most deadly that had been seen between 1928 and 1939, which bore witness to the revolutionary and indomitable spirit of the populations.[23]

The patriotic streak of the French characters appears only after they remember that the Germans haven't yet paid for the last delivery of supplies and that the bill likely won't be honored. The bar owner realizes that Germany is just next door, a few kilometers away. The Marseillaise is sung, and the sergeant of the colonial army reads legislative texts on the defense of the colonies and the mobilization that has been decided upon, along with the conquest of the neighboring German post, despite the humanist and realistic reservations on the part of the young geographer who is despised because of his lukewarm sentiments and his potential cowardice. He proposes that they negotiate with the enemy, recalling the latter's warrior qualities and strength in materiel, and pointing out that the colony is insufficiently prepared and that the plan of action has been improvised too quickly. However, these reservations are seen as treason. A few Africans of dubious health are recruited, encouraged by the gift of kitchen utensils that, along with other material objects and foodstuffs, are given as payment from the state, provided by the Réchampot brothers, these African "war profiteers." The missionaries educate these recruits by teaching them the words they'll need for the venture, such as "gun" and "bayonet," stressing the white man's superior competence and intelligence (witnessed in his ability to ride a bicycle) and the power of his god who makes this exploit possible. The traders arm the "volunteers" with some old rifles in questionable working order, and the expedition leaves after singing an out-of-tune "Chant du départ," as out of place as the "our ancestors the Gauls" that was printed in the school textbooks distributed in African schools. The sergeant is in the lead, accompanied by the colonists, men and women dressed elegantly for the event. When they reach a stream during their outing, this miniscule symbolic Rhine is designated as the border, which can be crossed by walking on a few stones. The colonists on a spree are traveling in palanquins, followed by several bearers carrying food and the things needed for a picnic that resembles as much Auguste and Jean Renoir's *Le Déjeuner sur l'herbe* (1959, 91 minutes, France, Jean Renoir) but without its charm as it does a stopping point on a cinematographic safari. The mood in the French community recalls that of *La Règle du jeu* (1939, 110 minutes, France, Jean Renoir), but without its dramatic form or its appealing sentimental character. The colonists get settled, eating and drinking while the African recruits are massacred by a German machine gun after having been primed with wine prior to their disorganized and suicidal attack. (On the European front, French soldiers received a ration of liquor before their attack, often Calvados, to bolster their nerve.) In panic, everyone rushes back in confusion.

The young geographer takes control of the situation with the consent of the sergeant, who is humiliated and disheartened, and the submission of the colonists, fearful of the Germans' arrival. "War is too important a thing to be entrusted to military men," said Georges Clemenceau, whom the geographer faintly recalls here. He takes an inventory of the provisions of the two trader brothers who try to hide some of them, hoping to profit a bit more from the events. He mounts a new campaign of forced recruitment with the village chief, assigning secondary missions to two of his fellow Frenchmen as a way to favor their naïve vanity and obtain their support for the situation. The two women in the colonial community sew some unorthodox-looking uniforms, while he seduces a pretty young black woman, which is in keeping with storybook colonial imagination. His servant becomes his secretary and an inevitable "state employee," ultimately making the geographer into a convert to supreme power and the best forms of its application. Thus he establishes the necessary traditional and psychological gap between citizens, on the one hand, and an authority considered infallible and requiring a blind obedience, on the other. He is confirmed in his position while making the presence of his mistress known during the ridiculous and protocol-ridden parade of a few dozen men. Shortly afterward the recruits receive real uniforms, mount another attack on impregnable German positions, bury themselves in rain-sodden trenches, and die as much from sickness as from the fire of fighting that leads to nothing,[24] despite the attention and paternalistic visits of the "commander in chief."

On the Western Front in Europe, the French army, with inadequate stocks of uniforms,[25] was subjected to German planning based on a strategic encircling from north to south passing through Belgium. This Schliefen plan was designed to surround and neutralize the French troops massed along the German border who were waiting for the end of Russian mobilization when the Tsar's armies would enter on the Eastern Front to reduce German pressure in the west. The French army succeeded in stopping the German advance in the battle of the Marne after disorganized or makeshift operations, whereas in Africa and in the film the initiative for the attacks depends on the (badly) organized "(badly) armed band." After 1915, with French soldiers in sky blue uniforms, the armies buried themselves in the trenches, lying elbow to elbow in mud, undergoing huge losses, in a development comparable to what is represented in *La Victoire en chantant*. The recent moment now belongs to the distant past when the young geographer, transformed into a war leader, reproached his compatriots for their inhumanity and their lack of respect for human beings.

The English troops soon arrive, led by an officer from India who announces the end of hostilities. The black soldiers of his majesty the king are ridiculed. Some are wearing kilts, their music and behavior seem anachronistic, outdated and out of place, and they don't speak French, unlike the Germans with whom the French community had dealings. Another reference is made here to Jean Renoir's *La Grande Illusion* (1937, 114 minutes, France, Jean Renoir) in which the aristocratic French officer and his German counterpart belong to the same world, speaking both their respective language and English. But the Frenchman does not communicate with the English officers who arrive at the camp, whereas his compatriots try to warn them of an escape tunnel that has been dug to which they no longer have access. In Annaud's film, the young German officer and the young geographer take to each other and leave the festivities to discuss German philosophy, recall their schooling at the École Normale Supérieure and the University of Heidelberg, and admit to each another that they were "socialists." Their initial antagonism is accidental; the deaths, the horror of mutilated bodies, and the blood that refer to European battles are like an avatar, a necessary but short-lived episode in the process of reconstructing society whose space, divided in preordained fashion according to two nationalities, offers no administrative, cultural or physical border. The stream that was the agreed-upon border apparently has no obvious political or geographical effectiveness, unlike the Rhine. When the film was being produced, French-German relations were the best possible, expressed in particular by the extensive and harmonious understanding between Chancellor Helmut Schmidt and President Valéry Giscard d'Estaing (1974–81), who were on a first-name basis.[26] Europe was being built by these two nations, at the expense of England. The way the film caricaturizes Great Britain is another reference to the present-day Europe the film promotes by intervening in the present in order to insert the past. In an analogous symbolic perspective, 18 years earlier on September 22, 1984, at Verdun, Helmut Kohl and François Mitterrand shook hands before meditating at the graves of German and French soldiers killed on the battlefields of World War I. The young administrator, level-headed, cultivated, and educated in the school of the Republic, begins the desired transformation following an initial humanistic attitude, according to a calculated and manipulative method involving a cold and austere rationality. He resembles his German counterpart, with whom he joins in an agreement to realize a new society.

Probably unconsciously, the film projects European concerns and perspectives onto the African continent in order to produce there a resolution of European conflicts, a latent and predominant cooperation in building a political Europe extending into the contemporary moment, and a vision that includes a current subjective perception of the social context. This set of historical representations is transformed and objectivized, bathed in an everyday that is too invasive and too constructed, in an effective and unavoidable emotionalism. These representations reproduce European political structures according to simplified and symbolic forms that are promoted by structuring the colonial world and its workings, which as a catalyst reveals the horror of war, human and ideological turpitude, and the randomness and evolution of viewpoints based on present interests and circumstances that are endured or created, staged and developed by cinematographic expression.

But colonialism, in its complex imaginary and historical outcomes, remains the primary subject of this film, which continues colonialism's history in a rich and original fashion, despite the still rather flagrant way it repeats colonialism's errors and prejudices, however attenuated, and despite the opinions some may have concerning this history's end. This history must still be done, continually, since history is interminable. Thus, by way of conclusion, the film joins the flow of history and the observation Paul Rivet makes:

> Our colonial past is far enough removed that we can finally establish a relation with it that is free from a complex of arrogance or guilt. We need not take on this legacy in order to savor its disquieting strangeness, as the mood of the times invites us to do, but rather in order to give it meaning. Otherwise, the colonial moment belongs to the part of our recent trajectory that is incomprehensible, poorly understood, and under-analyzed. As such, already in the nineteenth century it points to poorly understood parts of our century that henceforth historians must clarify. What must be done is to make sure that the temporal distance opening up between us and our colonial past does not become a dead moment without meaning, and that between us and the reality of colonization, which was an experience for some and an ordeal for many others, there takes place... "a transmission that is generative of meaning,"[27] to use the expression coined by Paul Ricœur.[28]

Notes

Chapter translated by Daniel Brewer.

1. Georges Balandier, "Il était une fois la colonie," in Georges Balandier and Marc Ferro, eds., *Au temps des colonies* (Paris: Seuil, 1985), 3.
2. *Le Président* (1961, 110 minutes, France, Henri Verneuil) is a notable French precursor, in which a head of the French cabinet, played by Jean Gabin, who is close to Georges Clemenceau, an eminent anti-colonialist, forcefully describes the damaging effects or the harmfulness of colonialism. To a deputy who notes the overseas locations that should be taken and France's duty to occupy them in order to establish new markets for her industry, a kind of test for her strength, he answers, "and a lesson in energy for her soldiers—I know the phrase. Personally, I find that this mission is unfounded, and its advantages are pathetic. Except of course for a few wheeler-dealers out for a buck, and a few missionaries who need souls to convert. I can see how the liabilities of these enterprises don't trouble a group whose members are just a bunch of self-interested managers."
3. See Colette Dubois, *Le Prix d'une guerre, A.E.F., 1911–1921* (Aix-en-Provence: Université de Provence, Institut d'histoire des Pays d'Outremer, 1989).
4. Marc Michel, *Les Africains et la Grande Guerre: l'appel à l'Afrique, 1914–1918* (Paris: Karthala, 2003).
5. Catherine Coquery-Vidrovitch and Henri Moniot, *L'Afrique noire de 1800 à nos jours*, rev. ed. (Paris: Presses Universitaires de France, 2005); Raoul Girardet, *L'Idée coloniale en France de 1871 à 1962* (Paris: Hachette Littératures, 2005); Til Gottheiner, *French Colonialism in Tropical Africa, 1900–1945* (New York: Pica Press, 1971).
6. Henri Brunschwig, *Mythes et réalités de l'impérialisme colonial français, 1871–1914* (Paris: Armand Colin, 1960); Yves Monnier, *L'Afrique dans l'imaginaire français, fin du 19e-début du 20e siècle* (Paris: L'Harmattan, 1999).
7. Pierre Boulanger, *Le Cinéma colonial de l'Atlantide à Lawrence d'Arabie* (Paris: Seghers, 1975), 7–8; Caroline Eades, *Le Cinéma post-colonial français* (Paris: Éditions du Cerf; Condé-sur-Noireau: Éditions Corlet, 2007).

8. Jacques Aumont, Alain Bergala, and Michel Marie, *Esthétique du film*, 3rd ed. (Paris: Armand Colin, 2004), 173–203.

9. A law to this effect was even passed, which gave rise to strong opposition, especially among historians; the law was since repealed. Law 2005–158, February 23, 2005; J.O. 46, February 24, 2005, text 2. Article 4 stated, "University research projects will give to the overseas French presence, especially in North Africa, the place it is due. Secondary school curricula will recognize in particular the positive role of the overseas French presence, especially in North Africa."

10. Jean-Jacques Annaud, *La Victoire en chantant* (Paris: Pathé Distribution, 2005).

11. For example *L'Attentat* (1972, 120 minutes, France, Yves Boisset); *R.A.S.* (1973, 110 minutes, France, Yves Boisset); *Section spéciale* (1975, 120 minutes, France, Constantin Costa-Gavras); *Bako, l'autre rive* (1978, 110 minutes, France, Jacques Champreux).

12. Marc Ferro, *Analyse de film, analyse de société: une source nouvelle pour l'histoire* (Paris: Hachette, 1976); Pierre Sorlin, *Sociologie du cinéma: ouverture pour l'histoire de demain* (Paris: Aubier-Montaigne, 1977); Jean-Pierre Esquenazi, "Penser le cinéma, penser le monde," *La Pensée* 300 (1994): 9–18.

13. Miguel Porter and Palmira Gonzalez, *Las Claves de la historia del cine* (Barcelona: Ariel, 1988), 3.

14. Marcel Oms, "Fonction idéologique de deux films historiques espagnols," *Cahiers de l'Université de Perpignan* 17 (1994): 7.

15. Georges Conchon, *L'État sauvage* (Paris: Albin Michel, 1964), adapted for film by Francis Girod in 1978 with the same title.

16. Jean-Joël Bregon, *Un Rêve d'Afrique: administrateurs en Oubangui-Chari, la Cendrillon de l'Empire* (Paris: Denoël, 1998).

17. Colette Dubois, *Le Prix d'une guerre*, 37–40.

18. Dubois, 16–17.

19. Henri Brunschwig, *Le Partage de l'Afrique noire*, 21st ed. (Paris: Flammarion, 1993).

20. Georges de Gironcourt, "Les Conquêtes franco-anglaises en Afrique, le Togo et le Cameroun," *Bulletin de la Société de géographie et d'études coloniales de Marseille* 40 (1916): 72–89; General Joseph Gauderique Aymerich, *La Conquête du Cameroun, 1 août 1914–20 février 1916* (Paris: Payot, 1933); Armand Annet, *En colonne dans le Cameroun: notes d'un commandant de compagnie, 1914–1916* (Paris: R. Debresse, 1949).

21. Pierre Mollion, *Sur les pistes de l'Oubangui-Chari au Tchad, 1890–1930: le drame du portage en Afrique centrale* (Paris: L'Harmattan, 1992).

22. Lieutenant-Colonel Charles Mangin, *La Force noire* (Paris: Hachette, 1910).

23. André Gide, *Voyage au Congo: carnet de route* (Paris: Gallimard, 1927).

24. Frédéric Rousseau, "Vivre et mourir au front: l'enfer des tranchées," *L'Histoire* 249 (2000): 60–65.

25. Their red pants made them easy targets for German riflemen.

26. Raymond Poidevin and Jacques Bariety, *Les Relations franco-allemandes, 1815–1975* (Paris: Armand Colin, 1977); Joseph Rovan, *Histoire de l'Allemagne des origines à nos jours* (Paris: Seuil, 1994).

27. Paul Ricœur, *Time and Narrative*, Kathleen McLaughlin and David Pellauer, trans., vol. 3 (Chicago: University of Chicago Press, 1988), 221.

28. Daniel Rivet, "Le Fait colonial et nous: histoire d'un éloignement," *Vingtième siècle: revue d'histoire* 33 (1992): 138. See Sophie Dulucq and Colette Zytnicki, "Penser le passé colonial français: entre perspectives historiographiques et résurgence des mémoires," *Vingtième siècle: revue d'histoire* 86:2 (2005): 59–69.

18

Making Space for War: *L'Action française, Je suis partout,* and French Right-Wing Understandings of the Spanish Civil War

Paul Schue

The Spanish Civil War (1936–39) was a crucial event in the course of French politics, both because of the kinship between the Popular Front regimes in France and Spain, and because many French observers saw Spain as a portent for France's future. Numerous French writers used their commentary on the Spanish conflict as a means of working through their own thoughts about class conflict and the political situation in France. In doing so, they often laid bare their own ideological preconceptions about modern society.[1] An example of this can be found in two right-wing French newspapers, the monarchist *L'Action française* and the fascist *Je suis partout.* Each newspaper envisioned the war as an ideological space and managed its representations of that space to suit its political preoccupations in France. *L'Action française* focused on Franco's ability to resolve class conflict by essentially removing the masses from the political process. The writers of *Je suis partout* displaced the class conflict of the war onto a geographical confrontation and so transformed class conflict into matters of space and territory. In construing the war in this fashion, both newspapers revealed how their ideologies served first to mediate their understanding of the war's events, and second to manage their political anxieties by envisioning a way that nationalism could neutralize class conflict in Spain.

The Chronicle of Fear: *L'Action française* Coverage of the War

By 1936, Charles Maurras' monarchist newspaper *L'Action française* was a venerable voice of right-wing ideology in France, although its parent organization, the Ligue de l'Action française, was banned by Léon Blum's Popular Front government in June.[2] The journalists at *L'Action française* had been critical of the Spanish Republic from the moment that it replaced the Spanish monarchy in 1931, saying that democracy was simply a manifestation of Spain's decadence. Thus, when the Spanish Civil War broke out, *L'Action française* immediately presented the conflict as the culmination of a process of social dissolution that had begun with the fall of the monarchy. *L'Action française* coverage of the war's first two months gave little attention to the ideologies of either Franco or the Spanish Popular Front. The dominant aspect of the war was the simple fact of social chaos.

Almost as soon as the first shots were fired, *L'Action française* columns began to describe "immense chaos."[3] One early article placed Franco's uprising at the end of a line of "nine revolutions in five years," lumping it together with the popular riots that helped topple the monarchy in 1931 and the anarchist-led Asturian revolt of 1934.[4] Another early article commented that a victory by either side would bring more killing and strife, for the Nationalists were not unified:

> Victory—quite improbable—by Madrid would be a catastrophe, as armed bands of Marxists would continue the sorry task already begun under the direction of emissaries from Moscow....If the Nationalists triumph, serious difficulties await them. Carlists, monarchists, fascists, and military men will agree on only one point: the execution, imprisonment or deportation of the Communist leaders who have not already fled; the dissolution of the Socialist syndicates and the Anarchist Federation.[5]

The most striking and terrifying manifestation of this disorder from the point of view of *L'Action française* was the arming and unleashing of the population:

> In no matter what country it is found, the human mass, left to its own devices, knows no limits. The crime of the Spanish Popular Front government in all of Spain, a crime that marks it forever with infamy, is to have armed it and launched it into the streets.[6]

This "human mass" should not be confused with the people of Spain, for *L'Action française* made a clear distinction between the two, based on political mobilization. The passive population was portrayed as good precisely because of its docility, while those who took up arms for a leftist political ideology were a mob, a human mass. An article on Republican Spain asserted that "whereas the population is unarmed, the most unsavory mob [*immonde canaille*] armed with rifles invades houses and reigns as mistress of the streets."[7]

If political mobilization was the key to creating the mob, then popular ideologies were a great danger, and in *L'Action française* coverage of the war, the *bêtes noires* were ultimately "the two poles of the Revolution: anarchy and Marxism."[8] The anarchists were "bands of veritable Spanish gangsters fresh from prison for the most part," and the communists were diabolically intent on creating tyranny.[9] Moreover, as Léon Daudet commented, "Marxism is a Muscovite import. Anarchy in Spain, and particularly in Barcelona, is autochthonous."[10] For the most part, however, the distinction between communists and anarchists was immaterial, for "From the point of view of violence, socialism, Sovietism, and anarchy are one."[11]

To prove this, *L'Action française* carried numerous articles that did little more than list murders, rapes, church burnings, and other crimes attributable to communists and anarchists. One reporter stated that "We have witnessed all sorts of crimes whose refinements in horror oblige us to repress our screams."[12] Another reporter denounced what he saw as "red satanism" given "a free rein."[13] There was even a semi-regular column entitled simply "The Red Terror."

Running throughout this coverage of the chaos in Spain was the metaphor of disease. Communism, anarchism, or any "red" ideology was a virus in the body politic, and so one article was entitled "The Communist Plague," and in another José le Boucher spoke of the "Bolshevik virus intentionally injected into the country by the propaganda services of Moscow."[14] The metaphor of disease implied that communism, or any ideology of popular power, was a form of illness or delirium. Popular mobilization was pathological. This assumption about popular participation led *Action française* thinkers to conclude that the infection of leftist ideology could succeed only in societies that had been weakened already by democracy. Maurras had long held that the creation of a republic always destroyed order and opened the way for truly dangerous demagogues to seize power.[15] Daudet reiterated this idea in 1936: "Democracy, which we also call the Republic, is only the first step on a staircase... of which the second step down is the cartel of the Popular Front, the third is socialism, and the forth terrorism." He then added:

> In Spain, where the indigenous blood is mixed with Moorish blood, a fact which makes the passions more intense, democracy, after several Republican pushes, has burned the socialist step entirely and gone straight to a communism as barbarous and unbridled as that of Russia or Hungary.[16]

To find evidence for this theory of democracy, *L'Action française* writers would continually return to the example of the First and Third French Republics. Charles Berlet commented that "even if the means of killing differ, the furor which animates the reds, be they from France or Spain, 1792 or 1936, is the same."[17] And Daudet opined that "Without going back to the deluge of blood of 1789–1793, it suffices to note that the Paris Commune of 18 March 1871 succeeded [the founding of the Third Republic on] 4 September 1870."[18] Daudet also said that

> The "Frente Popular" continues the Russian terror [*épouvante*] which itself continues the massacres of September, the drownings of Carrier, the guillotinings of Robespierre, of Danton and of Marat. All the atrocities, all the monstrosities, the issue of democracy, resemble and corroborate each other. But, faced with these things, there rise up the Chouans, [with their] absolute and vengeful acts of devotion.[19]

This extended comparison of Spain and France placed the Spanish conflict in a larger arena of ideological confrontation and made it part and parcel of the nationalist conception of the sacred memory of the nation, and even of civilization as a whole. The Maurrasian national memory was filled with moments of heroic resistance to danger, defense of the *patrie*, and beyond it, of civilization. Spain was simply inserted into this list.

Spain was therefore a moment of danger, especially to the democratically-enfeebled body politic of France. This was manifested in coverage of three issues: arms traffic across the border; Spanish communist and anarchist agitators in France; and Spanish refugees. These issues would rise and fall according to the tide of events in Spain, but all remained vibrant points of anxiety for *L'Action française* writers throughout the war.

The issue of arms traffic was articulated first. Less than a week into the war Maurice Pujo began a series of investigative articles on pending arms deals between the French and Spanish Popular Front governments.[20] France very quickly cancelled its shipments of airplanes and arms to Spain and eventually agreed to a nonintervention pact in Spain signed by several other nations, but *L'Action française* writers remained focused on clandestine arms traffic.[21] Over the next two years Pierre Héricourt wrote numerous articles for *L'Action française* on the continuing illicit Franco-Spanish arms traffic and then assembled these articles into a book.[22]

The issue of arms traffic soon bled into the issue of smugglers of all left-wing persuasions entering France. The smuggling routes were "well known to specialists from Perpignan, Marseille and Toulouse, in particular to the Soviet agent Roger Tolera, who has for some time held court at the Continental Bar in Perpignan."[23] Soon this fear of communist and anarchist arms traffickers had become a dread of leftist agents of all stripes infiltrating France. Daudet asserted that if Barcelona fell, "we will be infected by Hispano-Muscovite refugees who will stir up their Muscovite reactionism and their Civil War in France."[24] This anxiety about communist infiltration of France was at its most intense in 1936, and over the course of the war it often converged with concern about Spanish refugees in general.

The issue of refugees was a troubling one for the writers of *L'Action française*, ostensibly because of the fear of communists among them, but often simply because they were foreigners, the poor, the masses in general. After the battle of Irun in September 1936, *L'Action française* offered a short daily blurb on Spanish refugees entitled "Invasion of the foreigners [*métèques*]" in which the use of the ethnically charged and derogatory term *métèque* eloquently attests to how *L'Action française* viewed the situation.[25] In a similar vein, after the fall of Santander in the summer of 1937 an *Action française* reporter said of the refugees that "[Victor] Hugo himself would not have seen the 'ocean of people' [*le peuple océan*] but 'rabble scum' [*écume populace*]." He went on to claim that the crowd contained men who "specialized in executions, assassinations, and pillage," as well as "the dregs [*la lie*], common-law convicts set loose by the revolution, agitators, [and] rabble-rousers."[26] Although the reporters were often sympathetic to the suffering of women or children in the exodus, they saw the masses of refugees as disease vectors of revolution.

In the end, all of these issues—arms traffic, communist infiltration, and refugees—fed one another to create a single generalized fear of the revolutionary chaos coming to France and to reveal a fear of the masses on the part of *L'Action française*. For Maurras and the writers of *L'Action française*, the French people were to be lauded in the abstract, but never mobilized in reality. The great danger of the Spanish Civil War was precisely the fact that it allowed and encouraged the people to become politically active.

Thus, the dominant tone of *L'Action française* coverage of the war was an anxious one, but there was also a slowly growing appreciation of Franco. *L'Action française* writers never saw Franco's new nationalist movement as a complete solution to Spain's problems—for them only a monarchy was a complete solution—but they did comment increasingly favorably on Franco's abilities. The favorable image of Franco rested on his movement's apparent ability not only to restore order, but also to reconcile the warring classes, to instill a universal sense of identity with the Spanish nation, and above all to make the people passive once more. In mid-August of 1936 an unsigned article admitted that Franco "appears less and less like the author of some *pronunciamento* and more like the pacifier of the Iberian peninsula."[27] This message gradually became stronger over the course of the war, and Joseph Delest asserted in 1939:

> If it weren't for General Franco we would have been witnesses to millions of assassinations like Russia's, because the whole underworld was organized in Spain under the protection of Moscow. And we would not have escaped these revolutionary hordes who would then have set to creating disturbances in France.[28]

The key to this pacification, *L'Action française* writers asserted, was not just a good ideology, but also a strong sense of paternal discipline. Jean Dourec wrote in mid-1937 that "In spite of the iron hand of the master, and perhaps because of it, regenerated Spain, although still bloody and wounded, bruised [*meurtrie*] and in mourning for some time to come, vibrates with a new enthusiasm."[29]

In the end, the writers of *L'Action française* argued that by pacifying the country by force, Franco was literally creating the nation. Dourec noted that some captured governmental soldiers would "soon be 'new' Spaniards, 'new' Nationalists."[30] He also observed that a recaptured village was "Spanish again."[31] In describing the "force that Franco has," Héricourt said, "it is indeed a nation which has found its soul once again."[32] It should be pointed out, however, that Franco's success lay not in charisma, but in his willingness to enforce a strict discipline and his ability to embrace an ideal. For the ideal, not the bearer of it, was the key. As Maurras put it, "Neither 'technique' nor dexterity nor an innate sense of action is sufficient. One does nothing great, in politics in general, without the competition of opinion, and one does not lift up a nation without an idea."[33]

Thus Franco was seen as a disciplinarian who knew how to overcome class antagonism and social disorder. He gave the people a new ideal, and was literally recreating the nation and even saving civilization. Daudet would tell his readers that Franco "represents...civilization against barbarism."[34] Similarly, Maurras would write:

> The Spain of Franco, Mola, and Queipo de Llano, of the Requetes [monarchists] and the Falangists [fascists] is the right Spain: vigor, élan, skill, organization, all the noble civil and military forces which it has put in service to this just cause, have forged for it new titles to prosperity, happiness, peace and...Empire. The world is large, the Barbarians are everywhere, [and] the field open to Civilization is vast.[35]

But if Franco was an upholder of civilization and a unifier of the nation, was he not as good as a king? As much as Maurras, Daudet, and their fellow journalists lauded Franco, they never ceased asserting that only the monarchy, with its dynastic stability and its roots in the historical memory of the nation, could provide long-term stability and continuity. When Maurras surveyed Franco's cadre of nationalists, he asserted that "In order to guard against discord, they would be wise, as soon as possible, to establish amongst them, above them, this magistracy of Unity called the Monarchy, so that this organic work might be led with the maximum of precaution and artistry."[36] Yet this call for the monarchy did not represent a rejection of Franco himself, for the general was encouraged to find a role within a monarchist Spain. Maurras reasoned the king would either "exercise his authority personally or he will use an intermediary, such as Richelieu, Bismarck or Mussolini," and he cast Franco in such an advisory role.[37]

Overall, then, the vision of the Spanish Civil War that emerges from the pages of *L'Action française* was built upon the foundation of fear, and was focused on the issue of the border separating France and Spain. It was ultimately this fear that framed the picture that *L'Action française* drew of Franco and his Nationalist movement. Franco appeared as the disciplinarian of Spain; he had reawakened the nation's soul and saved Spain, and more importantly France, from chaos.

Yet if *L'Action française* remained decidedly royalist in its perceptions of Spain, some of its stepchildren did not. While Maurras, Daudet, and others were calling for the return of the king, Robert Brasillach, Lucien Rebatet, and other Action Française alumni at *Je suis partout* were lauding Franco as a permanent solution to Spain's difficulties.

The Fascist Vision: *Je suis partout* on the War

Eugen Weber describes the newspaper *Je suis partout* as "the home of an extremely active group of young Action Française *epigoni* gathered around Gaxotte and, later ... Brasillach."[38] Pierre Gaxotte, a former secretary of Maurras', had attracted to *Je suis partout* a whole stable of young journalists who had embraced Maurras' ideology while still at the *lycée*. By the late 1930s Gaxotte had stepped aside as editor to allow the younger Robert Brasillach to take over.[39] Under Brasillach, *Je suis partout* moved out of Maurras' shadow and took on a character of its own. Many of the writers at *Je suis partout*, including Brasillach, were happy to apply the fascist label to themselves, and in practice they replaced the dynastic monarchy at the center of Maurras' ideal nation with a dynamic, talented leader. Instead of a passive population, Brasillach and his group pictured a nation in action, bound together by an elite movement of willful, virile men. This fascist ideal would lead the *Je suis partout* crowd to embrace a whole host of foreign fascist movements, from the Belgian Rexists to the martyred Spanish fascist José Antonio Primo de Rivera and his Falange. It also led Brasillach to collaborate with the Germans during World War II.[40]

When the war in Spain erupted in July of 1936, the writers of *Je suis partout* immediately presented Spain as two distinct nations in conflict. One side was the Spain of chivalry, tradition, heroism, and valor, a glorious apparition from El Cid's *Reconquista*. The other side was the Spain of the rabble, of cowardice, brutality, and murder. More important, in the pages of *Je suis partout* a certain alchemy of space, time, and class was carried out. The front between the government and nationalist armies came to represent not only the dividing line between the two Spains, but also the ideological abyss between the *Reconquista* and communism, and the distinction between the "people" and the "rabble." Class conflict was remapped as geographical conflict.

In the first weeks of the war, while *L'Action française* was finding the chaos in Spain overwhelming, *Je suis partout* writers were establishing a clear order. In one of the first articles in *Je suis partout*, René Richard commented that "What is for sure, what is certain, is that a life and death struggle between the two Spains has begun. This time, the combat between order and anarchy should be decisive."[41] Even at this early date, when the names and goals of the nationalist leaders were still unknown, Richard was able to establish a spatial center to this conflict:

It must not be forgotten that in the history of Spain the northern front is the most important. There the population is the most energetic, and when it has the support of the army, instead, as in the Carlist wars, when the army was fighting against it, it [the population] represents the most powerful force in the country.[42]

Richard thus established Navarre, the traditional home of the Carlist monarchist movement, as the homeland of the rebellion.

Brasillach picked up this geographical graphing of ideologies two weeks later when he not only situated the heart of tradition and the rebellion in Navarre, but also identified Barcelona as the heart of anarchy. As Brasillach explained it,

> Barcelona sends [to Navarre] teams of killers in cars and trucks. In several hours they can easily nail a town's mayor, notables, and priest to the wall over the market, dig up the bodies of dead priests and nuns and display them in the market, and shoot all the young boys in red berets, because the red beret is not the sign of the falcons of M. Monnet [a Parisian Communist], but rather that of the Carlists. Then the teams of killers leave again in their trucks, at high speed, on darkened roads, and perhaps they sing the admirable old Catalan songs of revolt, or maybe their *Internationale*.[43]

By creating this image of arrival and return, Brasillach was constructing a geographical separation of social ideologies. Social revolutionaries did not live in Navarre, they had to arrive from Barcelona, and it was to Barcelona that they must return. Brasillach would articulate the converse of this a week later, when he described the Carlists as the "party of honor," and spoke of "peasants of the mountains who leave to seek out Fal Conde [leader of the Carlist party] with their priests."[44]

This geographical arrangement of ideology was in effect a redistribution of social struggle: class conflict was transformed into regional conflict. Navarre was symbolically cleansed of all indigenous class conflict and became a monolithic ideological block. Barcelona, Catalonia as a whole, and, to a lesser extent, Madrid were not the sites of an industrialized society with a full complement of political parties and social classes, but rather the homeland of a rabid communist or anarchist underclass. Most important, between these two forces were the bourgeoisie and the rest of the working class, indeed the rest of the country, uncertain in their convictions, or perhaps just inert, and waiting for leadership. Much of Spain was therefore ideologically empty space, waiting for definition, and any moderate political positions, such as the Catalonian nationalists, who were often essentially liberal, were erased.

To establish this graphing of social structure and ideology across the space of Spain, the *Je suis partout* writers would follow Brasillach's and Richard's lead and rely on a series of codings, shifting and uncertain at first, but solidifying as the coverage of the war continued. These codings consistently placed the underclass in Barcelona or Madrid and the valiant peasants in Navarre. Richard thus argued that "Madrid can only arm the dregs [*lie*] from the [largely working-class] suburbs, backed up by the prostitutes from the capital."[45] André Nicolas echoed this claim, arguing that "having been armed by the government of Madrid wherever it is in charge, the dregs [*toute la lie*] of the population have committed the most frightful common crimes."[46]

Nicolas immediately added, "Most notably in the north, there is a terrific morale, and, in particular, in Navarre, beyond the army and the police, it was possible to mobilize more than 40,000 men in a province with a population of only 800,000." Furthermore, Burgos, once it was made the nationalist capital, became a new center of order: "In Burgos, the headquarters of General Mola, the town 'in flames and defamed', the tranquility is absolute. If the activity is intense, it is in order and in joy."[47] In contrast, in Madrid at night, it was said, "The unleashed underworld [*pègre déchaînée*] profited from the darkness in order to assault the homes of the last bourgeois, and each night became a little Saint Bartholomew of 'suspects'."[48]

By thus dividing Spain into two regional forces, the dregs of the population and the true patriots, all uncertainty was expunged and the anxiety of class warfare, the terror of the servant rising up against the master or the employee against the boss, was transformed into a geographical division capable of separating the "good" workers from the "anarchist" ones. Thus an article in late August could observe that "Beyond Guadarrama, the front bends to the south, then climbs back north before Saragossa and Huesca. On one side patriotic and traditionalist Aragon, on the other autonomist and anarchist Catalonia."[49]

In this conception of the war, the front, the site of actual combat between armies, became a zone of honest confrontation, while the rear became the land of shadows and increasing anarchy:

> To be sure, there are in the ranks of the [governmental] militias men who are brave, disinterested, animated by great ideological motivations. But these men are all at the front. Those who remain in the rear and who terrorize the population are recruited from amongst the sadists, the savages [*apaches*] and the malefactors of every stripe.[50]

This image established the front as the only legitimate site for combat, and again negated the ideal of class warfare throughout society. The activities of the Barcelona anarchists and communists, because they did not occur at the front, became simply crimes, unrelated to the war. At the same time, Franco's mass executions of (now allegedly-criminal) prisoners became simply matters of justice not ideology: "In all equity, it is absolutely impossible to put on the same plane the condemnations to death decided on by Nationalists leaders, and the hideous unleashing of the Marxist mob [*canaille*]."[51]

To support this vision, *Je suis partout* presented two different, fluctuating visions of the bulk of the Spanish population: on the one hand, government regions had a Marxist minority tyrannically oppressing a passive but fundamentally traditionalist population that anxiously awaited liberation; on the other hand, the population was held to be an easily swayed crowd, a volatile creature with wild swings of mood, now ravenously Marxist behind Marxist leaders, now stalwartly nationalist behind nationalist leaders. As these images were developed, it became clear that the image of the masses as fundamentally traditionalist but inert and oppressed was often an image of bourgeois or petit-bourgeois masses. Similarly, the image of the masses as excitable but capable of being good or evil was usually associated with the working class. In this sense, the two images of the people in the middle were again a class coding of the conflict. In their deployment, however, these images were never clear-cut, and often the entire nation was pictured as inert and traditionalist, or the whole Spanish population was posited as being excitable but leadable. This two-faced nature of the masses, I would argue, reveals a profound schism in French fascist ideology itself, a duality that shaped all images of the masses.

The image of the people as good but inert, or at least voiceless, was a common one on the right in France during the Popular Front. Maurras and those who were in his intellectual orbit frequently made the distinction between the "legal" France of the leftist government and the "real" or "true" France of tradition. Whenever the left was in power, it simply meant that the "true" France was being oppressed. In *Je suis partout* commentary on Spain, this image of a fundamentally good people surfaced from time to time: "There are no 'cruel peoples', there are in every people certain dregs [*une lie*], which in times of trouble rise to the surface, and which are capable of the same excess, the same atrocities, the same savagery."[52] Even in Catalonia, the seat of anarchy, there resided "'apolitical' people," who had suffered "such horrors for two years that they now await with impatience their deliverance by Franco's troops."[53] As a rule, these images of the passive population were applied to the people as a whole, but from time to time the oppressed people became bourgeois. Thus when we are treated to eyewitness accounts of atrocities in Malaga, it was the bourgeoisie, those "wearing or *who have worn* a suit-coat, a tie or a hat," being "shot without pity by an underworld [*pègre*] whose ringleaders [*meneurs*] have only recently emerged from prison."[54]

Yet even this image of the solid, loyal—sometimes bourgeois—people being oppressed by the evil leftists often began to slip, and then the people lost their neutral passivity altogether. The crowd had moments of instability when docile submission became active support:

> Let us simply remember that the residents of Madrid, terrorized by a small, bloody anarchist and Bolshevik minority, await Franco as a liberator. In the beginning of the war, there was in Madrid an incontestable majority for the Popular Front. The crowd [*foule*] seized by a collective delirium, acclaimed the Republic, liberty, equality, fraternity and all the other items of the Masonic catechism.[55]

Thus we arrive at the second image of the people, as the mass of the population became a potentially dangerous weapon in the power struggle between the Marxist mob leaders and the stalwart nationalists: "Spain in a revolution can only be violently Marxist or violently anti-Marxist. There is no place over there for half-measures, for compromise, for retiring to the countryside [*chèvre et choutisme*]. It is a question of temperament."[56] If the nationalists could assert control, then the people would be good, but if the anarchists or communists asserted control, then the people would slide into a bloodthirsty frenzy. Thus, "the liberal Freemasons and the Marxists have made Madrid, Barcelona, and Valence accursed cities where the unleashed hordes know no impediments, and no law."[57] There was, it was argued, a permanent penchant for violence beneath the surface of the people, as was evident when Lucien Maulvault asserted that "Always, over the course of centuries, the association of the lubricious and macabre has been a profound trait of the Spanish character."[58]

In this conception of the crowd, if it was the Marxists demagogues who led the people astray and awakened their passions, then it was the nationalist leaders who reintegrated the nation and transformed the population from an unleashed mob to loyal Spaniards. The royalist Jean d'Elbée, when harkening back to the 1931 fall of the Spanish monarchy, commented,

> By removing the insignias of the royal house from monuments and uniforms one could say that men and things were stripped of this grace and nobility that so particularly characterize their race and country. In a few instants the rogue anonymity of the republic made its entrance.[59]

D'Elbée went on to add that "Since the renovation [by Franco] there is something in the air which, joined to an exquisite race, eliminates all vulgarity."[60]

Furthermore, it was the personal aspects of the personalities of the leaders of the nationalist cause that would effect this transformation: "But Marxism will only be repulsed by the people if the people find in a national and social party sincere acts of ambition, a will to struggle, a sentiment of generosity and comprehension, and an absolute devotion to the grandeur of the fatherland."[61] Or, as André Nicolas argued, "the Spanish are ready to follow, even to death, a leader [*caudillo*] who inspires confidence in them." Thus, "Only the authority of the generalissimo, loved and respected by all, could have obtained such a result...: the union of traditionalists and Falangists."[62] In this vision of the leader, *Je suis partout* went well beyond Maurras. The heroic leadership of an elite would not just pacify and discipline the people, but arouse them to feats of great nationalist vigor and renewal. The passive people had been rendered spontaneously active, their awakened energy channeled into the proper paths by the will of the leader. However, there are important ways in which the presentation of leadership in *Je suis partout* belied this sense of a purely voluntary discipline of attraction. In many cases, the newspaper dwelt on the aspect of command, and even indulged in a fantasy of control of the people.

When Simionesco reported on General Queipo de Llano's radio addresses to Nationalist Spain, he indulged in this fantasy of the power of the leader over the working class: "the microphone of Radio-Seville is not simply an instrument of propaganda and enthusiasm: it is also an infallible means of command."[63] This fantasy of the power of the commanding voice revealed an abstract vision of the people as intrinsically excitable, but also controllable:

> In fact, democracy, which is agreeable to no people, is particularly disastrous for meridional temperaments, which are more easily excited than others by demagogic propaganda. However, under other regimes we have seen Spain, faithful to its traditions, follow leaders who have the grandeur of the nation and the good of the people in mind.[64]

Thus the people, it was argued, were molded by the hands of a leader, but they were incapable of democracy, because that gave them over to excitable emotions and exposed them to demagoguery.

In many ways the vision of the crowd as the potentially dangerous mob was simply another iteration of the tradition in France and Europe of seeing the crowd as female and emotionally unstable, enraged or tamed by the virile male speaker. Taine and Zola had depicted the crowd as mercurial and emotional, for instance, and Hitler presented the crowd as a female to be mastered by the virility of the speaker.[65] However, the fantasy of the voice's power over the crowd was perhaps specific to fascism. Alice Yaeger Kaplan notes that many French fascists were fascinated by the idea of the reproduced or amplified voice.[66] Brasillach certainly mentioned frequently his delight in listening to fascist leaders on the radio, and he had a vaunted idea of the power of the transmitted, or amplified, voice. During the Czech crisis in 1938, Brasillach said,

> we were able to see again what the Spanish War had shown us several months earlier: the role which radio plays today. The evening of Hitler's speech, the evening when they announced the conference at Munich, in a total silence, all the world listened.[67]

It should be noted that this power of command, the magic of the voice, was considered legitimate and distinct from demagoguery. The power of the communists over the crowd was considered to be illegitimate, based only on lies. The communists, it was asserted, used the rule that "'the credibility of a lie depends on its size.' This technique of lying is based on a judicious psychology of crowds [*foules*]."[68] Thus the legitimate leading of crowds was through the voice of command, the illegitimate manner was through the big lie.

It was precisely this reading of the crowd as entirely subject to the character of its leaders that allowed the clear geographical mapping of social stress and ideology. Spain, in this conception, was clearly divided between regions of total anarchy and regions of total order and calm, and in each region that Franco conquered the anarchic and frenzied people were magically transformed into loyal subjects by the power of the nationalist leaders' will, conveyed through their voices; they ceased being beasts and became Spaniards.

Thus it seems that there were two competing visions of the people. The vision of the crowd as being molded and shaped by leaders of the left and right was always in tension with the other image of the passive people, awaiting salvation from communist oppression. If it was the latter conception that was the foundation of the moral justification of Franco's coup in the pages of *Je suis partout* (i.e., the people really wanted Franco but had no voice), it was the former vision that was the basis of the *Je suis partout* group's aversion to democracy. In a larger sense, both views displayed a profound contempt for the population at large, be it a helpless mass or a powerful force with no intrinsic direction or intelligence. Underlying these characterizations was often a class distinction, and the image of the passive traditionalist people was often coded as an image of the bourgeoisie, while the image of the volatile mob subject to leadership was often coded as an image of the working class. However, these distinctions and codings are not absolute and varied with events.

The presence of these distinctions between classes, coupled with the instability and variability of the codings, reveals a great deal about the motivating force behind fascist ideology. Fascist anticommunism resolved itself into a fear of an angry working class, but at the same time it revealed contempt for the docile bourgeoisie. Fascist ideology's acute consciousness of the apparently intrinsic differences between these two classes was in tension with the contradictory desire for class conflict to simply disappear. Thus the virile leader's power of command became a fantasy to hold this nightmare at bay, and claims of fascists to unite the entire nation in perfect harmony revealed their foundations in class fear.

Some *Je suis partout* writers displayed a greater tendency toward this fear than others. Lucien Maulvault articulated a near-hysterical vision of the crowd that left little room for the picture of a fundamentally good people oppressed by a small minority. Others, like Jean d'Elbée, were far more likely to attribute fine intentions but imperfections to the people. Brasillach, for his part, was always willing to idealize the simplicity of the working class, albeit condescendingly.[69] Thus the tension in the image of the crowd in *Je suis partout* reflected the opinions of the multitude of writers gazing across the Pyrenees, but it can be argued that fascism in France needed the idea of the silent, traditionalist majority to justify its conception of the nation, while it also needed the mechanism of the power of command to transcend class conflict.

Taken as a whole, coverage of the war in *Je suis partout* transformed social oppositions into geographical ones and made nationalist leaders into the unifying map-readers who could magically cleanse Spain of class conflict by sweeping across its territory in a great scouring arc. The removal of class conflict required the destruction not of classes of people within Spanish society, but simply a few strongholds wherein the dregs of society, the Marxists, the criminals, the Freemasons, and the Russians had managed to turn the people into monsters, unleashing the volatile emotion of the mob. This was part of a larger unstable vision of the people as either the traditionalist, inert, and bourgeois population or the excitable, commandable mass, awaiting emotional direction. In this latter vision, the difference between calm order and bloodthirsty frenzy was leadership: Nationalist leaders had the calming power of command while Marxist leaders had the insidious power of silver-tongued demagoguery.

The presentation of the Spanish Civil War in *L'Action française* and *Je suis partout* was, in many respects, quite consistent. In both periodicals Franco's nationalist movement was presented as saving Spain from the bloody terror of communism and the fratricidal nightmare of anarchism and civil war. Franco was restoring order and his Nationalist movement captured something of a "true" Spain, a Spain that was threatened or submerged by the rising tide of democracy, communism, and anarchism.

The major difference between the images of Spain in the *L'Action française* and the pages of *Je suis partout* lay in the power of the leader and his relationship to the masses. In the eyes of Maurras' followers, the masses, when aroused, were always dangerous, even pathological, but Franco's dedication to a strict discipline and nationalism would save Spain by pacifying it and excluding the masses from politics. In *Je suis partout*, on the other hand, Franco's movement represented a heroic, personal power, exercised through command and leadership, and it required a dynamic, motivated, politicized population. Both ideological positions therefore reveal a profound fear of the revolutionary population, but while Maurras' vision remains fixed in fear, the vision of Brasillach and the *Je suis partout* writers was to harness this powerful, mobilized population to the fascist cause through the power of the leader's charisma.

Notes

1. Surprisingly little has been written about French press' treatment of the war. Lars Peterson's dissertation on French weeklies dedicates a chapter to this issue but focuses mostly on how images of the painter Goya were used to interpret the war. See Lars Peterson, *The Age of the Weekly: Contesting Culture in the Hebdomadaires Politico-Littéraires (1933–1940)*, dissertation, University of Iowa, 2003.

2. See Eugen Weber, *The Action Française: Royalism and Reaction in Twentieth-Century France* (Stanford: Stanford University Press, 1962), 363. Following Weber, I will use the "Action Française" to denote the political movement, and "*L'Action française*" to render the movement's eponymous newspaper.

3. José le Boucher, "L'Espagne dans le chaos," *L'Action française*, 18 August 1936, 2.

4. Unsigned, "Soulèvement militaire en Espagne," *L'Action française*, 19 July 1936, 2.

5. Unsigned, "La Guerre civile en Espagne: la liaison Franco-Mola est effectuée," *L'Action française*, 12 August 1936, 1.

6. Inigo Bernoville, "Le Front populaire à Barcelone," *L'Action française*, 21 August 1936, 3.

7. Unsigned, "La Guerre civile en Espagne: la terreur à Madrid," *L'Action française*, 5 August 1936, 3.

8. Léon Daudet, "À Barcelone, les révolutionnaires se battent entre eux," *L'Action française*, 6 May 1937, 1.

9. Unsigned, "*L'Action française* chez les rouges d'Espagne," *L'Action française*, 30 April 1937, 3. "Gangsters" is in English in the original.

10. Léon Daudet, "À Barcelone, les révolutionnaires se battent entre eux," *L'Action française*, 6 May 1937, 1.

11. Léon Daudet, "La Virulence de la démocratie," *L'Action française*, 5 August 1936, 1.

12. Unsigned, "'L'Action française' chez les rouges d'Espagne," *L'Action française*, 17 May 1937, 1. Similar articles appear under the same title 30 April and 1, 3, and 18 May 1937.

13. Unsigned, "À Barcelone sous le règne de la Tchéka," *L'Action française*, 16 December 1937, 3; and Jean Dourec, "En Espagne: le satanisme des rouges," *L'Action française*, 30 March 1938, 2.

14. Léon Daudet, "La Peste communiste," *L'Action française*, 18 August 1936, 1; and José le Boucher, "Révolutions espagnoles," *L'Action française*, 20 July 1936, 2.

15. See Charles Maurras, *Enquête sur la monarchie, 1900–1909*, 2nd ed. (Paris: Nouvelle Librairie Nationale, 1909), 544.

16. Léon Daudet, "La Peste communiste," *L'Action française*, 18 August 1936, 1.

17. Charles Berlet, "Démocraties sanglantes: France-Espagne," *L'Action française*, 20 September 1936, 3.

18. Léon Daudet, "Après Unamuno: Maranon," *L'Action française*, 23 February 1937, 1.

19. Léon Daudet, "Corneille en Espagne, Shylock en France," *L'Action française*, 3 October 1936, 1. See also Léon Daudet, "L'Agonie de la révolution espagnole," *L'Action française*, 25 September 1936, 1.

20. See articles by Maurice Pujo, *L'Action française*, 1, most days from 24 July 1936 through early September of that year.

21. For a larger discussion of France's aid to the Spanish Republic see Michael Alpert, *A New International History of the Spanish Civil War*, 2nd ed. (New York: Palgrave Macmillan, 2004), and Gerald Howson, *Arms for Spain: The Untold Story of the Spanish Civil War* (New York: St. Martin's Press, 1998).

22. The book is made up of articles in *L'Action française* that appeared between 23 May 1937 and 2 November 1937. See Pierre Héricourt, *Pourquoi mentir? l'aide franco-soviétique à l'Espagne rouge* (Paris: Éditions Baudinière, 1937). The book was published in November of 1937 and was translated into English as *Arms for Red Spain* (London: Burns, Oates and Washburn, 1937).

23. Héricourt, *Pourquoi mentir?*, 22–23.

24. Léon Daudet, "Avant la prise de Madrid: fuite d'Azana à Barcelone," *L'Action française*, 23 October 1936, 1.

25. Unsigned, "L'Invasion de métèques," *L'Action française*, 8–12 September 1936.

26. Joseph Delest, "L'Invasion rouge: déserteurs, assassins, et cambrioleurs," *L'Action française*, 12 September 1937, 2.

27. Unsigned, "Les Nationaux progressent lentement," *L'Action française*, 13 August 1936, 5.

28. Police report of a 17 March 1939 meeting of the Dames Royalistes. Archives de la Préfecture de Police de Paris, BA1895, 340.700–28-B, 18 March 1939.

29. Jean Dourec, "Vers la fin de la guerre," *L'Action française*, 15 September 1937, 2.

30. Jean Dourec, "Le Moral des rouges sur le front de Santander," *L'Action française*, 20 August 1937, 1.

31. Jean Dourec, "La Prise de Torrelavega vue par notre envoyé spécial," *L'Action française*, 28 August 1937, 3.

32. Pierre Héricourt, "Circuits franco-espagnols," *L'Action française*, 17 October 1936, 3; and Pierre Héricourt, *Pourquoi Franco vaincra* (Paris: Éditions Baudinière, 1936), 101. Like *Pourquoi mentir?*, Héricourt's *Pourquoi Franco vaincra* was composed of articles by Héricourt which had already appeared in *L'Action française* over the course of 1936.

33. Charles Maurras, *Vers l'Espagne de Franco* (Paris: Éditions du Livre Moderne, 1943), 28. This book, completed in 1940, was composed of articles by Maurras on Franco during the Spanish Civil War.

34. Léon Daudet, "Avant la prise de Madrid, fuite d'Azana à Barcelone," *L'Action française*, 23 October 1936, 1.

35. Charles Maurras, "'Pourquoi Franco vaincra,'" *L'Action française*, 7 December 1936, 1. This article was later published as the preface to Héricourt's book of the same title and then used once more in Maurras' own book on Spain. See Héricourt, *Pourquoi Franco vaincra*, 11–12; and Maurras, *Vers l'Espagne de Franco*, 136.

36. Charles Maurras, "'Pourquoi Franco vaincra,'" *L'Action française*, 7 December 1936, 1; and Héricourt, *Pourquoi Franco vaincra*, 14; Maurras, *Vers l'Espagne de Franco*, 137. The passage in *Vers l'Espagne de Franco* omits the word "organic."

37. Maurras, *Vers l'Espagne de Franco*, 190–91.

38. Weber, *Action Française*, 506.

39. See Annie Brassié, *Robert Brasillach ou encore un instant du bonheur* (Paris: Robert Laffont, 1987), 179–80.

40. See Pierre-Marie Dioudonnat, *Je Suis Partout, 1930–1944: les maurrassiens devant la tentation fasciste* (Paris: La Table Ronde, 1973), 133–71 for an extended discussion of some of these various attractions for the *Je suis partout* group in the late 1930s.

41. René Richard, "La Révolte d'un peuple: l'Espagne entre la vie et la mort," *Je suis partout*, 25 July 1936, 1.

42. René Richard, "La Révolte d'un peuple: l'Espagne entre la vie et la mort," *Je suis partout*, 25 July 1936, 3.

43. Robert Brasillach, "En attendant les camions de tueurs," *Je suis partout*, 8 August 1936, 2.

44. Robert Brasillach, "Le Parti de l'honneur," *Je suis partout*, 15 August 1936, 2.

45. René Richard, "Le 'Frente popular' en route vers la défaite," *Je suis partout*, 15 August 1936, 5.

46. André Nicolas, "Réflexions sur le soulèvement," *Je suis partout*, 15 August 1936, 6.

47. Unsigned appendix to André Nicolas, "Non, les nationaux ne sont pas les ennemis de la France!" *Je suis partout*, 22 August 1936, 6.

48. Unsigned, "Les Nationaux ont repris l'offensive," *Je suis partout*, 29 August 1936, 5.

49. Unsigned, "Les Nationaux ont repris l'offensive," *Je suis partout*, 29 August 1936, 5.

50. Unsigned, probably André Nicolas, "Là comme ailleurs, les Soviets reculent," *Je suis partout*, 19 September 1936, 9.

51. Unsigned, but probably René Richard, "La Guerre de libération espagnole," *Je suis partout*, 22 August 1936, 5.

52. Jean d'Elbée, "De Burgos à Guernica," *Je suis partout*, 17 September 1937, 6.

53. André Nicolas, "Avec les troupes nationales en Aragon," *Je suis partout*, 18 April 1938, 4.

54. Charles Kunstler, "Les Atrocités rouges de Malaga," *Je suis partout*, 17 October 1936, 5. Emphasis in the original.

55. Unsigned, "Tolède délivrée — Madrid en plein désarroi. Franco chef suprême des nationaux. Le Frente popular perd le contrôle de la mer," *Je suis partout*, 3 October 1936, 5.

56. Dorsay, "La Pavé espagnol dans la mare aux grenouilles parlementaires," *Je suis partout*, 20 January 1939, 2.

57. Pierre-Antoine Cousteau, "Saragosse by Night," *Je suis partout*, 29 July 1938, 5.

58. Lucien Maulvault, "Le Vertige de la mort," *Je suis partout*, 9 July 1937, 8.

59. Jean d'Elbée, "Comment on vit en Espagne rénovée," *Je suis partout*, 1 May 1937, 5.

60. Jean d'Elbée, "Comment on vit en Espagne rénovée," *Je suis partout*, 1 May 1937, 5.

61. Pierre Gaxotte, "'L'Insurrection, dernier des devoirs,'" *Je suis partout*, 8 August 1936, 1.

62. André Nicolas, "M. Fernandez Cuesta nous dit ce qu'est la 'Phalange Traditionaliste'," *Je suis partout*, 22 April 1938, 5.

63. Simionesco, "En écoutant le général Queipo de Llano," *Je suis partout*, 8 August 1936, 5.

64. André Nicolas, "Réflexions sur le soulèvement," *Je suis partout*, 15 August 1936, 6.

65. Hitler's treatment of the crowd as female is well known. See Adolph Hitler, *Mein Kampf*, Ralph Manheim, trans. (Boston: Houghton Mifflin, 1943). For Taine and Zola see Susanna Barrows, *Distorting Mirrors: Visions of the Crowd in Late Nineteenth-Century France* (New Haven: Yale University Press, 1981).

66. See Alice Yaeger Kaplan, *Reproductions of Banality: Fascism, Literature, and French Intellectual Life* (Minneapolis: University of Minnesota Press, 1986), 134–35.

67. Robert Brasillach, *Une Génération dans l'orage: mémoires; notre avant-guerre, journal d'un homme occupé* (Paris: Plon, 1968), 254.

68. Saint-Aulaire, "Le Mythe basque: Guernica," *Je suis partout*, 7 January 1938, 4.

69. See Mary Jean Green, "Toward an Analysis of Fascist Fiction: The Contemptuous Narrator in the Works of Brasillach, Céline, and Drieu la Rochelle," *Studies in Twentieth-Century Literature* 10:1 (1985): 81–97.

19

American Audiovisual Propaganda in France, 1948–1955

Angélique Durand

In the wake of research in the area of politics and economics, research in cultural diplomacy offers new perspectives to historians of the Cold War. It reveals the conflict's psychological dimension, which stems from ideological stratagems organized by the two protagonists, the United States and the Soviet Union. Both countries sought to disseminate collective social models by making use of forms of mass culture such as film. More so than radio and photography, film offers a representation of society and of life styles that feeds forms of the social imaginary, dreams, and aspirations. Because of these stratagems' institutional character, they are meant to be obvious to the man in the street as well as to leaders, leading both to define themselves in relation to them. Thus, the activities of American cultural policy toward France are of particular interest. France was considered to be the "keystone" of Western Europe, and for the East as well as the West it represented an important part of what was at stake politically as well as economically.

At the end of World War II, cultural policy represented an object of recent interest for political leaders. It had gone through a period of considerable change before reaching a level of institutionalized organization after the war, attracting increasing interest on the national and international level during the 1930s. In the face of the social and financial crisis of the time, certain filmmakers felt the need to speak to their fellows by circulating images describing the poverty-stricken living conditions of European and American populations. These documentary films were designed to educate the masses and foster socioeconomic reconstruction. In the United States, private organizations such as Films and Photo Leagues and Frontier Films developed an active social agenda. Their mission was to spread ideas, contest forms of social oppression, and fight fascism. Robert Gessner, a writer and a member of this movement, stated in the *New Theater Magazine* in June 1935, "The red meat of social reality is a better diet than Hollywood cream puffs," and "the American working class is our casting list."[1] Receptive to this form of expression and the private initiatives supporting it, and hoping for government involvement in this dissemination of information, Franklin Delano Roosevelt decided initially to provide financial encouragement before becoming involved personally by supporting and commissioning works charged with promoting the aims of the New Deal. American authorities came to understand the advantage of using forms of mass communication to inform, educate, and persuade the people. This technique had also been shown to be effective in fascist countries, as can be seen most readily in such Leni Riefensthal films as *The Triumph of the Will* (1934) and *The Gods of the Stadium* (1936).

In reaction to this kind of propaganda and cultural control on the part of fascist countries, American authorities decided to introduce a cultural dimension to their domestic and foreign policies. In 1938 they established "an organization that would represent the American tradition of intellectual freedom and integrity in the area of education."[2] This was the origin of the Interdepartmental Committee on Scientific and Cultural Cooperation with the American Republics, later the Department of State's Division of Cultural Relations. At first, this policy was improvised and halting, but over time the pressures of war caused it to be organized in such a way as to meet the growing need for information. In this context, aspects that strictly speaking were cultural and social yielded to political propaganda, despite the resulting conflict with American democratic values and a paradox that would continue throughout the Cold War. This decision led to requisitioning all the means of supporting information that henceforth would be used for political propaganda and the war effort. The political documentary film was favored over films with a social theme: "The documentary film won out over others, establishing the standard of the 16 mm. format. Projectors of the 'light' medium spread throughout the world. Filmmakers dedicated their practical knowledge to the struggle against Nazism, either by filming in the field or by working with archives."[3] These films multiplied and came to occupy an important place in American cultural policy, either in the education of young soldiers or by showing the horrors of war following the liberation of the Nazi concentration camps. Filmmakers were commissioned predominately by the Office of War Information (OWI), created in 1942. Along with the Office of Strategic Services, it was charged with centralizing the American information services and explaining governmental policy to the media and American and foreign public opinion during the Second World War.

Following the Second World War, while a new world order was taking shape, uncertainty concerning U.S.-Soviet relations led the United States to question how best to pursue its international cultural policy. What political choices would be made, and what role would they play in East-West ideological power relations? What impact would they have on France's position on the world chessboard? Examining these questions will allow us to show the causes and the activities of American cultural and international policy in France and to observe the place allotted to the institutions that commissioned a cinematographic propaganda aimed at the French population. Thus we can describe the audiovisual themes employed, their evolution, and their spread, and we can assess their impact on the French people and French politics during the initial years of the Cold War, from 1948 (the date of the creation of the U.S. Public Information and Educational Exchange Act of 1948, Public Law 402, 80th Congress and of the Foreign Assistance Act) to 1955 (the date of the closing of the Film Unit in Paris in October).

Cultural Aspects of American International Policy since 1945

Aware of the advantages of an international cultural policy, the American government decided in December 1944 to strengthen the administration of information services by creating the position, within the Department of State, of Assistant Secretary of State in Charge of Public and Cultural Affairs, a position first held by Archibald MacLeish, assistant director of the OWI since 1942. On August 31, 1945, President Truman, convinced of the need to maintain an information service during peace time, transferred the activities of the OWI to the Interim International Information Service (IIIS) in the State Department. Its mission was to insure "that other people have a full and exact picture of American life and of the goals and policies of the American government."[4] However, the means and activities of this service did not offer sufficient security against Soviet propaganda. Increasing East-West tensions led the Department of State to redirect its international and cultural policy. In the future, it planned to communicate directly to various populations, "people to people," and it increased its use of all mass culture media such as radio, the press, publications, photography, and film. The diplomats connected to the previous system were shifted aside in favor of personalities belonging to the world of communication, business, and higher education. They met on March 1, 1946, at the Office of International Information and Cultural Affairs (OIC). These goals were presented to the Department of State on October 1, 1945, by OCI Director William Benton, the Assistant Secretary of State in Charge of Public and Cultural Affairs. "All the programs on the ground involving so-called 'cultural relations' must support American foreign policy in its long-term goals and must serve as an instrument of this policy." He also stated that "the basic goal of the program is to advance the cause of peace by improving mutual understanding between the people of the United States and those of other nations."[5] Cultural relations thus became a way of mobilizing foreign public opinion in favor of the United States by spreading an image of the country that was more human and supposedly more "exact," and by lessening the aversion of European public opinion, which was encouraged by local communist parties. Cultural relations transmitted the values of a liberal democracy for economic prosperity and an improved quality of life in the countries affected by the war. However, East-West tensions, caused by issues pertaining to the reconstruction of Europe, as well as the American policy of containment announced by President Truman on March 12, 1947, brought a new ideological dimension to the policy.

This ideological power struggle caused the educational aspect of American cultural relations to be developed, leading to the replacement of the Office of International Information in fall 1947 by the Office of International Information and Educational Exchange (OIE). Subsequently, a bill was proposed in Congress in 1947 to confirm the existence and the missions of these different services.[6] This bill drew in part on surveys conducted involving the French population and describing the image of the United States in France. This image appeared to have suffered from the war itself and its political and moral consequences during the post-Liberation period. The criticisms frequently made included the following: the American life style, which was considered to be materialistic and pleasure seeking; distrust of American political and economic motives, considered to be self-interested; the denigration of French cultural accomplishments and intellectual traditions; and the perversion of American social and moral values. For the people conducting the survey, these opinions seemed to be heavily influenced by the anti-American writings of prewar French intellectuals and by communist and nationalist French political cells. These surveys also pointed out American supporters in France, who included the French who were grateful for the role the United States played during the Liberation and members of certain industrial and financial circles. This bill also presented the main objectives of the cultural relations and information policy, orchestrated by the future Smith-Mundt bill (named for its sponsors, Representative Karl Mundt and Senator Alexander Smith), the "U. S. Public Information and Educational Exchange Act of 1948, Public Law 402, 80th Congress." These goals included: providing better information concerning the real conditions of American life so that the French people could understand them more accurately; countering the errors and combating the forms of harmful propaganda directed against the United States; attempting to create a current of opinion that would be friendly and understanding toward the United States; explaining the methods, motives, and goals of American foreign policy; and facilitating the flow of information designed to publicize achievements in the area of science and culture. Cultural relations were thus incorporated legally into American international policy, whose goals involved military security, the protection of European democracy, and the containment of Soviet influence.

In terms of organization, the OIE was assisted in foreign embassies by cultural affairs attachés and information supervisors connected to the U.S. Information and Educational Exchange Program (USIE).[7] Through this international presence, Americans could come closer to the people and enter into direct communication with individuals. Thus they could react more responsively to local communist propaganda. This strategy was pursued very actively in France, where the French Communist Party garnered between 22.6 percent and 28.3 percent of the votes in legislative elections between 1945 and 1951. Moreover, in the eyes of American authorities, France occupied a strategic position in Western Europe concerning issues of political security and influence:

> It is in our interest to do everything in our power to help France morally and physically to regain her former strength and influence. The United States wishes to see a strong and friendly France, serving as a shield for our security, our view of democracy, and our material interests on the European continent, and sharing in the responsibility, through the intermediary of the United Nations in particular, for maintaining peace. In return, we must try to prevent the development of a France that could possibly become the western stronghold of a hostile "continental system."[8]

These fears prove that France had not been won over totally to the American cause. Thus, a 1947 Department of State memorandum reveals that "the funds and limited personnel available to the OIC require agents to focus on important objectives: informing the French people about America, counteracting propaganda that is false or hostile towards America, and creating a friendly and understanding opinion towards America."

In addition to its political dimension, an economic dimension of American cultural propaganda existed as well, and which was linked to the Marshall Plan. On June 5, 1947, Secretary of State George Marshall chose Harvard University as the place where he would present the basic elements of a European economic recovery plan. This plan was to serve as a catalyst for rebuilding European economies. It was fully defined in the Foreign Assistance Act signed by President Truman on April 3, 1948. Responsibility for carrying out the program for European recovery fell to a new federal organization, the Economic Cooperation Administration (ECA), which was placed under the authority of Congress. Its mission in France consisted of developing economic and financial policy, in collaboration with the French government and other European governments receiving economic recovery aid. The 1948 act assigned three functions to the ECA: promoting industrial and agricultural production in the participating countries; pursuing the restoration and stabilization of European currencies, budgets, and finances; and facilitating and stimulating trade development between participating countries through appropriate measures, in particular by eliminating the obstacles to trade.

To succeed, however, this program required assistance, as well as popular consensus in Europe and the United States. A policy of cultural propaganda was established within the ECA, the Office of the Special Representative-Information Division (OSR-ID), whose goal was to present the Marshall Plan as a form of friendly aid and not a threat to national sovereignty. To obtain American popular support, this propaganda highlighted the financial and economic situation in Europe that could promote support for communist influence, in opposition to liberal American values. The mission of the ECA thus was concentrated on political modernization and integration.

Wishing to win over European populations by spreading information concerning the underlying aspects of American policy, social values, and economic models, the USIS and the ECA used one of the most extensive forms of cultural propaganda ever organized by a democratic government during peace time. Because of the size of their sphere of activity and the variety of the subjects involved, they could reach a greater population and a variety of social groups. This they hoped would increase their chances of winning the war of ideas against the USSR. Film and documentary film in particular occupied an important place in this ideological war, and their production was organized within well-defined structures. Film became a significant tool in the psychological stratagems introduced in American and Soviet policy during the Cold War.

The Organization of the USIS and ECA Information Services

Within the French Embassy and its regional branches, the USIS represented the authority and coordinated the objectives of the Office of International Information and Education Exchange, which derived its authority from the Department of State. It received its directives by diplomatic pouch or telegram. But operations policy was directed in Washington by an administrator named by the president, the Assistant Secretary of State in Charge of Public and Cultural Affairs. He determined the policy line that was dictated in the long run by the will to maintain the alliance with France within the American protective shield, and in the short run by the will to make clear to the French how the Soviets threatened security and freedom.

To ensure the effectiveness of cultural policy in France, the USIS was organized into five sections: cultural affairs, regional branches (Bordeaux, Lille, Lyon, Marseille, Strasbourg, and Tours), information, the office of the press attaché, and an administrative unit. These sections in turn were split up into divisions. The information section, which in 1954 was housed in Paris at 41 rue du Faubourg Saint-Honoré, comprised four divisions, including press, exhibitions, radio, and film. It was responsible, through these various means of communication, for the diffusion of information about the American way of life, which between 1948 and 1950 took a social perspective, and later a more politicized one during the truth campaign initiated by President Truman.

Moreover, during this time the USIS staff in Paris grew from 149 to 174 employees. The unit was supervised by an office of information, which was responsible for transmitting activity reports to the head of the USIS public affairs agency and to other information agencies in France. It worked in close cooperation with other European embassies to coordinate information and maintain a liaison with the ECA.

Within the ECA, the dissemination of information was orchestrated by the Office of the Special Representative-Information Division (OSR-ID), created in 1948. It grew out of the bilateral agreements and economic cooperation agreements between France and the United States signed by U.S. ambassador Jefferson Caffery and French Foreign Minister Georges Bidault. These agreements provided for establishing an exchange of information concerning the economic decisions and advances brought about by the Marshall Plan in France. According to the signatories, these relations were to provide the material needed to diffuse information in Europe and the United States concerning the economic recovery underway.

The OSR-ID soon began "internal propaganda [aimed at] Europe, which also required a strong information campaign, in other words an active propaganda campaign."[9] Personnel in this office quickly became aware that more was needed than simple governmental reports to convince the American Congress to make the necessary financial means available and to rally European public opinion. In France in particular, public opinion was widely influenced by communist arguments denouncing American imperialism and warning of the social and economic impact of such a plan. The goals of this propaganda were thus in line with the speech given by George Marshall at Harvard on June 5, 1947, and they allowed for presenting aid as a proof of friendship in accord with the decisions of the participating governments. Traditional media were employed in this propaganda, including photography, exhibitions, the print media, radio, and film, which were organized in three services: the Visual Information Unit (publications, posters, and exhibitions), the Film Unit (news, documentary films, and technical films), and the Radio Unit (news bulletins, documentary programs, and entertainment programs). Within these three, the film unit represented a significant source of information for the personnel working there. Film was a rich medium, one that was often more explicit than a list of figures detailing the credits allocated by the Marshall Plan. Directing films and shooting them on the sites where Europe's reconstruction and modernization were taking place struck a better chord with social imagination. Moreover, as Stuart Schulberg, one of the heads of the film unit, wrote in 1951, "ERP ads could do as much for Europe's mental depression as ERP shipments could do for Europe's economic ills. An important tenet of ECA philosophy was fashioned in a slogan: 'The Marshall Plan—helping people to help each other.'"[10]

In 1948, both government agencies, the USIS and the ECA, had the structural and financial means necessary to spread audiovisual propaganda aimed at improving the image of the United States and promoting a consensus favoring that country's integration in the politics and economy of France. Nevertheless, although similarities can be observed in the motives behind the creation and organization of the information and propaganda agencies, differences exist as well. These differences are largely related to their specific hierarchical connections, to the Department of State and Congress, respectively, which determined the choice of news content, as well as that of film production.

USIS and ECA Film Production

The production of USIS films was shaped by the wish to show the American life style to the French, along with technological, scientific, and agricultural advances. These information programs were designed not only to persuade but also to encourage social and economic modernization. The latter was presented as the indispensable means of improving each person's living and working conditions. There lay the subtlety of American propaganda. It developed an imaginary reality and created new needs in a population for some of whom basic modernization was still unknown, and whose daily life was marked by the bombing of the Second World War. To enhance these films' realism, they were filmed in America with the support of U.S. production teams. Two examples of such films are *The United States Military Academy* (Columbia Pictures Corporation) and *Small Town, USA* (Pathescope Company of America).

The production of Marshall Plan films was organized differently. Its goal was to promote the achievements of the European economic recovery plan in Europe in order to increase the French public's awareness of the advantages of American aid. Marshall Plan slogans were just about everywhere and left little room for interpretation. As it was not enough to show American intervention, these films explained and demonstrated its necessity. Filming took place in countries receiving aid, in factories, fields, ports, and cities being rebuilt—everywhere that Marshall Plan assistance was given. These propaganda films took their inspiration from the progress made in Western European economic reconstruction. Their basic direction came from the reports submitted by the recipient countries. This documentation provided the ECA with a large source of inspiration in commissioning the production of a particular film. This documentation was also supported by the office of news and communications in the French Ministry of Foreign Affairs, which submitted ideas for particular proposals when it was determined that producing a particular film would benefit the use of Marshall Plan aid. The most representative example of this is *The Gold of the Rhône*, whose function the ministry defined as follows: "This production presents the role that this river played from a cultural point of view and the significance it is destined to have now in the French economy."[11] Made in 1951, this documentary described the economic development of the Rhône-Alpes region following the Second World War. This development was financed and promoted by means made available by the Monnet Plan and by Marshall Plan assistance, which made it possible to construct the hydroelectric infrastructure in the region. The film stresses the building of the Donzère and Mondragon dams, which provided electricity for city dwellers within French borders and beyond (particularly in Italy). In addition, during the conception and writing of these film projects, the ECA accepted the assistance of private production companies through the French Ministry of Foreign Affairs. However, beneath this arrangement often lay a simple attempt to gain financing. This was this case with *La Fugue de Mahmoud*, made in 1950 by Roger Leenhardt with Films Compas. This film presented the adventures of a young Moroccan and his meeting with an agricultural engineer with the ERP who teaches him the basics of modern agricultural techniques. This kind of collaboration with the private sector of European film professionals expanded to include technicians. People in charge of the ECA film office preferred to hire local directors and technicians because they knew the local people and their habits. The message expressed by these films would be much better received, they felt, if it came from a team that was familiar with the culture of the film's audience. One such example is *Jetons les filets* (*Het schots is te boord*), directed by Herman Van den Horst. This film relates in poetic fashion a Dutch trawler's trip to the Dogger Bank during herring fishing season. This beautifully-crafted documentary won the grand prize for shorts at the Cannes Film Festival in 1952. More representative of the travel film genre, *Liberté* presents the planning and building of the German steamship *Europa*, which was renamed *Liberté* when it was acquired by France. This documentary film was made in 1950 by Jean Mitry, the well-known French director of the 1940s and 1950s.

Contextual factors also played a role in the choices of scripts and topics in the ECA audiovisual propaganda. Situated in the daily life of Europeans, these films also had to reflect the international situation. Thus, increased concerns relating to national security and military rearmament generated new ideas for films. These concerns were expressed in American international policy as early as 1949. Alarmed by the coup d'état in Prague in February 1948, the countries allied with the United States sought to prevent a potential Soviet threat by creating the North American Treaty Alliance (NATO) in 1949. Under the pressure of international events, which reached their high point with the Korean War in 1950, military questions gradually took precedence over social and economic ones. On January 1, 1952, the ECA was replaced by the Mutual Security Agency (MSA). As defined in the mutual security bill, this agency was to allow "North Atlantic defense plans to be carried out, while maintaining economic stability in the countries in the region so as to allow them to confront their defense responsibilities, and to encourage a greater economic unification and the political union of Europe."[12] Although it maintained certain economic aspects, the MSA audiovisual information agency gradually became closer to the USIS in order to produce a form of propaganda better able to meet new goals, until they both were discontinued in 1953. The film laboratory of the Films Unit would not be closed until 1955.

Films and Their Topics

Marshall Plan films produced as ECA-MSA propaganda represented a wide variety of topics and techniques. The name change of the organization in charge of European economic recovery brought with it a redefinition of goals and thus a change in topics. The period from 1948 to 1950 was marked by films designed to show how the Marshall Plan had contributed to solving economic, industrial, energy and agricultural problems, as well as to ensuring European support for American democratic and liberal values. These films also highlighted problems specific to World War II such as housing and refugees. They did not compare countries, selecting instead one nationality as background to the film. These films were thus given the generic title of "one-country films." They reflected the first stage of the American aid program, support for reconstruction and recovery in the area of technology and industry. John Ferno's 1950 film *Return from the Valley* offers a good illustration of this first period. It describes the return from exile of a group of 700,000 Greek refugees after the Second World War and the civil war in 1949. This documentary film makes reference to the reconstruction and the new agricultural techniques transmitted by the Marshall Plan; through its geographical localization it expressed the close connection between economic aid and political doctrine. The only Balkan country to receive American aid was Greece, situated at the center of American-Soviet competition.

This period was followed by a transition between 1950 and 1951. The international situation and the Korean War set a new tone for Marshall Plan films. Topics shifted from Europe to focus on the international context. They explored the idea of need for military and international political cooperation. The MSA worked selectively with NATO and Supreme Headquarters Allied Powers Europe (SHAPE) in coproducing films that supported the role of the organization within the "Atlantic community." The film *Alliance for Peace*, produced in 1951, is one of the best examples of this partnership because it describes NATO's international command structure and military force. Moreover, to make the film available to everyone, it was translated into the languages of all the allied countries.

Although the films' economic conception focused between 1952 and 1953 on topics of technical support and innovation, it nonetheless remained closely connected to questions of joint military security. "This reduction was above all the result of the burden that the Korean War and the building of NATO imposed on still very delicate European economies. Europe had to adopt this approach in order to produce as much as possible with a minimum of means in the area of capital, primary goods, and workers."[13] For this reason, NATO films such as *Soldiers of Freedom* and *Promotion atlantique* (*School for Colonels*) were produced during the same period as were films on the functional organization of industrial production, such as *Enseignement technique en Holland* (*Training for the Job*) or *La Marche du travail* (*Work Flow*). The increase in productivity that these films described was supposed to lead to increased exports. Thus, during this period it was possible to see films on commerce and free trade. The 1951 film *Sans visa, ni frontière* (*Clearing the Lines*) is one of the most typical because it advocated the removal of borders that merely blocked economic development. *Nous et les autres* (*We and the Others*), produced in 1951, echoes this theme. It presents Europe through the eyes of a man from the nineteenth century who is astounded by progress that appears useless to him because of the purpose to which is it put. For him, this progress has merely brought about the separation of people. This documentary film shows the need to strengthen international bonds and broaden trade agreements between the countries of Western Europe.

The variety of techniques employed in these films is not so much related to historical developments as by the wish for films that match various topics. The corpus of Marshall Plan films comprises approximately 180 documents in French, divided into three distinct genres: documentary films, news films, and animated films. The documentary films in this collection were produced according to different cinematographic styles that met the requirements of particular forms of propaganda. These various procedures echo the methods used during the Second World War: the "classic documentary," "the "docu-drama," the "fictional documentary," and the "educational documentary." On the one hand, use of the educational genre recalls the techniques of audiovisual propaganda employed by the Vichy government. One example of a film illustrating these characteristics is *Les Aventures extraordinares d'un litre de lait* (*The Extraordinary Adventures of a Quart of Milk*), a film that explains the making of powdered milk, from milking to packaging. Both technical and precise in its description, it uses a child-like off-screen voice aimed at a variety of publics. On the other hand, the fictional documentary is often used to portray the rebuilding of European cities and villages financed by American aid. The film *The Village That Wouldn't Die*, produced in 1950, describes the rebuilding of a village in Normandy through a guided tour given by a village inhabitant who is actually the wife of the doctor-mayor. The theme of building also provides an opportunity for films in the docu-drama genre. The film *Et c'est arrivé ainsi* (*Dunkerque*) shows viewers the anguish of the people of Dunkerque who received assistance from their counterparts in the American city of Dunkirk. These latter two film genres mix reality and fiction, corresponding to propaganda in the use that is made of the important process of identification. Through his or her own life experience as a worker or a victim of the war, the viewer identifies with these actors or characters who were selected during filming for their believability. The receptive viewer is carried along by them and guided into believing in their hope for renewal, made possible by the Marshall Plan. Finally, the last genre corresponds to classic documentaries that take the form of a simple and well-defined scenario, designed to provide a realist view of things. These films show different versions of the Marshall Plan, from the perspective of observers of the reconstruction and modernization of French and European industries, agriculture, and economy, two results made possible by American aid. It is also important to point out the decision to group certain documentary films in a series. Six of these films representing a variety of genres were put together beginning in the 1950s in a series entitled *Le Grand Espoir* (*The Changing Face of Europe*). In technicolor and produced by Wessex Film Productions, this series was designed to give the people of the countries receiving aid from the Marshall Plan an idea of its results and the economic and social condition of Western Europe.

This organization in series led to another genre represented in this collection, the current events films. These films were collected in two groups, *Les Nouvelles illustrées* (*The News in Pictures*), which were 11 minutes in length, and *Les Revues filmées* (*Film Magazines*), which were 7 minutes long. Differing from the presentation and classic form of news films, these documents were closer to news reports, filmed and edited using rushes from documentary films. The animated film represented the last filmic genre in Marshall Plan films. These films explored three topics commonly seen in American propaganda: free trade, industry, and anti-communism.

Like Marshall Plan films, USIS films, which numbered about 450, included documentary, fictional, and technical films, current event films, and animated films. Putting little or no emphasis on the international situation, they highlighted the history of the people of essentially one country, the United States. These films transmitted the stereotypes of the American Way of Life and offered the way to achieve the American dream. These stereotypes were embodied in important political and artistic figures. Three films were devoted to the inauguration of President Eisenhower: *General Dwight Eisenhower; Dwight Eisenhower, 34th President;* and *The Second Inauguration of President Eisenhower.* The first film, which was edited using a collection of personal and official archives, traced his childhood, his family, his college education, his military career and his election as president of Columbia University, and finally his assuming the position of supreme commander of NATO forces in 1950. The second film was more neutral, focusing on the inauguration ceremony on January 20, 1953, at the Capitol in Washington, D.C. It highlighted the international role of the United States, and it presented the country and its new president as the guarantors of world peace. This film is comparable to the one that recounted his second election in 1957.

This latter film employed a glorifying and patriotic tone, with the national anthem in the background. This tone was perfectly in keeping with Eisenhower's offensive and militaristic pledge, which was emphasized by a parade displaying the armada of America's modern armed forces. These portraits also stressed individuals from American minorities. The film *Edith Sampson, Ambassador to the United Nations* recounted the various stages in this politician's career, her college degrees, her family life, her career as a lawyer, and her social activities in the Afro-American community in Chicago. Professional and social success was shown to be possible for anyone who acquired the means to achieve it. Political and social values made up the thematic cornerstone of the film collection, which required a rigorous selection of situations, places, and human anecdotes. To provide a nuanced portrayal of American industry without promoting liberal ideas that might displease the French public, these films instead stressed the human side of things. The film *With These Hands* recounted the daily life of textile industry workers in the United States, highlighting their relations with employers and the solidarity among workers against the backdrop of the history of trade unions in the United States. It also showed how this kind of collaboration involving efforts and skills enabled greater productivity. Upward social mobility and personal fulfillment were important themes in USIS films. *American Women at Work* presented the portraits of four women: Esther, Joan, Anne, and Martha. They all have jobs allowing them to feel fulfilled in their family and providing them with social recognition. These films also showed how new technologies led to improvements in daily life, whether in the area of agriculture, industry or health care. The film *Breast Self-Examination* informed women about advances in cancer research and preventive techniques for detecting a possible tumor. Culture and leisure were also important ideas for making the American lifestyle appealing. They echo the consumer society and the improvement in lifestyle. These films showed the organization of American free time through images displaying citizens engaged in activities involving play and culture such as sculpture, painting, visits to museums, and sports. Examples of such films are *Loisirs bien employés* (*Leisure Time Well Spent*) and *Sports étrangers pratiqués aux États-Unis* (*Foreign Sports Played in the United States*), both titles expressive of the films' content. This collection is much richer than these few examples, and it offers carefully selected information concerning various aspects of rural and urban life in the United States.

Film Distribution in France

Despite the difference in their topics, these two collections regularly used the same distribution networks, which were organized by the USIS. Visitors to the film section offices could obtain distribution plans and film catalogs. In addition, before each screening, the USIS undertook to place an article in the daily paper. For regional screenings, the USIS was often assisted by regional offices that, besides placing articles in the press, distributed fliers in the local markets and placed posters in the city and surrounding areas. Everyone who had taken part in Franco-American cultural exchanges was also invited to these screenings, according to the specific topic. These invitations were aimed mainly at teachers who were invited along with their classes. Plans for specialized screenings were carefully organized. Officials were responsible for the appropriate selection of locale and audience members. This was the case, for example, for the distribution of three films that dealt with motorized transportation: *Remorque 201*, *Panorama*, and the third installment of *Nouvelles illustrées*, *La Jeep en civil* (4 mn). After the war, the jeep was used as a farm and utility vehicle because of its endurance and sturdiness, as shown in *L'Autocar de l'avenir* (1 mn). *Les Trotteurs* (5 mn) dealt with the taming and training of race horses for equestrian competitions. These films were shown in Marseille to 300 truck drivers. In addition, to facilitate access to these screenings, the USIS also organized buses for people who were isolated or lived too far from cities.

The distribution of American films received financial aid from French authorities, both as tangible support and as a way to ensure the message would be communicated precisely when the film's topic corresponded to the information disseminated by the government. In January 1950, six films concerning the medical and social problems relating to children were organized by the USIS/Paris Film Unit, the French Ministry of Labor (social security branch) and the Caisse nationale de sécurité sociale. Eighty people attended the screenings of these films.[14] The USIS also used pro-American distribution networks, including private associations under the patronage of the American Embassy such as the France-United States Association. This organization announced its co-sponsorship of certain screenings with a political theme so that they would be better received by a sometimes hostile public. In fact, some screenings gave rise to virulent opposition on the part of communist-directed political groups. On September 19, 1950, a symposium in Roanne on the theme of American policy toward Korea led to the arrest of 18 communist protesters.

Although French authorities were not opposed to the distribution of American audiovisual propaganda, often for economic reasons, this was not the case for the French people. American activity sometimes met strong popular opposition led by the French Communist Party. This group was a significant element in the French post-war political landscape, as it received on average 25 percent of the votes in the national elections between 1945 and 1951. These figures prove the political influence this party had on the French population. Thus, when the Communist Party was expelled from the Ramadier government in May 1947, and when it announced on October 30, 1947, that it would join the opposition to a government it claimed had been "sold" to Washington, many people followed the party's belief that the Marshall Plan represented French dependence on the United States. For this reason, protest operations against forms of American imperialism multiplied. The headline in the October 31, 1947, issue of *L'Humanité* read, "Following Washington's orders, will M. Schuman propose a devaluation today in the council of ministers?" The November 9 headline read, "The Americans are going to choke French railroads." This context calls into question the influence that Marshall Plan propaganda had in France. Moreover, according to the opinion poll conducted by the ECA in mid-1950, only 55 percent of the French people interviewed approved the plan, as opposed to 75 percent in Norway, Denmark, Holland, Austria, and Italy.[15] In addition, according to the same survey, workers and peasants were the least convinced by American aid. These figures affirm the observations of an upper-level administrator in ECA headquarters in Paris in 1949: "The European worker listens listlessly while we tell him we are saving Europe, unconvinced that it is his Europe we are saving."[16]

In conclusion, numerous parameters must be taken into consideration in assessing the influence of American audiovisual propaganda in France. The weight of other components of cultural policy such as the press, photography, radio, exhibitions, and cultural exchanges must be included in any overall assessment. Yet the importance of private media must not be overlooked. Their influence is unquestionable because they achieved a much larger distribution than institutional films. Moreover, this distribution was encouraged by the Blum-Byrnes Agreement signed in Washington on May 28, 1946. In return for a final settlement of debts and war claims amounting to $720 million payable over five years, as well as for an agreement on a freer international trade policy and on the opening of additional credits allocated by the Export-Import Bank, the French government was required to allow American films easier entry into France. Theaters were required to project French films for four weeks a quarter, during the other nine of which they would present foreign films selected competitively. Given the quantitative superiority of American films on the world market, a new French-American film agreement was signed September 16, 1948, by Robert Schuman, French Minister of Foreign Affairs, Robert Lacoste, French Minister of Industry and Commerce, and Jefferson Caffery. The agreement set "the import limit at 121, the number of American films that can be brought into France each year, and the new screening quota at five weeks per quarter."[17]

Compared with the competition of private media, the influence of USIS and Marshall Plan films is difficult to assess. The hypothesis of their contribution to the Americanization of France and Europe between 1948 and 1955 thus cannot be determined with certainty.

Notes

Chapter translated by Daniel Brewer.

1. Jean-Paul Colleyn, *Le Regard documentaire* (Paris: Édition du Centre Georges Pompidou, 1993), 38.
2. Yves-Henri Nouailhat, "Aspects de la politique culturelle des États-Unis à l'égard de la France de 1945 à 1950," *Relations internationales* 25 (1981): 88.
3. Colleyn, *Le Regard documentaire*, 41.
4. Extract from the Department of State Bulletin, 13:323 (September 2, 1945).
5. Extract from the Department of State Bulletin, 13:330 (October 21, 1945), 590.
6. Department of State, Appropriation Bill for 1947. Hearings before the Subcommittee of the Committee on Appropriations. House of Representatives, 79th Congress, Second session.
7. From 1945 to 1948, the main agency for carrying out the American policy of public affairs was the United States Information Service (USIS). In 1948 it became the United States Information and Educational Exchange with the Fulbright program. From 1950 to 1952 it was the international information agency, and from 1953 to 1958 the United States Information Agency (USIA) or the agency of American information.
8. Extract from N.A., R.G. 59, 711.51/9 – 1546. Department of State, France, Policy and Information Statement. September 15, 1946; Secret. 26 pp.
9. Noël Van Rens, "Les Films du plan Marshall: 1948–1953," in Roger Odin, ed., *L'Âge d'or du documentaire, Europe: années cinquante* (Paris: L'Harmattan, 1998), 199.
10. Stuart Schulberg, "Making Marshall Plan Movies," *Film News*, Septembre 1951.
11. Archives of the French Ministry of Foreign Affairs, cultural relations agency, cultural exchanges service, office of information and distribution, articles 225, 1950.

12. Archives of the French Ministry of Foreign Affairs, report on a bill for mutual security, United States, foreign policy, article 11, 1944–52.
13. Van Rens, "Les Films du plan Marshall," 204.
14. William Tyler, "Report of Motion Picture," January 30, 1950. Record group 59. Department of State. Decimal File 1950 to 1954. Box 2383. NARA, Washington, DC.
15. David Ellwood, "You Too Can Be Like Us: Selling the Marshall Plan (American Propaganda During the European Recovery Program)," *History Today*, 48 (October 1998).
16. Ellwood, "You Too."
17. Archives of the MAE, article 277, "nouvel accord cinématographique franco-américain, Paris, le 16 septembre 1948, circulaire no. 215-IP."

20

Two Films and Two Wars in the Public Sphere

Steven Ungar

"Attention: un train peut en cacher un autre!" Some railroad crossings in France still display a sign warning pedestrians to look both ways because a train going one way on one track often hides a second train coming from the opposite direction on a second track. I take the sign as a cautionary reminder with reference to what I have seen while drafting this paper, which I began a year ago as a study of *La Trahison* (The Betrayal), a 2005 feature-length film about a French military unit stationed in an isolated part of southeastern Algeria in March 1960. Director Philippe Faucon adapted the film with journalist Claude Sales from the latter's 1999 memoir based on an incident during Sales' tour of duty as an officer in the French army. I first saw *La Trahison* in January 2006 and was drawn to the economy with which it conveyed complexities affecting all parties to the incident in question. I considered the film's mix of drama and historical account to support its status as a possible commentary on and allegory of the Algerian war close to 50 years after the fact. I was also curious how the film's release in early 2006 might be understood in light of a February 2005 proposal by then President Jacques Chirac for the national school system to present in a positive light the contributions of colonization in overseas territories formerly occupied by France. (Chirac withdrew the proposal in January 2006.) *La Trahison* was by all measures a small-scale film—low-budget and marketed mainly on the independent and art-house circuits—whose perspective on a large and complex topic I found worthy of further scrutiny.

Attempts to access documentation on the critical reception of *La Trahison* during the summer of 2006 invariably led me to coverage of another war film with a colonial slant, *Indigènes*, whose major award at the Cannes Film Festival had heightened anticipation surrounding its forthcoming release. Wherever I looked in the French press and online, *Indigènes* was so unavoidable (*incontournable*) that I found myself in the dubious position of feeling obliged to write—at least tentatively—about a film I had not yet seen in conjunction with a film I had seen and about which I had strong convictions. What follows is an account of my evolving views on *La Trahison* and on *Indigènes*, the latter of which I have since seen on screen and on DVD. After presenting a draft of the paper at the October 2006 "Spaces of War" conference, a June 2007 return to Paris allowed me to recast my earlier remarks on both films in light of debate surrounding the extended aftereffects of France's colonial history. In so doing, I tried whenever possible to avoid comparisons between the two films that, following the railroad warning sign, might hide one film from the other and thereby fail to do justice to their respective specificities.

War in the Public Eye

Unprecedented and thus all the more symbolic, the award for best actor at the May 2006 Cannes Festival went to the five male leads of *Indigènes*, Rachid Bouchareb's epic account of four North Africans recruited by the French army out of their home villages in 1943 to fight against German occupying forces in Italy and France.[1] Critics of the film have consistently lauded the earnestness with which the film honors the 130,000 "native" veterans from Morocco, Algeria, and Tunisia whose contributions to the World War II liberation of France remain under acknowledged some 65 years after the fact. Director Bouchareb has stated that he saw film as a vehicle for encounters and emotions, even if this initial response did little to contend with the serious issues the film raised. "Only in this way could I carry out the story and create a tie with the audience. I didn't want to be didactic, which leads to nothing."[2]

Bouchareb's remarks express a personal commitment to salvage a neglected aspect of France's recent past by transmitting it in the form of a major-release feature film targeting spectators many of whom are too young to have direct experience of the events it depicts. His remarks also point to the symbolic significance of the film for communities of North African and sub-Saharan ethnicity whose collective voice in current debate surrounding national identity conservative politicians consider threatening to a certain idea of France and French life. Finally, Bouchareb's recognition of a necessary balance between emotion and pedagogy touches on historiography. In particular, his remarks point to the kind and degree of understanding that films like *Indigènes* promote concerning France's colonial past. Moreover, they do this within a public sphere whose breadth extends beyond those of scholarly discourse and the independent film market in which Bouchareb's earlier films, *Bâton Rouge* (1985) and *Cheb* (1991), tended to circulate. In this sense, I find it crucial to consider *Indigènes* as a major-release film with which at least part of its target audience may strongly identify. (See figure 20.1.)

Indigènes draws on conventions of the buddy film grounded in male bonding for which wartime is a prime setting and for which the characters played by Jean Gabin, Marcel Dalio, and Pierre Fesnay in Jean Renoir's *La Grande Illusion* (1937) mark a notable precedent. The historical basis of Bouchareb's film is Operation Anvil-Dragoon, launched by the August 1944 landing in southern France by troops of whom some 85 percent were recruited in North and sub-Saharan Africa under orders from Generals De Gaulle and Giraud to constitute an African army headquartered in Algiers. These African troops initially fought in Tunisia, Sicily, and Italy alongside American soldiers. On August 16, they entered the Mediterranean port city of Toulon where they joined some 120,000 members of De Gaulle's Forces Françaises de l'Intérieur, with whom they moved north to Lyon, Belfort, Strasbourg,

Figure 20.1   Rachid Bouchareb (director), *Indigènes* (2006).

and Germany.[3] Despite the crucial role these African troops played in the eventual liberation of France from the south, De Gaulle withdrew some 20,000 Senegalese soldiers from service in early 1945 in order to "whiten" the southern surge and to emphasize the northern liberation initiated by the June 1944 allied landing in Normandy.[4]

Bouchareb tells and shows his account of this operation by means of wide-screen images, explosive surround sound, and period details of dress and music intended to convey historic authenticity. The sensory effect is dazzling as an enhanced version of *mode rétro* films such as *The Conformist* (Bernardo Bertolucci, 1970), *Stavisky* (Alain Resnais, 1974), and *Lacombe Lucien* (Louis Malle, 1974) whose evocation of interwar and wartime France likewise relied on production details to promote a critical "backward turn" toward the period in question.[5] *Indigènes* follows these earlier films by evoking World War II through special effects including black-and-white to color wipes at the start of each of the film's sections that recall the scrolling texts of silent and early sound-era colonial films such as Jean-Benoît Lévy and Marie Epstein's *Itto* (1933). Sound and music tracks sustain an emotional pitch coded to convey dramatic sacrifice. The effect promotes a degree of identification among all spectators in conjunction with an episode whose contemporary resonance specifically addresses (interpellates) descendants of the North Africans depicted in the film. Spectator identification is crucial to the impact *Indigènes* has fashioned in light of debate surrounding government policies related to immigration and employment. Because the scope of this debate extends in public sectors of the press, mass media, and broad social constituencies throughout France, *Indigènes* has taken on exceptional visibility in a public sphere that few feature films attain. Accordingly, it is appropriate to consider exactly what convergence of elements and concerns—aesthetic, commercial, and political—make up this cinematic object. What was at stake—for its makers, spectators, and critics—in the fall 2006 release of this epic account of World War II heroism? What, in particular, might spectators of *Indigènes* learn about World War II, about the role of North Africans "natives" recruited to fight for France, and perhaps about themselves with reference to the story it tells?

The high level of media visibility enjoyed by *Indigènes* between May and September 2006 did little to allay two misgivings I had concerning this film before I saw it. A first misgiving pertained to its title. My students—especially those from North and sub-Saharan Africa—continually remind me that the French term *indigène* has a pejorative ring that is far less present in the synonym *autochtone*. The *Petit Larousse* concurs when it lists one meaning of the term as "originaire d'un pays d'outre-mer, avant la décolonisation" (coming from an overseas country prior to decoloniza-tion). At most, I was ready to consider that Bouchareb chose the title as an ironic provocation to transform its negative resonance into a contemporary assertion of difference and pride. Such an interpretation is in line with the film's English title, *Days of Glory*, which echoes the epic nostalgia depicted in *Glory*, the 1989 Edward Zwick film (with Matthew Broderick, Morgan Freeman, and Denzel Washington) that chronicled the Massachusetts 54th, the first black volunteer infantry unit in the Civil War. A second U.S. film that comes to mind is *Saving Private Ryan* (Steven Spielberg, 1998), which follows *Indigènes* by ending with an update sequence to bring its depiction of a World War II episode toward the present. As with usage of the English term "nigger" by black militants and others since the 1960s, much depends on utterance; that is, on who is using the term and to whom the term is addressed.

A second misgiving concerns the choice of World War II to address the histor-ical record of France's recent past in the summer and fall of 2006. In this sense, the tense aftermath of October and November 2005 riots in communities on the northern perimeter of Paris makes it all the more unfortunate that *Indigènes* sub-sumed the political, cultural, and ethnic hegemony associated with the long history of French colonization within a primary reference to national identity fashioned through a primordial antagonism with Germany that framed the start and end of the French Third Republic between 1871 and 1945. Yes, Bouchareb portrays the four Algerian soldiers as colonial subjects fighting in the name of a France that has occupied their homeland for more than century. And yes, the role of such sol-diers in World Wars I and II has long warranted due recognition. And yes again, injustices inflicted by the French army and French government on these soldiers, on their families, and on their descendants remain in place. Yet at a moment when debate surrounding French national identity increasingly engages minority pop-ulations, notably descendants of Algerians who came to France during the two decades following Algeria's independence in 1962, the choice of World War II in place of the 1954–62 war in Algeria can be seen at least initially to mark a signifi-cant compromise.

Coming to Terms with the Algerian War

These initial remarks on *Indigènes* support my choice to study *La Trahison*'s por-trait of a French military unit in Algeria whose 30 include 4 Algerians, whom the film describes alternately as Muslims (*musulmans*, a shortened version of *Français musulman d'Algérie* or FMA) and French of North African stock (Français de souche nord-africaine, henceforth FSNA). Un-heroic in scope and restrained in tone, the film recounts events in the unit during a period of 13 days after a senior intelligence officer, Captain Franchet, informs Lieutenant Roque (the name given in the film to Sales' surrogate character) of plans involving the 4 Algerians to murder him and other French soldiers in his unit, allegedly on orders from regional command of the Armée de Libération Nationale (ALN).[6] Having worked on a daily basis with the 4 men over the previous 15 months, Roque is at first unwilling to accept the allegations, especially since they implicate his personal interpreter, Corporal Taïeb, whom he has grown to trust. Yet after reflecting on Franchet's allegations, Roque recalls exchanges and actions to which, in retrospect, he ascribes an ambiguity he is unable to deny. This change, which Faucon picks up in faithful detail from Sales' memoir, is nothing less than semiotic, as Roque detects infinitesimal signs of a forthcoming threat in what he had taken earlier as nothing more than "an expression of their distress and confusion" (Sales, 21).

But exactly who are *they*? Who are these 4 men whose distress and confusion Roque comes to see as suspect? What is their status in conjunction with the French troops stationed in the remote post? These are among the questions the film raises in conjunction with the 1954–62 Algerian "war without a name" around which historical inquiry has yet to evolve toward the type of critical understanding of France between 1940 and 1944 revised over the past 35 years on the basis of the efforts by Robert Paxton, Marcel Ophuls, Henry Rousso, and others.[7] Anne Donadey has tested the viability of what she has termed an Algeria syndrome, with reference to an emphasis on World War II in French public life, the media, education, and literature during the 1970s and 1980s that she saw as obscuring critical engagement with events in Algeria between 1954 and 1962.[8] Films such as *La Trahison* and *Indigènes* can serve to address the extent to which the shattering of established myths concerning World War II since the early 1970s can be seen today in conjunction with the Algerian conflict and its extended aftermaths on both sides of the Mediterranean.

In what ways might a film such as *La Trahison* contribute to a critical historiography of the Algerian war modeled on revisions of received understanding concerning Vichy France? As with *Indigènes*, a number of historical details depicted in the film warrant mention. Because French Algeria in 1960 comprised three administrative departments, it is important to note that the four Muslims serving under Roque were completing their military service as FSNA serving alongside their fellow *Français de souche européenne* (henceforth FSE). After 1957, the French army depended heavily on FSNA recruits. By 1962, 50,000 FSNA were in the French army in Algeria, serving mainly as interpreters mediating relations with local populations. The four soldiers portrayed in the film are thus *not* in a strict sense among the 180,000 North African civilians known as *harkis*, who contracted to work as auxiliary troops (*forces supplétives*) or in other roles to help occupying French military units.[9] At the same time, they were subject to the same doubts and questions—were they mercenary or committed? faithful comrades or double agents?—that made the *harkis* objects of suspicion among all parties involved in the Algerian conflict. For the record, the four FSNA portrayed in Faucon's film are draftees whose presence in the unit is less a matter of choice than one of compliance with recruitment practices instituted in French Algeria.

La Trahison forcefully stages tensions surrounding the four Algerians who find themselves increasingly isolated from both the white FSE in the unit and the local Algerian population. When children in the village taunt Taïeb by calling him *harki*, the term is meant as an insult directed toward one who has seemingly abandoned the cause of Algerian independence by siding with the French occupiers. When a fifth FSNA member, Lachène, returns to the outpost after briefly deserting during a military action, Roque reassures the other four that this changes nothing concerning their role in the unit. Taïeb asks Roque for a gun to kill Lachène, adding "He is not on our side. He is neither with the *fellaghas* [Algerians favoring national independence] nor with the French; he has betrayed twice" (Sales, 21).

Sales and Faucon portray betrayal as a multiple phenomenon that evolves from individual to group. Taïeb embodies personal betrayal after the allegations made by Franchet undermine a trust and friendship that Roque had taken to be reciprocal following months of daily interaction. The allegations project outward from Roque and Taïeb toward all members of the unit with whom the FSNA had worked. At the same time, it might be argued that Roque betrays Taïeb when he follows Franchet's orders without confronting Taïeb. From the perspective of the local villagers, the FSNA working with the French army betray the cause of independence represented by the FLN and its military branch, the Armée de Libération Nationale (ALN). In retrospect, the film can be seen as staging conditions that led to the deaths of thousands of FSNA, *harkis*, and *pieds noirs* abandoned (betrayed?) by France, and by De Gaulle, following the 1962 independence of Algeria some two years after the incident recounted in the film.

Working with Philippe Faucon to adapt his memoir for the screen allowed Claude Sales to deepen his personal sense of a betrayal he had lived 40 years earlier by recasting it as a series of glances and silences that his 1999 memoir had only approximated. Accordingly, he asked in a recent interview:

> Does the word "betrayal" still have a meaning? It is always the one betrayed who defines the traitor. At the time, it is a shock. Afterward, you question yourself. By accepting to do his military service, was [Taïeb] not already in a state of betrayal from the point of view of other Algerians [*vis-à-vis des siens*]? Perhaps my death was for him the only means to reassert his identity. To a degree, I had booby-trapped him and his comrades. The village for which we were responsible was not continually hostile toward us, which delighted Taïeb. But the FLN was also present. After the publication of the book, I received numerous letters from Algerian veterans who had lived similar stories.[10]

Prominent among changes between the written memoir and film versions of *La Trahison* was that of period, which Faucon set in 1960, or up to twenty months later than Sales' tour of duty in Algeria. The effect of this change is perhaps coincidental. Yet the interim was marked by increased polarization resulting in phases from De Gaulle's June 1958 "Je vous ai compris" statement in Algiers, his September 16, 1959, call for a vote on self-determination for Algeria, and General Maurice Challe's Operation Jumelle that systematically destroyed much of the armed Algerian resistance. The March 1960 moment depicted in Faucon's film is cast openly within the context of the eventual French withdrawal and the prospect of reprisals against FSNA and *harkis*. By dramatizing this moment, Faucon portrays the plight of both the FSNA and those FSE veterans of the Algerian conflict whose experiences Sales' memoir had retrieved.

I initially chose to cast *La Trahison* against *Indigènes* because what I first learned about it illustrated a distinction I found crucial in light of ongoing debate on the extended afterlives of these two major "spaces of war." The distinction I have in mind sets narrative fetishism against the work of mourning. As formulated by Eric L. Santner in a 1992 article on the representation of trauma, narrative fetishism is an effort to construct and deploy a narrative simulating a condition of intactness that releases one from having to constitute one's self-identity under posttraumatic conditions.[11] In essence, it is a form of conscious or unconscious denial. By contrast, what Freud called *Trauerarbeit* ("work of mourning") integrates the reality of loss or traumatic shock "by remembering and repeating it in symbolically and dialogically mediated doses" (Santner, 144). *Indigènes* would appear to subsume the specific trauma of loss associated with colonial troops fighting to liberate France in 1943 by equating all victims within a memory of patriotic sacrifice. In so doing, it repositions the World War II episode within a broader fashioning of national identity under the French flag in ways to which the Algerian war seen in *La Trahison* does not (not yet?) lend it itself.

The release of *Indigènes* during the same year as the opening of the Musée du Quai Branly prompts me to see them both as promoting narrative fetishism by setting France's long history of colonization within institutions and discourses of national identity whose goal is marked by consensus. By contrast, *La Trahison* forsakes the tone and scale of epic heroism for an interplay of individual glances and gestures whose meaning remains unstable As a result, its narrative strength comes especially from ambient or peripheral movement in the back alleys of the rural village and in the thoughts of young French and Algerians waiting to act. In terms of critical historiography, the depiction of this movement in Sales' written memoir and Faucon's adaptation extends to the contested memory of events a half-century ago whose implications remain central to debate related to identity formation and constructions of belonging in contemporary France.[12]

In thinking how best to address this wartime displacement from 1954–62 to 1940–44, I take my cue from a distinction Henry Rousso has made concerning the late twentieth-century culture of memory directed in France first toward the Shoah and more recently toward the Algerian war.[13] While he asserts that consensus over a national representation of the Algerian war is unlikely, Rousso argues that whereas debates of the 1980s between surviving resistance group members and deported Jews targeted a common adversary embodied mainly in the occupying German forces, ongoing debate surrounding the Algerian war sustains oppositions between yesterday's adversaries that remain unreconciled. Rousso sees the impossibility of true consensus on a national representation of the events in Algeria in the ongoing failure to agree on a single date—such as the March 19 date of the 1962 Evian agreements—to satisfy those for whom a fuller and more dutiful commemoration (*devoir de mémoire*) ought also to evoke the October 17, 1961, Paris demonstrations that resulted in hundreds of unrecorded deaths, the February 8, 1962, killings at the Charonne metro station, the May 26, 1962, gunfire at the rue d'Isly, and the July 5, 1962, murders in Oran.

Rousso notes two types of effort on the part of the French government to do justice to these and other episodes related to the 1954–62 period. The first involves the systematic recourse to raising public awareness of the 1954–62 period resulting from a 1999 parliamentary vote changing the former "events of Algeria" to the status of a declared war. Rousso notes that the vote merely ratifies in legal terms a semantic shift over the past 40 years and is an anachronism because if France had admitted on November 1, 1954, that a war had broken out in Algeria, the admission would have implied recognition of Algeria's independence as a nation, which was precisely what proponents of French Algeria were unwilling to recognize.

As legal actions seek to set the historical record straight with reference to crimes against humanity linked to violence and torture during the 1954–62 period, Rousso rightly concluded that what was at stake was the putting into place of strategies of collective action that attribute to justice a formidable vector of memory. What has occurred with reference to World War II starting with the 1945–49 Nuremberg trials and Adolf Eichmann's 1961 trial in Jerusalem through the 1997–98 trial of Maurice Papon needed to continue toward fuller disclosure of similar abuses and atrocities related to the period of the Algerian war. The structure of *Indigènes* asserts a teleology of closure that commemorates France between 1940 and 1994 in the format of epic narrative whose focus on World War II elides the more complex and unresolved nature of debate surrounding French Algeria between 1954 and 1962. By contrast, the markedly unheroic atmosphere in *La Trahison* suits the very kind of non-resolution that I see as promoting the fuller coming to terms that remains in process concerning the Algerian war.

In all fairness, Bouchareb offsets the displacement away from the Algerian war in a closing sequence in which the sole survivor among the five North Africans recruited to fight in 1943 returns to Alsace in a sequence entitled "Alsace, 60 years later" (*Alsace, 60 ans plus tard*). The survivor, Abdelkader (played by Sami Bouajila), is now an elderly man with a slow gait, eyeglasses, and a white beard. He is first shown walking over the crest of a hill in a military cemetery dotted with crosses marking the graves of Christian (presumably European) soldiers. A French flag is visible in the distance against a windless white sky.[14] The man crouches opposite a cross before a counter-shot shows a plaque with the name of the group's ranking officer, Sergeant Roger Martinez. He walks next among graves marked with stone slabs covered in Arabic calligraphy and locates the graves of two fallen comrades, each with a bronze marker with the same phrase in French used for all military victims: "Mort pour la France" (died for France). He returns to an urban setting and to a single room in a housing development where he sits alone on his bed. A fade to black is followed by three passages in prose, shown on screen at intervals of ten to twelve seconds:

> In 1959, a law was passed freezing the pensions of infantrymen from countries of the colonial empire that were attaining independence.
>
> In January 2002 following lengthy legal action, the Government Council ordered the French government to pay these pensions in full.
>
> But successive governments have succeeded in pushing back the due date in question.

Indigènes addresses this failure to honor an outstanding debt by staging preludes to it throughout the film in a series of betrayals on the part of military officers who fail to keep stated promises to give the North Africans their due in salary, rank, food, and leave on a par with their (white) European peers. Rumor has it that when former French President Jacques Chirac saw the film in the spring of 2006, his wife Bernadette urged him to resolve the matter without further delay. Whether or not the anecdote was true, the call for action in response to the film's final sequence soon expanded through concerted efforts on the part of Bouchareb and others connected to the film to address issues raised by the film in public venues that featured communities of North Africans portrayed in the film in urban settings of Paris, Lyon, and Marseille.

Two Films and Two Wars

La Trahison and *Indigènes* disclose markedly distinct approaches to contending with France's colonial past in the public sphere of commercial film. It is notable that websites for both provide additional information on the films as sites for inquiry surrounding the periods and conflicts depicted in the films as well as their relevance to contemporary debate linked to national identity. I have asserted my preference for the understatement and narrative irresolution in *La Trahison* over the epic account of heroism that structures *Indigènes* until the final sequence that links consequences of this World War II episode to the present. This preference derives, in part, from the irresolution with which *La Trahison* engages an episode within a conflict about which irresolution continues to be divisive among historians and in the larger public sphere of which film culture is only one sector.

By contrast, the awards and recognition received by *Indigènes* mark what I see as an unstable point between institutional engagement and monumental consecration, as though the film itself constituted the kind of recognition that might allow the soldiers involved and their families to come to terms (and be done) with the episode in question. While this is clearly not what the concluding sequence of *Indigènes* stages, much depends on the extent to which the closing sequence's call for reparations evolves from symbolic to material action. The film's September 2006 release provided additional opportunities to channel media visibility toward activism in the cause of local communities such as those in Paris affected by demonstrations and disorder in October and November 2005.

Additional statements by Bouchareb have grounded his work on the film in family history: "Members of my father's family died in combat in the war of 1914. One of my uncles fought in Indochina, the Algerian war. I had wanted to recount these elements through my skills as a filmmaker."[15] Jamel Debbouze was already well known in France as a comic actor before his role as the produce seller assistant Lucien in Jean-Pierre Jeunet's *Amélie* (2001) drew worldwide attention. In *Indigènes*, he exploited his star status to portray the character of Saïd Otmari and to co-produce the film. Along with other members of the cast, he spoke about the film to youth groups in urban neighborhoods of Paris, Lyon, and Marseille. Having earned credibility as a role model who refused to serve as symbolic flag waver or social worker, he offered a message in public appearances and press interviews that was one of reconciliation and appeasement to turn hatred and humiliation toward something positive.[16]

From a jaded perspective, such statements by Bouchareb and Debbouze might be seen as a marketing ploy looking to capitalize on social unrest to promote a film. But evidence to the contrary suggests that these personal appearances are designed primarily to mobilize populist support to press the government to honor lapsed promises of retirement and related benefits to veterans. A handout I received in Paris in June 2007 urged individuals to write directly to President Jacques Chirac in order to end the scandal of non-coverage affecting "native" veterans (*anciens combatants "indigènes"*) in twenty-five countries of France's former colonial empire.[17]

Interviews with Claude Sales and Philippe Faucon reveal that while personal links inflect their work on *La Trahison*, these are secondary to what they both see as the film's engagement with the contemporaneity of the drama the film stages on a human scale. Repeated visual cues remind spectators of *Indigènes* that they are watching a historical film of epic proportions. By contrast, *La Trahison*'s directness and sobriety internalize the historical frame within a human drama centered on choices made by individuals—especially by the characters Lieutenant Roque and Taïeb—whose consequences they may not fully understand, even after the fact. (Faucon notes that at one point, he and Sales had considered titling the film *Le Choix* [The Choice].)[18] This emphasis enhances the dramatic force by which the incident passes toward the present by recording actions, words, and objects in a quasi-documentary style.[19] Faucon has stated that the film simulates a visual chronicle whose transposition of the war into human drama casts conflict through the evident estrangement of the soldiers in military unit from the physical setting of the rural village. A double discrepancy involving the Algerians serving in the military unit complicated their relations with villagers who saw them as having betrayed the cause of national revolution and independence and with white members of the military unit who came to question their loyalty to the French army.[20]

A potential risk for both films is to fall victim to the force of recent debate surrounding communalism (*communautarisme*) as a closing in of ethnically defined communities on themselves, an action that threatens political consensus at the level of nation when citizens value their affiliation to these communities over their collective participation in the nation.[21] The risk derives in part from narrative structure. To the extent that *Indigènes* suggests that all military victims are equal in their respective deaths for France, the postulate that all citizens subscribe to the same values may prove to be inadequate to the full degree of material and symbolic reparation that the film promotes. By contrast, the irresolution at the end of *La Trahison* supports an openness that is likely to spawn inquiry and debate whose scope appears more circumscribed than that afforded to *Indigènes*. But instead of prioritizing one film over the other on the basis of the respective public spheres in which debate over them occurs, I want to conclude by setting both films within what the historian Benjamin Stora has termed a war of competing memories related to France's colonial past. Like Henry Rousso on World War II, Stora argues for heightened attention to the role played by the transmission of France's colonial past for which the history of Algeria is preeminent. More pointedly than Rousso and Anne Donadey, he asserts that if France in the person of President Jacques Chirac is ready to recognize the responsibility of the Vichy government in the World War II deportation of Jews, his desire to close the page on Vichy has left matters concerning its role in Algeria unopened.[22]

In response to a question from Thierry Leclère concerning the critical success of *Indigènes*, Stora noted that its success was ambiguous because even if it jumped from January 1945 to 2006, the consensus it asserted around the French flag failed to account for what happened in overseas territories occupied by France following World War II such as the May 1945 massacres in Sétif and those of March 1947 in Madagascar (Stora, 55). Looking at a counterexample of accounts of the 1954–62 Algerian war as it is taught in contemporary Algeria, Stora noted that textbooks (*manuels scolaires*) remained essential tools to facilitate the forging of national mythologies based on historical accounts produced in terms of state interests (Stora, 59). One may wonder—as Roland Barthes did some 50 years ago in "Le Mythe, aujourd'hui"—if any myths escape ideological compromise of one kind or another on the political left and/or right.[23] Even so, the questions raised by and surrounding *Indigènes* and *La Trahison* hold the prospect of ongoing debate as a potential means of breaking with received representations of a recent past whose inadequacy these films link to present and ongoing debate. Whether these films draw on personal ties (as is the case for Bouchareb, Faucon, Sales, and Debbouze) or on issues linked to the writing of history (as is the case for Rousso, Stora, and others), they can promote the fuller disclosure of France's colonial past in a broad public sphere that films such as Ophuls' *Le Chagrin et la pitié* and books such as Robert O. Paxton's *Vichy France* and Russo's *Vichy Synd*rome have promoted with reference to Vichy France.

Notes

1. The five actors are Jamel Debbouze, Samy Nacéri, Roschdy Zem, Sami Bouajila, and Bernard Blancan.
2. See http://www.festival-cannes.fr/films/fiche_film.php?langue=6002&id_film=4352781 (accessed February 1, 2007).
3. Laurent Lemire, "C'est nous les Africains...", *Nouvel Observateur* (September 28, 2006): 6–7.
4. See Myron J. Echenberg, *Colonial Conscripts: The Tirailleurs Sénégalais in French West Africa (1857–1960)* (Portsmouth, NH: Heinemann, 1991).
5. On the phenomenon of the *mode rétro*, see Lynn A. Higgins, *New Novel, New Wave, New Politics* (Lincoln: University of Nebraska Press, 1996), 189–90 and 214–15, her entry on it in Bertram M. Gordon, ed., *Historical Dictionary of World War II France* (Westport, CT: Greenwood, 1998), 245–46, and Naomi Greene, *Landscapes of Loss: The National Past in the Postwar French Cinema* (Princeton: Princeton University Press, 1999).
6. My comments on *La Trahison* draw heavily on a pedagogical file prepared by Valérie Marcon and Hélène Chauvineau in conjunction with the film's January 2006 release. The file is available at http://www.agence-cinema-education.fr/latrahison-dossierpeda.pdf.
7. In English, see John Talbott, *The War Without a Name: France in Algeria, 1954–1962* (New York: Knopf, 1980); in French, see Benjamin Stora, *La Gangrène et l'oubli: la mémoire de la guerre d'Algérie* (Paris: La Découverte, 1991), Patrick Rotman and Bertrand Tavernier, *La Guerre sans nom: les appelés d'Algérie 54–62* (Paris: Seuil, 1992), and Alain-Gérard Slama, *La Guerre d'Algérie: histiore d'une déchirure* (Paris: Gallimard, 1996). With reference to Vichy France, see *Le Chagrin et la pitié (The Sorrow and the Pity)*, Marcel Ophuls, 1973; Robert O. Paxton, *Vichy France: Old Guard and New Order, 1940–1944* (New York: Knopf, 1972); and Henry Russo, *The Vichy Syndrome, History and Memory in France Since 1944*, Arthur Goldhammer, trans. (Cambridge, MA: Harvard University Press, 1991).
8. Anne Donadey, "'Une Certaine Idée de la France': The Algeria Syndrome and Struggles over 'French' Identity," in Steven Ungar and Tom Conley, eds., *Identity Papers: Scenes of Contested Nationhood in Twentieth-Century France* (Minneapolis: University of Minnesota Press, 1996), 217.
9. See http://www.agence-cinema-education.fr/latrahison-dossierpeda.pdf. For an analysis of semantic and historical distinctions among *harkis* and FSNA, see Tom Charbit, *Les Harkis* (Paris: La Découverte, 2006).
10. See http://www.agence-cinema-education.fr/latrahison-dossierpeda.pdf.
11. Eric L. Santner, "History Beyond the Pleasure Principle: Some Thoughts on the Representation of Trauma," in Saul Friedlander, ed., *Probing the Limits of Representation: Nazism and the "Final Solution"* (Cambridge, MA: Harvard University Press, 1992), 144.
12. Jean-Michel Frodon, "*La Trahison* de Philippe Faucon: ce qui bouge dans le creux de l'Histoire," *Cahiers du Cinéma* 608 (January 2006): 26.
13. Henry Rousso, "La Guerre d'Algérie et la culture de la mémoire," *Le Monde* (April 5, 2002): 17.
14. The distinction between Christian Europeans and Muslim North Africans does not hold in every case. At one point in the film, Saïd (Jamel Debbouze) discovers an old photo in

the pocket of a shirt belonging to his superior, Sergeant Martinez (Bernard Blancan). The photo shows a North African woman with a child. Its inscription, "Mom and me" (*maman et moi*), suggests that Martinez , who passes throughout the film as a European and repeatedly sets himself above the North Africans whom he commands, is the offspring of a mixed couple.

15. Rachid Bouchareb, "*Indigènes* arrive au bon moment: entretien avec Dominique Widemann," *L'Humanité* (September 27, 2006).

16. See, for example, statements by Debbouze in Marie-France Etchegoin, "Jamel, nouveau soldat de la République," and in Fabrice Pliskin, "Pourquoi j'aime la France." Both texts appear in a dossier devoted to *Indigènes* published in the September 28, 2006, issue of *Le Nouvel Observateur*. Similar statements by Debbouze and by other actors in the film appear in the September 27, 2006, issue of the Communist daily, *L'Humanité*. For an early American take on the film, see A. O. Scott, "Yes, Soldiers of France in All but Name," *New York Times* (December 6, 2006).

17. The document is "Appel pour l'égalité des droits entre les anciens combattants français et coloniaux" (Call for equal rights between French and Colonial Veterans).

18. Pascal Mérigeau, "De la guerre d'Algérie à la crise des banlieues: un entretien avec Philippe Faucon," *Nouvel Observateur* (January 26, 2006).

19. Antoine de Baecque, "Conflits intérieurs," *Libération* (January 25, 2006).

20. Jacques Mandelbaum, "Tragédie discrète dans le bled," *Le Monde* (January 25, 2006).

21. John R. Bowen: *Why the French Don't Like Headscarves: Islam, the State, and Public Space* (Princeton: Princeton University Press, 2006), 156.

22. Benjamin Stora, *La Guerre des mémoires: la France face à son passé colonial* (La Tour d'Aigues: L'Aube, 2007), 21.

23. Roland Barthes, "Le Mythe, aujourd'hui," *Mythologies* (Paris: Seuil, 1957). I have written on this essay and Barthes's mythologies in the context of the Algerian war in "From Event to Memory Site: Thoughts on Rereading Mythologies," *Nottingham French Studies* 36:1 (1997): 24–33.

Contributors

Patrice Arnaud, a former student at the École Normale Supérieure and *agrégé* in history, received his doctorate in 2006. He is the author of numerous articles and is preparing for publication his doctoral thesis on French civilian workers in Germany during World War II. He teaches at a *lycée* in Fontenay-sous-Bois, France.

Annette Becker, Professor of Contemporary History at Paris Ouest Nanterre La Défense has written extensively on the two World Wars and their memories, particularly on intellectuals and artists (Halbwachs, Bloch, Edith Wharton, Dadaists). Her books, *Obus-Roi, Apollinaire, la guerre, les avant-gardes* (Tallandier) and *Les Occupations, 1914–1918, laboratoires du XXème siècle* (Perrin), will appear in 2009.

Daniel Brewer is Professor of French Studies at the University of Minnesota. His areas of teaching and research include the literature and culture of ancien régime France and contemporary cultural theory. His books include *The Discourse of Enlightenment* and *The Enlightenment Past: Reconstructing Eighteenth-Century French Thought*. He has co-edited volumes on the *Encyclopédie* and is co-editor of *L'Esprit Créateur*.

Nicole Dombowski Risser is Associate Professor of European History at Towson University and Director of the Master of Science Program in Social Science and Global Analysis. She is editor of *Women and War in the 20th Century: Enlisted with or without Consent* (New York: Routledge, 2004). She earned her Ph.D. in History from New York University and her BA from the University of Wisconsin-Madison.

Angélique Durand is a graduate student in cultural history at the Université de Versailles Saint-Quentin en Yvelines. She is completing a doctorate in contemporary history at the Centre d'histoire culturelle des sociétés contemporaines (CHCSC).

Christopher J. Fischer is Assistant Professor of History at Indiana State University. His *Alsace to the Alsatians? Visions and Divisions of Alsatian Regionalism, 1870–1939*, is based on his Fritz Stern Prize winning dissertation and will be published in 2009 by Berghahn Books.

Julien Fragnon holds a Ph. D. in Political Science and is an Assistant Professor at the University of Lyon. His works concern the French political discourses on terrorism and the uses of past in the political field.

Kirrily Freeman is Assistant Professor of History at Saint Mary's University in Halifax, Nova Scotia. Her first book, *Bronzes to Bullets: Vichy and the Destruction of French Public Statuary, 1941–1944* (Stanford, 2008), investigates the campaign undertaken by the French collaborationist government to melt down bronze statuary throughout France.

Richard Golsan is Professor of French and Head of the Department of European and Classical Languages at Texas A&M University. He is the author of *French Writers and the Politics of Complicity* (Hopkins, 2006) and *Vichy's Afterlife: History and Counterhistory in Postwar France* (Nebraska, 2000), among other works. He is editor of *South Central Review*.

Floréal Jiménez Aguilera obtained his *doctorat* in history and civilizations from the École des Hautes Études en Sciences Sociales. He is an independent researcher and a documentation officer for the Archives Nationales d'Outre-Mer in Aix en Provence. Among his many articles on the social and political aspects of film and cinematography in both Francophone and Anglophone journals are "L'Homme du Niger, The Man from Niger: A Cinematographic Construction of Colonialist Ideology in the 1930s," *Studies in French Cinema*, and "Ces garçons qui venaient du Brésil: l'homme prévu et l'homme imprévu," *Cinéma: rites et mythes contemporains*.

Martha Hanna is Professor of History at the University of Colorado, Boulder. She is author of two books on France during the First World War: *The Mobilization of Intellect: French Scholars and Writers during the Great War* (Harvard University Press, 1996) and *Your Death Would be Mine: Paul and Marie Pireaud in the Great War* (Harvard University Press, 2006). *Your Death Would Be Mine* received the J. Russell Major Prize from the American Historical Association in 2007 and the Distinguished Book Award in Biography from the Society for Military History in 2008.

Bronson Long is a tenure-track Instructor of History at Georgia Highlands College. As a doctoral student at Indiana University, he received the Chateaubriand Fellowship from the French government and the Dr. Richard M. Hunt Fellowship from the American Council on Germany. In October 2007, he successfully defended his dissertation entitled "The Saar Dispute in Franco-German Relations and European Integration: French Diplomacy, Cultural Politics and the Construction of European Identity in the Saar 1944–1957." He has also published articles on the Saar question in the *Atlantic Affairs Journal* and *Football Studies*.

Patricia M. E. Lorcin is Associate Professor of History at the University of Minnesota-Twin Cities and editor of *French Historical Studies*. Her publications include: *Imperial Identities: Stereotyping, Prejudice and Race in Colonial Algeria*; *Algeria and France 1800–2000: Identity, Memory and Nostalgia*; numerous book chapters and articles on French colonialism; and two co-edited volumes.

Libby Murphy is Assistant Professor of French at Oberlin College. Her current book project develops the picaresque as a conceptual framework for understanding the ways in which French novelists, journalists, cultural critics, and graphic artists attempted to make sense of the Great War.

Alexandre Niess received his doctorate in history at the Université d'Orléans. He is a member of the research laboratory SAVOURS (Contemporary Knowledge and Powers of Antiquity) and CHPP (Committee of Parliamentary and Political History) and copyeditor of the review *Parlement(s)*. His research focuses on memory and war memorials after World War I and on the politicians of the Third Republic. Author of a work on military cemeteries and war memorials, *First World War Military Cemeteries and Memorials—Marne*, with Maurice Vaïsse he co-edited *Léon Bourgeois (1851–1925): du solidarisme à la Société des Nations* (2006).

Christy Pichichero is a postdoctoral fellow in the Introduction to the Humanities Program at Stanford University. She completed her Ph.D. in Early Modern French Studies at Stanford (2008), an A.B. in Comparative Literature (French and Italian) at Princeton University (1998), and a B.M. in Applied Music (Voice—mezzosoprano) at the Eastman School of Music (2000).

Paul Schue holds a doctorate in European History from the University of California, Irvine, and is an Associate Professor of History at Northland College. He has published articles in *French Historical Studies, The Intellectual History Review*, and *National Identities*. His current research focuses on how ideology influenced French intellectuals' responses to the Spanish Civil War, and how these responses point to a larger crisis of conceptions of modernity in the late 1930s.

Jennifer E. Sessions is Assistant Professor of History at the University of Iowa, where she teaches French and European history. Her research focuses on nineteenth-century French empire, colonial cultures, and the comparative history of settler colonialism. Her current book project is an interdisciplinary study of the conquest and colonization of Algeria, which combines visual, literary, and archival sources to trace the origins of French Algeria in post Revolutionary French political culture.

Steven Ungar is Professor of French and Chair of Cinema and Comparative Literature at the University of Iowa. His book-length publications include *Roland Barthes: The Professor of Desire* (1983), *Scandal and Aftereffect: Maurice Blanchot and France Since 1930* (1985), *Popular Front Paris and the Poetics of Culture* (2005, with Dudley Andrew), and *Cléo de 5 à 7* (2008).

Béatrice Vernier-Larochette is Assistant Professor at Lakehead University (Canada) where she teaches French language and literature. Her primary field of interest is art and literature, the writings of letters, of autobiographies and the expression of oneself in the letters of war. Her most recent publications include articles on Paul Gaugin, Fernand Léger, and Émile Zola.

Michael Wolfe is Professor of History at St. John's University in Queens, New York. His six books include *The Conversion of Henri IV: Politics, Power and Religious Belief in Early Modern France, Changing Identities in Early Modern France*, and his forthcoming study *Walled Towns and the Shaping of France*. He is currently completing a companion study on building and living with fortifications in urban France between 1450 and 1650.

Index